Composition

Chap. 7

Bonding Between Atoms

Chap. 3

Microstructure and Macrostructure

Chap. 4   Chap. 9   Chap. 10   Chap. 11

Cold-Work-Anneal Cycle

Chap. 6   Chap. 9

n Hardening

9

er Hardening

9

tion Hardening

p. 9

ecomposition

Chap. 10

eatments

Chap. 11

ects

ap. 13

Practical Applications

Chap. 2   Chap. 4   Chap. 15

THE SCIENCE OF
ENGINEERING MATERIALS

THE SCIENCE OF ENGINEERING MATERIALS

Second Edition

CHARLES O. SMITH

Professor of Engineering
University of Nebraska at Omaha

Prentice-Hall, Inc., *Englewood Cliffs*, New Jersey *07632*

Library of Congress Cataloging in Publication Data

Smith, Charles O   1920–
    The science of engineering materials, 2nd Ed

Bibliography: p.
    Includes index.
    1. Materials. I. Title.
TA403.S593   1977   620.1'1   76-7357
ISBN 0-13-794990-1

© 1977 by Prentice-Hall, Inc.
Englewood Cliffs, New Jersey 07632

10 9 8 7 6 5 4 3 2 1

Printed in the United States of America

Prentice-Hall International, Inc., *London*
Prentice-Hall of Australia Pty. Limited, *Sydney*
Prentice-Hall of Canada, Ltd., *Toronto*
Prentice-Hall of India Private Limited, *New Delhi*
Prentice-Hall of Japan, Inc., *Tokyo*
Prentice-Hall of Southeast Asia Pte. Ltd., *Singapore*
Whitehall Books, Limited, *Wellington, New Zealand*

CONTENTS

PREFACE
TO SECOND EDITION

This second edition has the same focus as the first edition, i.e., attention to properties as responses to demands of the immediate environment, within a practical context through discussion of materials selection and failure analysis.

This edition is an enlargement of the first edition with some reorganization and addition of much discussion of ceramics and polymers. As a consequence, I believe it has better balance than the first edition. A number of examples have been added throughout the text in the hope of better illustrating various concepts. Chapter 15, dealing with failure analysis, has been enlarged by providing two more cases of failures, one of which deals with polymers.

COS

PREFACE
TO FIRST EDITION

This book has been written in terms of properties which are regarded as responses to the demands of the immediate environment. The properties (responses) are functions of the kinds of atoms present (chemistry and composition), type of bonding among them, geometrical arrangement of large numbers of atoms, microstructure, and macrostructure. The last items are, in turn, functions of fabrication procedures and various treatments. An attempt has been made to put this in a practical context through a discussion of selection of materials and failure analysis.

The book is primarily intended for use in a first course in Science of Materials. At the same time, I believe it can be of substantial value to any engineer. Much of the text was developed and used at the Oak Ridge School of Reactor Technology at the Oak Ridge National Laboratory. It thus has the merit of having been rather thoroughly and critically scrutinized by many of the illustrious graduates of that school as well as by many undergraduate engineers at the University of Detroit.

The book contains more than I believe can be successfully mastered in a one semester course. This has been done for the sake of providing reasonably complete coverage of all major aspects of materials science. In addition, it allows any teacher more flexibility in selecting the areas of most interest for his particular context.

Credit for assistance is due to many people with whom I have been associated over several years. I would like to list all of them, but it would be a very long list. The absence of such listing in no way reflects any lack of appreciation on my part, as I am indeed most grateful for all the help I have received. I would like to especially thank Dr. J. V. Cathcart of the Oak Ridge National Laboratory for his comments on oxidation and Mr. R. C. Wilson of Ex-Cell-O Corporation for his comments on failure analysis. I further appreciate the many hours of expert typing by Mrs. Gwen Wicker of the Oak Ridge National Laboratory and Mrs. Imogene Marshall of the University of Detroit. Last, and above all, I express my thanks to my good spouse for her understanding and forebearance over the years this book has been in the making.

CHARLES O. SMITH

ACKNOWLEDGMENTS

The author gratefully acknowledges the courtesy of: Addison-Wesley Publishing Company, Inc., for permission to use Figs. 3-10, 6-10, 8-6, 8-8, 8-9, 9-11, 10-4, 10-5, and Table 8-1 (from A. G. Guy, *Elements of Physical Metallurgy, 2nd Ed.*, 1959), Fig. 3-10 (from A. G. Guy, *Physical Metallurgy for Engineers*, 1961), and Fig. 4-8 (from B. D. Cullity, *Elements of X-ray Diffraction*, 1956); American Society for Metals, for permission to use Figs. 7-24 and 7-25; American Welding Society, for permission to use Fig. 12-5; Consolidated Western Steel Division (U. S. Steel Corp.), for permission to use Fig. 10-9; Engineering Case Library, Stanford University, and Dr. Geza Kardos, Carleton University, for permission to use all of Section 15-5; McGraw-Hill Book Co., for permission to use Figs. 4-23 and 4-27, and Tables 4-6 and 4-7 (from A. T. DiBenedetto, *Structures and Properties of Materials*, 1967) and Fig. 11-4 (from A. G. Guy, *Introduction to Materials Science*, 1972); "Metal Progress," for permission to use Figs. 10-2 and 13-8 and Table 10-1; Oak Ridge National Laboratory, for permission to use Figs. 4-34 and 9-4; Prentice-Hall, Inc., for permission to use Figs. 4-13, 5-1, 5-12, 5-13, 5-14, 5-16, 5-17, 5-18, 5-19, 5-20, 5-21, 5-22, 5-23, 5-25, 5-26, and 5-27 (from Allen Nussbaum, *Electronic and Magnetic Behavior of Materials*, 1967), Figs. 6-10 and 6-13 (from N. H. Polakowski and E. J. Ripling, *Strength and Structure of Engineering Materials*, 1966), Appendix A (from A. L. Ruoff, *Introduction to Materials Science*, 1972), and Table 11-1 (from J. Schultz, *Polymer Materials Science*, 1974); Reinhold Publishing Corp., for permission to use Fig. 1-1; Dr. Ray B. Richards, Imperial Chemical Industries, Ltd., for permission to use Fig. 4-32; U. S. Steel Corp., for permission to use Fig. 15-2; John Wiley & Sons, Inc., for permission to use Fig. 6-11 (from W. C. Dash, *Dislocations and Mechanical Properties of Crystals*, 1957), Figs. 4-1, 4-20, 4-21, 4-22, 7-1, 7-2, 8-1, 11-1 and 11-2 (from W. D. Kingery, *Introduction to*

Ceramics, 1960), Fig. 12-6 (from J. Wulff, H. F. Taylor, and A. J. Shaler, *Metallurgy for Engineers*, 1952), and Fig. 11-3 (from J. Wulff, et al, *Structure and Properties of Materials*, 1964).

PROLOGUE

(TO THE STUDENT)

Some knowledge of materials is necessary for all engineers since use of materials is all-pervasive. It may be trite, although nonetheless true, but we can make nothing unless materials are used in some form. The intent of this book is to provide an introduction to materials (stuff that things are made with) whose usefulness is determined by properties (responses to demand of the environment) not position in the periodic table. The book presents concepts of basic phenomena and mechanisms which, while interesting and important in themselves, are highly important for design and application. This viewpoint of use in real situations underlies the entire text.

In this context, it may be worth noting that 10×10^6 in materials represents 250×10^9 in gross national product. Some who tend to view materials aspects as rather unimportant may have been influenced by this ratio of 1 to 25,000. May you not be trapped!

May you read and study this book and profit therefrom!

THE SCIENCE OF

ENGINEERING MATERIALS

1

INTRODUCTION

1-1 INTRODUCTION

Ever since Adam and Eve felt a need for some sort of apparel in the Garden of Eden, mankind has been concerned with problems of materials. While Adam and Eve and their immediate descendants had only relatively simple materials problems such as clothing and the construction of shelter, we in our modern civilization have far more numerous problems. We still have the age-old needs for clothing and shelter, but, in our intense modern life, we have become extremely dependent upon a great variety of materials. In fact, materials constitute the basic sinews of our modern civilization.

The rise of materials from a relatively dormant role to one of foremost importance is recent. For several centuries the few metals and alloys (such as iron, copper, brass, tin, and zinc), the few ceramics (such as pottery), and the natural polymers (such as wool, cotton, asbestos, cellulose, etc.) were sufficient to meet most of man's needs. The birth of the new age of materials took place in the late 19th and early 20th centuries. Steel became the major engineering material, aluminum became a commercial metal, Hyatt invented Celluloid, and Baekeland developed Bakelite. This modest start spawned a virtual revolution in materials. Only routine progress had been made for centuries, but now new materials were being developed at a greatly increased rate. Technical ceramics and polymers, almost unknown about 1900, have increased several hundredfold in quantity and variety. Even metals, with a slower rate of development, have expanded greatly with hundreds of new alloys each year.

Not only has the number of materials increased but the combinations have also increased. Clad metals, reinforced polymers, laminates, honey-

1

combs, organometallics, and composites of many types have introduced the systems concept to material usage.

The tremendous increase in available materials, coupled with demands from new applications and more severe service requirements, has brought about many changes in attitudes and viewpoints. In effect, many and diverse factors have forced a renaissance in materials. Coupled with advances in fundamental science, they have led to the development of a new technical area which we might call a unified science and technology of materials.

1-2 CLASSES OF MATERIALS

In earlier, less complicated days, we were content to treat metals, ceramics, and polymers as rather separate entities with relatively little in common. This is not completely realistic today, but we still find it useful to make such a subdivision. Therefore some definitions seem in order.

Metals are chemical elements which form substances that are opaque, lustrous, good conductors of electricity and heat, and, when polished, good reflectors of light. Most metals (and alloys) are strong, ductile, and malleable, and, in general, are heavier than most other substances. We are familiar with steels, aluminum, copper, zinc, and many other metals in the tremendous variety of applications which we use every day.

Ceramics are any inorganic, nonmetallic solids (or supercooled liquids) processed or used at high temperatures. We immediately think of such things as pottery, sanitary whiteware, tiles, table china, etc. We often overlook the oxides, carbides, nitrides, etc., which are of great industrial interest. Urania (UO_2), beryllia (BeO), and alumina (Al_2O_3), for example, are of prime importance in the nuclear reactor field. Ceramics also include materials such as glass, graphite, and Portland cement (concrete).

Polymers (also called *plastics* or *resins*) are materials formed from large numbers of comparatively low molecular weight units (monomers or mers), which are bonded together by primary valence bonds in a repetitive manner to form independent large molecules (macromolecules) in some sort of a "chain." These macromolecules, in turn, are bonded together by additional bonds to form the final material. Natural polymers (starch, cellulose, wool, various proteins, etc.) have always been part of our normal environment and have been used extensively. In the last forty years, synthetic organic and inorganic polymers have become increasingly important. In the United States alone, the "plastics," fiber, rubber, and paint industries account for over 5×10^9 kg of synthetic polymers annually. These materials, although basically made from the same types of molecules, range from very thin liquids (thinners, release agents, coating materials, etc.) to very strong crystalline solids (constructional materials).

1-3   STATES OF MATTER

Traditionally, engineers have been concerned with three states of matter: vapor, liquid, and solid. Today we often hear about a fourth state, i.e., plasma, which, however, is of no specific concern in this context. The other three are of interest, but we will focus almost exclusively on the solid state.

We all have a good working concept of the differences among these three states without having made any great technical distinctions. We are aware that a vapor or a gas requires a completely enclosed container, that a liquid can be open to the atmosphere, and that a solid needs no container. We are also aware that all three can exist simultaneously with one material; e.g., H_2O can exist as vapor, liquid, and solid all at the same time. Obviously, there is a distinction among the states. Part of this distinction lies in the geometrical arrangement of atoms and molecules in each of the states. In the case of ice, the hydrogen and oxygen atoms are arranged in a regular pattern which is repeated "indefinitely" in three dimensions. In water vapor, the hydrogen and oxygen atoms are joined in water molecules, but the molecules are free of one another in a completely random array. Good comprehensive theories have been developed to explain the solid and vapor states.

There is, however, no satisfactory theory in terms of structure which can tell us why a liquid is a liquid. Most approaches have treated a liquid as a solid with many imperfections or as a gas which is crowded together. J. D. Bernal of the University of London has developed a theory which states that molecules in a liquid are coherently packed, but without any regularity. It has been observed that at high pressures there is a perceptible interval, marked by high specific heat, that separates liquid from gas. Bernal interprets this interval as a transition from a loose but coherent arrangement of molecules in the liquid to an incoherent arrangement of clumps of molecules in the gas. Bernal's theory considers the structure of liquids as organized irregularity. We might extend this theory to regard solids as organized regularity and gases as disorganized irregularity.

1-4   BASES OF MATERIALS PROPERTIES

Foremost in the unified approach to materials is the realization that the properties of all materials arise from their structure, i.e., from the manner in which their atoms aggregate into hierarchies of molecular or crystalline order or into disordered amorphous structures. Moreover, the properties of bulk matter of all kinds depend strongly on the nature and distribution of imperfections, either chemical or architectural, in the main array. Most of the properties observed and exploited in materials are cooperative properties of the aggregate rather than of the constituent atoms and molecules. The arrangements of the outer electrons of the atoms are of primary importance, and these are strongly modified by the configuration of neighboring ones.

The principles that may be formulated in the science of materials give the engineer a basis for understanding the nature and behavior of a wide variety of materials. With such a basic background, the engineer should have the potential to anticipate the properties of materials not yet studied, or, for that matter, not yet developed.

As shown in Fig. 1-1, properties of materials are directly related to the structures found within the materials and to the conditions under which the materials are used.

In studying materials, it is perhaps helpful to draw an analogy from preclinical medicine and informally classify the subject into anatomy, physiology, and pathology. In this context, anatomy means the study of atomic structure, electronic configuration, bonding forces, and architectural arrangements of aggregates of atoms. Physiology refers to the macroscopic properties of materials, especially the structure-insensitive properties such as electrical conductivity, magnetism, etc., which can be explained in terms of the anatomy. Pathology refers to the macroscopic properties of materials, especially the structure-sensitive properties such as strength, which can be explained in terms of chemical impurities and/or imperfections of the structure, in effect, properties which can be explained in terms of departure from anatomy. Pathology also refers to the various procedures which exist for altering properties in materials.

If there is a materials science and engineering, it is partly due to materials users' preoccupation with properties rather than with chemical constitution. But it is due even more to the unity in the rules by which atoms and molecules join into groups and into groups of groups. Once the principles of interaction and the possible geometrical aggregations are thoroughly understood, they appear to be almost universally applicable. Such understanding is not only effective in guiding development of new and useful materials but it also makes the scientist or engineer feel at home in an unfamiliar setting. A science based on structure seems most able to unite the microscopic and the macroscopic, theory and practice, intuition and logic, beauty and utility.

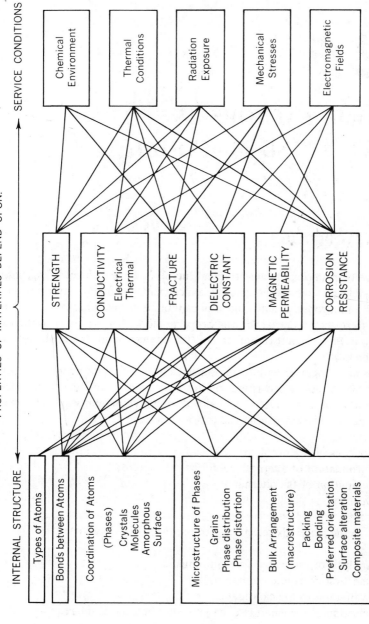

Fig. 1-1. Properties of materials are related to internal structure and to service conditions.

5

2

SOME FACTORS IN
SELECTING MATERIALS*

2-1 INTRODUCTION

Every use of materials, however trivial, involves some element of selection. It is possible through understanding, experiment, or manipulation to maximize any one property, but—and this cannot be overemphasized—it is not possible to select a material for one property alone. It is precisely in the balance of one factor against another that the engineer finds his challenge and his satisfaction.

Like most engineering efforts, materials selection is a problem-solving process. Much has been written on problem solving, and the major steps involved have been expressed and defined in many different ways. However, there is general agreement that the major steps are:

1. Analysis of the problem
2. Formulation of alternate solutions
3. Evaluation of the alternates
4. Decision

Applied to the materials selection process, these steps become:

1. Analysis of the materials requirements
2. Selection of candidate materials
3. Evaluation of the candidates
4. Selection of the candidate material that best meets the requirements

*It is recognized that many concepts and terms are used without definition in this chapter. These will be discussed in the following chapters.

Too many engineers select a material for two incorrect reasons: (1) "It's always been used for this application," and (2) "Its properties." With regard to the first incorrect reason, the time when each application had its preferred material and each material had its secure markets has gone. In other words, tradition and good technology do not necessarily mix.

The second reason is incorrect because it implies that properties are something mysteriously, almost metaphysically, connected with the material, something a material *has* without reference to the specific use of the material. Almost any type of behavior is possible in any material, since behavior is the response of the material to a given set of conditions and circumstances. In this context, strength, for example, is not a property of a material, but rather the response of the material to a given set of conditions.

A test is simply the process of changing the environment of the material and describing its reaction to the change. The results of tests are a significant description of the material. The only true definition of a property includes a description of the test by which the definition is obtained.

In this context, the four-step procedure indicated above might be revised to read: (1) the materials engineer carefully lists the conditions of service and environment the product must endure (analysis of requirements); (2) he lists the responses needed to withstand the conditions and changes of conditions in service; (3) he compares the needed responses with the properties (read responses) of the more than 70,000 engineering materials in today's materials spectrum; (4) he chooses the material which gives the best match or represents the best compromise possible in light of common sense, experience, and judgment.

2-2 GENERAL FACTORS

In selecting materials, there are general characteristics (see Table 2-1) to consider. Obviously, many applications do not involve all of these factors. There may be other applications where circumstances require consideration of additional factors.

Table 2-1. General considerations for selection of materials

Mechanical Strength
Ductility
Stability
Fabricability
Availability
Corrosion Resistance
Properties of Unique Interest
Cost

Behavior of materials is determined by composition, structure, service conditions, and the interactions among them. All materials have limitations within which they perform well but beyond which they cannot be used satisfactorily. Standard testing procedures will measure service behavior only in exceptional cases, for behavior of materials is an integrated interaction of many characteristics, attributes, and service conditions, many of which can often be specified only with a low degree of accuracy.

In selecting materials for a given application, the first step is evaluation of service conditions. The characteristics (responses) of various materials are next evaluated to select the most promising materials. These sometimes need further appraisal through various simulated service tests. Actual service is, of course, the ultimate test. Knowledge of the behavior of materials in other applications should help in exercising sound judgment for a new application, whether similar to or different from a case of known performance.

Unless we know and understand the conditions of fabrication and service, including temperature, specific environment, operating stress level, and other factors, we have little hope of selecting the proper material. If we have a clear knowledge of service conditions and properties (responses) of various materials, past experience indicates we can usually "engineer" around the difficulties and pitfalls to obtain satisfactory service life.

2-3   STRENGTH

One question usually asked in selecting materials is whether strength is adequate to withstand the stresses imposed by service loading. While the primary selection criterion is often strength, it may also be toughness, corrosion resistance, electrical conductivity, magnetic characteristics, thermal conductivity, specific gravity, strength-weight ratio, or other properties. For example, in household usage with relatively low water pressure, weaker and more expensive copper tubing may actually be a better choice than stronger steel pipe. One major difference lies in installation, since steel pipe comes in sections and is joined by threaded connections with elbows at corners, whereas soft copper can be obtained in coils and can be threaded around corners. The lower installation cost of copper could overcome its higher material cost. Also, since copper has adequate strength, the greater strength of steel is not necessary. Furthermore, in the event of freezing, copper tends to yield instead of burst. In general, the usual criterion in selection is not just one property, e.g., strength, but some combination of chemical, physical, and mechanical properties plus economic factors.

2-4 DUCTILITY

The question of ductility is related to the question of strength. Considerable ductility is generally obtained at a sacrifice of strength. During cold working, for example, there is gain in strength and loss in ductility. It may seem that some ductility is always required, and the more ductility obtainable without great loss in strength, the better. This is often true, but, at the same time, many metals and alloys have little ductility and may not need much. Railroad rails, despite their severe service, actually have little ductility. In fact, the amount of ductility actually usable under service conditions is a controversial subject. There is, however, considerable agreement on 1 to 2% elongation. Studies of brittle behavior in large sections of steel indicate that approximately 1 to 3% deformation is the useful limit. Other work indicates a permanent deformation of about $1\frac{1}{2}\%$ in an aircraft wing substantially destroys its aerodynamic efficiency. In some cases, brittleness may be an asset, for example, the use of readily replaceable frangible members that are intended to fail first and protect the rest of the system. At the same time appreciable ductility (or plasticity) is required for fabrication by rolling, drawing, extrusion, and other mechanical working processes.

2-5 DESIGN

Design is closely related to strength and ductility in selection of materials. It is also quite widely recognized that a large portion of service failures are due to fatigue. J. O. Almen,* a prominent automotive engineer, has said:

> "Fully 90 per cent of all fatigue failures occurring in service or during laboratory and road tests are traceable to design and production defects, and only the remaining 10 per cent are primarily the responsibility of the metallurgist as defects in material, material specification, or heat treatment.
> Study of fatigue of materials is the joint duty of metallurgical, engineering, and production departments. There is no definite line between mechanical and metallurgical factors that contribute to fatigue. This overlapping of responsibility is not sufficiently understood.
> Hence, the engineers are constantly demanding new metallurgical miracles instead of correcting their own faults. Until metallurgists are less willing to look for metallurgical causes of fatigue and insist that equally competent examination for mechanical causes be made, we cannot hope to make full use of our engineering material."

*J. O. Almen, "Probe Failures by Fatigue to Unmask Mechanical Causes," *SAE Journal*, Vol. 51, May 1943. (Makes a point which is still as highly valid as in 1943.)

This quotation deals specifically with fatigue, but the philosophy is certainly applicable on a far broader basis. This does not excuse the materials engineer when he makes a mistake, nor does it allow him to slacken his efforts to find a better material for a given application. It does insist, however, that all aspects be considered, since even the seemingly most insignificant factors often have far-reaching effects. There is at least one case in which a too-heavy hammer blow on an inspection stamp led directly to the failure of an aircraft in flight. There are many cases in which the search for a substitute material led to feasible design modifications which were much more advantageous than a change in alloy composition.

2-6 STABILITY

Stability of material in service is intimately related to temperature, fluctuations in temperature, and length of time at temperature. Exposure to radiation may also be an important condition in some applications. Temperature not only directly affects strength and creep, but it can also produce changes in the microstructure of the material, e.g., tempering of martensitic steels and overaging in precipitation-hardening alloys. Obviously, time is important in determining the extent to which these phenomena occur and, consequently, in the stringency of stability requirements. A rocket motor, for instance, may be required to operate only briefly, whereas a steam turbine is expected to operate for many years. In many components it is desirable to have characteristics which preclude shutting-down for repairs. In other components, especially those subject to mechanical wear, replacement at regular intervals is anticipated, and the part is made to be readily detachable. In nuclear reactors, for example, the problem of stability is far less drastic in a reactor that is to be operated for some months to test design feasibility than it is in a central power station reactor that is to be operated for several years.

Another aspect of stability is the question of seriousness of failure. For instance, a leak in a tea kettle may have only nuisance value but a leak in a vessel containing an inflammable or radioactive fluid is an entirely different matter. We should note that any design for long-time operation may be an extrapolated or educated guess, since the best available data are often for times much shorter than anticipated in long-term operation.

We must note that an "educated guess" can be a treacherous proposition. It has been pointed out by C. F. Kettering* that engineering judgment may be based on too-limited experience and that the simple way to widen experience is to make simulated service tests. He cites the example of running a variety of pistons, of both orthodox and unorthodox design, in a single-

---

*C. F. Kettering, "Get Off Route 25, Young Man," *Collier's*, Dec. 3, 1949, p. 15.

cylinder engine "to find out which the engine liked best." Its choice was not the anticipated one as "the engine was smarter than the engineers."

2-7 AVAILABILITY

Regardless of the merits of a material, if it is unavailable it is not reasonable to base a design on it. This question involves availability of material at an appropriate cost and availability in the desired form. A material obtainable only in castings obviously cannot be used in applications requiring tubing, wire cloth, or various other fabricated shapes.

2-8 FABRICABILITY

Fabricability is closely related to availability. A material may not be commercially available in the desired state of fabrication, but it may be possible, with relatively small-scale development-type operations, to produce it in the desired form. This, of course, entails considerable expense, but circumstances may justify the extra cost. The development of production and fabrication procedures of beryllium and zirconium for nuclear reactor use provides two specific examples. In general, however, we prefer to use materials which can be fabricated by standard methods of forming, joining, etc., without special precautions. If fabricating operations can also be successfully performed under field conditions, this is often desirable.

Directly related to fabricability is the number of pieces required. If several thousand duplicate parts are necessary, the high cost of dies or other specialized equipment for quantity production may be economically justified. If few pieces are required, hand production from relatively expensive materials in stock may be less expensive overall than more elaborate methods using less expensive materials.

2-9 CORROSION RESISTANCE

A material may or may not be regarded as corrosion resistant depending on the particular service requirement. The criteria for corrosion resistance can be considered in three degrees: (1) avoiding contamination (e.g., food products); (2) preventing leaks of closed containers or conduits; (3) maintaining strength and other properties during corrosive attack. In certain situations, it may be sound engineering to use an inferior material that requires periodic replacement, since this may be more economical than use of an initially more expensive material. In other situations, however, this approach may not be acceptable because of high potential risk to life or for other reasons.

There are other factors which should not be overlooked when dealing with corrosive media, since neglect of these may lead to erroneous interpretations of corrosion tests and handbook data (responses): (1) a sample in a simple static immersion test at a given temperature may corrode in a manner and at a rate significantly different from the way it would if the material were simultaneously transferring heat to the solution; (2) the test solution may change its corrosiveness as a result of contamination; (3) lack of correlation between laboratory tests and operating conditions is often due to a different ratio of surface area to solution volume; (4) pressure of gas or vapor above a corrosive liquid may appreciably affect corrosion by its influence on the amounts of oxygen or other gases which may be forced into, or withdrawn from, the solution; (5) alloys that owe their corrosion resistance to development of passive films are particularly susceptible to the development of concentration cells.

The possibility of corrosion should always be considered in any design. We may obtain acceptable resistance to corrosive attack under a particular set of conditions only to discover that some change in conditions gives a new or modified problem.

2-10 PROPERTIES OF UNIQUE INTEREST

The foregoing discussion is in very broad terms since it comments on factors which should be considered in almost all problems involving selection of materials for a given application. In a number of cases, however, perhaps one or two properties are of overriding importance. These properties include: (1) density (e.g., one desires low density—light weight—simultaneously with strength for aircraft use); (2) melting point (e.g., a "low" temperature for fusible heads in fire sprinklers); (3) volume change on melting or freezing (e.g., expansion on freezing for type metal); (4) ablative behavior (e.g., reentry vehicles for space missions); (5) antibacterial characteristics; (6) thermal expansion; (7) thermal conductivity; (8) electrical conductivity (including superconductivity); and so on to include a wide variety of properties.

If high thermal conductivity is required, a number of metals and alloys are of particular interest. If thermal insulation is necessary, metals are not appropriate but a number of polymers and ceramics, especially the latter, should be considered. If electrical insulation is required, many of the polymers are very useful since they combine mechanical flexibility with lack of electrical conductivity. Sometimes there are materials with unusual combinations of properties such as beryllia (BeO) which has about the same thermal conductivity as aluminum but is an electrical insulator. The form of the material may have a very significant effect on its characteristics. Magnesia (MgO) used for thermal insulation in a more-or-less powdered form has a conductivity of about 0.3–0.4 Btu/hr-ft°F. In dense form for structural use, it has a thermal conductivity of about 27 Btu/hr-ft°F.

Developments in polymer chemistry in the design of giant molecules have led to creation of systems with unique properties. For example, fluorocarbon polymers have been developed with versatile and useful characteristics. Fluorine, extremely active as a chemical element, is equally inert in the combined state. The fluoro-carbon polymers, e.g., du Pont's Teflon, will not conduct electricity. They are immune to attack by fungi, molds or pests, chemical reagents including strong acids or solvents, and are stable up to about 600°F. They have widespread use in the electrical and electronic industry, as bearings, in processing industries, and for such household items as steam irons, frying pans, pots, etc., where nonstick qualities are desired.

2-11 RESPONSIBILITY FOR SELECTION

Presumably, the designer is basically responsible for selecting materials. He should not, however, rely exclusively on handbook data but should consult with materials specialists. These, in turn, must know the complete story of intended use, sources of supply, fabricating facilities, and costs. Thus, it often happens that wise selection of materials involves the cooperation not only of design and materials engineers but also production engineers and purchasing agents.

Materials producers often have valuable information which has not yet become widely available. A designer should tactfully tap this source when he has a need. This direction can be very profitable if the producer is cooperative and there is a minimum communication gap in contrast with a situation in which a user needed material for a rocket nozzle throat insert, and was considering a nylon fiber-filled phenolic. In a material of this sort, type of resin, resin content, type of filler, and amount of filler have effects on the properties. While the producer had made tests to determine the type and amount of resin and the effects of various fillers on the properties, he would not release this information to the potential user. He would only say that the constituents of the composite material were standard materials. The net result was that the user tried an improperly constituted material, there was an unsuccessful firing, little was learned about the reason for the failure, and a promising candidate material was lost.

2-12   COST

In every case, final selection of a material for a specific application depends on a compromise. In some applications, there are specialized requirements which restrict us to a choice from among relatively few materials. Even then, there is a compromise among the contending factors previously discussed. In nearly all instances, the compromise and final selection involves economic considerations. The initial cost of a piece of equipment involves raw material, fabrication, and installation costs.

Once operation has started, however, there may be additional costs in the form of replacements due to failure, shutdown expenses while undergoing repair or replacement, and the economic damage of production losses.

The tests for value (Table 2-2) developed by the U.S. Navy are directly pertinent to economic considerations. If the answer to any one of the items in Table 2-2 (and perhaps other similar questions) is yes, the selection task is incomplete.

Table 2-2.   Tests for value

Every material, every part, every operation should pass these tests

1. Can we do without it?
2. Does it do more than is required?
3. Does it cost more than it's worth?
4. Is there something better to do the job?
5. Can it be made by a less costly method?
6. Can a standard item be used?
7. Considering the quantities used, could a less costly tooling method be used?
8. Does it cost more than the total of reasonable labor, overhead, material, and profit?
9. Can someone else provide it at less cost without affecting dependability?
10. If it were your money, would you refuse to buy the item because it cost too much?

No industry is immune to savings through more effective application of materials. At least three major approaches may be taken to reduce cost through better use: (1) reconsider the material selected; (2) reconsider the form of the material; and (3) redesign to take full advantage of properties. While these apply primarily to production line parts rather than tailor-made parts, the philosophy is pertinent in general.

In many situations, definite savings can be realized by the simple expedient of changing from one material to another without substantial change in form or processing procedure. There are other applications for which two or more materials can be considered as alternates with the choice at any given time dictated by current market prices. This, in turn, is dependent on variations in supply and demand. Another aspect of alternative materials relates to the adage of not putting all the eggs in one basket.

Great savings can often be realized by changing fabrication procedure or the form in which material is used. For example, some relatively complex shapes which can be made directly by expensive machining or forming methods can also be made by joining simple components readily fabricated from sheet, plate, castings, or wrought forms. On the other hand, assembly costs can often be minimized or eliminated by using a casting or an extrusion. One prominent example is the automobile engine crankshaft. For many years crankshafts were machined from forged steel which had to meet rather stringent specifications of strength and toughness. It was eventually realized that a bent crankshaft is just as useless as a broken one and high toughness was not needed since impact was not severe. Today there are millions of cast crankshafts, incapable of appreciable bending but with the proper rigidity and wear resistance, in successful operation. The cost of a cast and machined crankshaft is significantly less than of a forged and machined one.

With changes from one fabrication method to another, there are often concurrent design changes. There are opportunities for cost reduction in some cases, however, even while retaining the same basic design. Designers have often used appreciably heavier sections than necessary, primarily due to lack of knowledge. As more information about materials becomes available and nondestructive testing techniques are perfected, it is possible to approach maximum efficiency in utilizing materials in a given design. Miniaturization is another possibility. Although used principally in electrical and electronic components, the same idea might be applied to other equipment to obtain smaller, more compact, and more powerful apparatus with appreciable reduction in overall cost.

2-13 SUMMARY OF PRINCIPLES

The foregoing discussion is obviously too brief and too limited to provide adequate selection for any and every application. Proper selection can be made only on the bases of service requirements, knowledge of properties of alternative materials, experience, and engineering judgment. It is hoped, however, that the discussion provides: (1) an indication of the philosophy of materials selection; (2) some idea of the complex interrelationships among materials characteristics; and (3) a perspective for specific problems.

2-14 SANDWICHES FOR 10 CENTS, 25 CENTS, AND 50 CENTS

As a specific application of the principles outlined above, let us consider a problem to which we all know the answer, i.e., the coinage currently being produced and used in the United States. There were many technical requirements to be met but there were also a number of other considerations which were equally important in influencing the final decision. The interplay of considerations and the compromises made in reaching a solution are typical.

Alexander Hamilton recommended the original 90% Ag–10% Cu coinage alloy to Congress in 1792. The new coinage materials, authorized by Congress and signed into law by President Johnson in 1965, represented a drastic departure from the original alloy. Behind the change lay a complex history of coin shortages, increasing industrial demand for silver, and declining sources of silver. When the price of silver rose to $1.29 per troy ounce in 1963, the U.S. Treasury became a seller instead of a purchaser of silver to prevent a further rise in price. At $1.38 per troy ounce, the value of the silver in the old coinage would equal the face value of the coin. Clearly, no coinage material should be in danger of approaching this "economic melt point."

The Treasury engaged Battelle Memorial Institute to make a study of possible coinage materials and recommend suitable alternatives to the Ag-Cu alloy. In an effort to obtain a balanced perspective of possible approaches, a broad spectrum of metals, alloys, and nonmetallic materials was considered and evaluated. Certainly many characteristics were desired but five were considered outstanding.

For coinage, a material should be in good supply at a reasonable price; it should be acceptable to the American people who have certain traditions and prejudices regarding their coinage; the material should be one easily processed by the mint; the coins made from it should have a minimum disruptive effect on coin-operated devices; and, lastly, it must be unique enough to discourage counterfeiting.

The availability and price criterion had to be used for coarse, preliminary screening. Any choice had to be available at a reasonable price in sufficient quantities for years to come. The importance of this is sharply focused by considering that it was the supply and price of silver that created the problem.

The bases for public acceptability are highly subjective. Intrinsic value is not itself a critical element in our coinage. This is obvious from the fact that copper "pennies," cupronickel "nickels," and paper currency all circulate freely. A summation of all the subjective aspects indicates the following: "If the public is confident that a 25-cent coin will buy 25-cents' worth of goods, then the actual material from which the coin is made is not important."

This obviously is not a technical question but it is certainly vital to the problem.

A more technical question is that of processing by the mint. Making a finished coin involves melting, casting into ingots, rolling into strip, blanking, and coining. While the mint prefers to perform all the operations itself, all but the last step could be performed by subcontractors. The final coining step, however, had to be performed by the mint with existing equipment which was geared to relatively malleable materials with relatively low melting points.

Coin-operated machines pose a different set of problems. Those that perform services, e.g., coin laundries, normally have limited protection against slugs, usually a sizing slot and a magnet. Vending machines, on the other hand, usually have an eddy-current selector which sizes the coin, weighs it, makes sure it is nonmagnetic, and accepts or rejects it on the basis of eddy-current response.

Counterfeiting is easily discouraged with a silver coinage alloy because of the high price of raw materials. Cheap imitations of silver coinage are readily detected. Coping with slugs in coin-operated devices is basically the task of the manufacturer. In any event, it is clear that new coinage material must not make any counterfeiting method easier.

Economic studies indicated that, in addition to silver, there were eleven metals (Al, Cr, Cu, Mn, Mo, Nb, Ni, Ti, W, Zn, and Zr) which would satisfy the supply-price criterion up to the year 2000. Certain ceramics and polymers might have been added to the list but represented too radical a departure at the time.

Translating the coinage criteria into technical terms, the coin had to be white in color (public acceptance aspect), have a density of 8–12 g/cc and an electrical conductivity of about 2.1 ohm-cm. It would be nonmagnetic, easy to manufacture in the mint but not elsewhere, and it would resist wear and corrosion.

Only the Ag-Cu alloys satisfy the combination of white color, conductivity, and density requirements. A coin having an acceptable white color with sufficient corrosion resistance requires an alloy containing at least 72% Ag. Thus a perfect fit is not possible, especially since it would be practically impossible to duplicate all the characteristic features of the Ag-Cu coinage while maintaining the various specified conditions.

Many possible compromises were considered before the final decision was made. A sandwich with cupronickel (75% Cu–25% Ni) on the outer layers and copper in the core gives a coin with the required conductivity and color. It was not a perfect solution since the edge shows the copper core when the coin is blanked from the three-layer sheet. At the same time, the copper edge may be sufficiently distinctive to discourage counterfeiting. This solution was accepted for the dime and the quarter. For "prestige" purposes, the Kennedy half dollar contains silver. It is also a sandwich having Ag-Cu (80-20) outer layers with a copper-rich Ag-Cu core. This eliminates the copper edge found in the dime and quarter and reduces the silver content from 90% to about 40%. This leads to a minimum consumption of silver while retaining some silver in coinage.

The composite coin has the virtue of satisfying the criteria of cost and supply, mint processing, and public acceptance to a high degree. Moreover, this solution—really an across-the-board compromise—has the potentiality of meeting coinage needs for years to come.

QUESTIONS AND PROBLEMS

1. What are the basic steps in selecting materials for a given application?

2. Table 2-1 lists the item, "properties of unique interest." How many such properties can you list that may enter into the selection of materials?

3. Can you think of any other items which should be listed in Table 2-1?

4. What is the relationship between strength and ductility, and what effect does this have on the selection of materials?

5. What role does design play in selecting materials (and vice versa)?

6. What is meant by stability? Of what importance is it in selecting materials? What factors have an effect on stability? How do these factors operate with regard to stability?

7. What is meant by compatibility? What role does it play in selection of materials? What factors affect compatibility? How?

8. The final selection of materials for any given application is a compromise among several factors. What factors generally have major influence?

9. Discuss the applicability of the items in Table 2-2 to selection of materials. Can you think of any other items which should be listed in this table?

10. Examine an automobile. List all the ceramic materials (not only parts) you can see. Do the same for polymeric materials.

11. Take apart any one (or several) of a number of discarded products such as a ball point pen, alarm clock, lawn mower, flashlight, transistor radio, incandescent light bulb, or other item in common use. List the types of materials used and indicate why each was selected. Handle each material, trying to break it, etc., to get some "feeling" for its characteristics.

12. Twelve metals were initially considered as possibilities for the new coinage. What reasons can you give why most of them (or alloys based on them) would not be acceptable in the context of the five major criteria?

3

ATOMIC STRUCTURE
AND BONDING

3-1 INTRODUCTION

Before we can adequately determine the best material for a given application or how a material may be modified to make it suitable, we must understand the fundamental nature and structure of materials. In our search for understanding, we either start with, or ultimately come back to, the primary assumptions and axioms upon which science is based. All experimental evidence leads to the conclusion that everything is composed of energy. Matter and materials are, therefore, a form of energy. We postulate that matter (M) and energy (E) are interchangeable, the conversion factor being the square of the velocity of light (c); that is

$$E = Mc^2 \tag{3-1}$$

3-2 STRUCTURE OF THE ATOM

The building blocks of matter with which we are concerned are electrons, protons, and neutrons (see Table 3-1). The dual nature of matter sometimes causes considerable difficulty. In some instances, it is desirable to think of matter as particles, and in other cases it is more helpful to consider it as quanta of energy which obey the laws of wave mechanics rather than the laws of particle mechanics. For the materials engineer, it is usually sufficient to use the particle concept, and this will, in general, be our approach.

The electron is the unit quantity of electrical energy. It can be considered as a minute particle or as an energy wave of negative charge. The electrical energy which lights an electric light or runs a motor is a flow of electrons

Table 3-1.   Significant properties of basic parts of matter

	Charge (esu)	Charge (coulomb)	Rest mass (gram)	"Effective" radius (cm)	Rest mass equivalent energy (Mev)
Electron	-4.8×10^{-10}	-1.69×10^{-19}	9.1×10^{-28}	4.6×10^{-13}	0.51
Proton	4.8×10^{-10}	1.69×10^{-19}	1.67×10^{-24}	1.4×10^{-13}	938
Neutron	0	0	1.67×10^{-24}	1.4×10^{-13}	938

through a conductor. The charge of the electron is 4.8×10^{-10} esu. If we think of a free electron as a particle, then its rest mass is 9.1×10^{-28} gm, with an effective radius of 4.6×10^{-13} cm. Considered as a packet of energy, the electron is worth 0.51 Mev.

The proton is a heavy particle of matter, 1836 times as heavy as the electron. It carries a positive unit charge; that is, the charge on a proton is equal in quantity but opposite in sign to the charge on an electron. If a free proton is pictured as a particle, it has a rest mass of 1.67×10^{-24} gm with an effective radius of 1.4×10^{-13} cm. A free proton can also be considered as a packet of energy of 938 Mev.

The neutron is also a heavy particle of matter, 1838 times as heavy as the electron, i.e., just slightly heavier than the proton. This particle is electrically neutral and sometimes can be treated as though composed of one proton and one electron. Since it has no charge, it can readily penetrate atoms when used as a projectile for transmutation experiments, whereas the electron and the proton, being electrically charged, do not penetrate as readily due to electrostatic repulsion by the target nuclei. In the normal atom, the neutron has no electrical or chemical effect, but contributes only mass. This results in subtly differing atoms called *isotopes*. For all practical purposes, the neutron can be considered either as a particle having the same mass and radius as a proton or as an energy packet with the same amount of energy as a proton.

In materials engineering, we often think of the atom as having a very small size and a shape something like a ball, and it is frequently treated as a tiny, hard sphere. Although this concept is acceptable for many purposes, such as the study of crystal structure, it is a very superficial view. A more realistic concept is that the atom is an aggregate and that the electrons, protons, and neutrons within it are arranged in a configuration which resembles a miniature solar system. Thus, the atom is, in fact, mostly space. For example, in hydrogen the most probable distance of the electron from the nucleus is 0.53×10^{-8} cm, whereas the diameter of the nucleus is only about 1.4×10^{-13} cm.

The small, positively charged nucleus at the center of the atom is composed of all the protons and neutrons in the atom. Since these constitute about 99.98 % of the weight of the atom, the nucleus is an extremely dense structure. Obviously, some stabilizing or interattractive force must be present, since, according to electrostatic theory, the protons, having like charges, should repel each other. The nature of this interattractive force is not clearly understood, and for our purposes it is not important, although much research is being done on this problem. It might be noted that it is within the nucleus that the neutron contributes to the existence of isotopes.

The electrons (considered as particles) are attracted to the nucleus, but repel each other and are, in addition, kept at some distance from the nucleus by a kind of centrifugal force. They may be thought of as revolving around the nucleus in a series of orbits or shells which are analogous to the planetary orbits of our solar system. This visualization of the atom as an extremely small solar system is appropriate only if we think of the electrons as particles. It is somewhat more realistic, however, to consider the electron as being a "smeared-out glob" to be dealt with in terms of probability. The treatment is that of quantum mechanics or wave mechanics, which applies to the motion of all particles, regardless of size. Classical mechanics, a simplified approximation to quantum mechanics, can only be successfully applied to large-scale events.

Apart from the mathematics, the main difficulty in understanding wave mechanics (or its implications) is abandoning some of the traditional notions of classical mechanics, in which it is assumed that we know or can measure *both* position and momentum of a given particle. Actually, this assumption is false, for rigorous analysis shows that, in principle, it is impossible to determine position and momentum simultaneously. Either quantity can be measured as accurately as desired, but the experimental refinements necessary to do so are precisely the kind which make measurement of the other quantity less accurate. This phenomenon is known as *Heisenberg's uncertainty principle*. This states, in effect, that

$$\Delta p \, \Delta x \geq \frac{h}{2\pi}$$

where Δp is the uncertainty in momentum, Δx is the uncertainty in position and h is Planck's constant. The principle becomes trivial when applied to macroscopic bodies, but it is extremely important when applied to nuclear particles such as electrons.

On the basis of wave mechanics, we can only state the probability that an electron will be found in a given region of an atom. This probability can be represented by a wave function, the square of whose amplitude at a given point is a measure of the probability of finding an electron at that point. For our discussion, the various basic parts of matter, especially electrons, will be treated principally as minute discrete particles located at the most probable positions. On this basis, the electrons in an atom constitute electron clouds in a series of shells which form the outer regions of the atom. Electrons in other atoms are so completely and elastically repelled by this electron cloud that the atom behaves much like a tiny sphere in its physicochemical reactions.

The electron in an atom can occupy different energy states. Each electronic state gives rise to a specific electron cloud pattern and is characterized by a definite energy. The question then arises as to the division of the electrons among the main energy groups (shells). The state of an electron in an atom is described by four quantum numbers: n, l, m_l, and m_s. In a general way, the quantum number n is a measure of the energy of the electron, and l is a measure of its angular momentum. The quantum number l may have any integral value from 0 to $(n - 1)$, but a state where $l = 0$ is not to be regarded as one in which the electron is at rest; rather, it is the state in which the motion of the electron does not give rise to an angular momentum.

The quantum number m_l is a measure of the component of the angular momentum in a particular direction. This quantum number may have any integral value from $+l$ to $-l$, including 0. The fourth quantum number m_s is introduced because an electron behaves in some ways as if it were spinning about its own axis as well as in orbit about the nucleus. Magnetic properties are associated with this spin, and it has been found experimentally that two opposite spin directions are possible. These spin directions are specified by the fourth quantum number, which has values of $+\frac{1}{2}$ or $-\frac{1}{2}$.

By specifying the four quantum numbers associated with an electron we give all the information required to define its state in the atom. In any given atom, each electronic state has a definite, specific energy. The energy of the atom as a whole depends on the way in which its electrons are arranged in the different possible electronic states. Each arrangement gives rise to a state of the atom. The most stable state is that for which the energy is a minimum, and states of higher energy are called excited states.

We might think at first that the ground state of an atom is the one in which all electrons occupy the lowest quantum state, but this is not so. According to the *Pauli exclusion principle*, no two electrons in an atom can be in exactly the same state as defined by all four quantum numbers. The fact that l varies from 0 to $(n - 1)$, m_l varies from $+l$ to $-l$ including 0, and $m_s = \pm\frac{1}{2}$ implies that the maximum possible number of electrons with a principal quantum number n is $2n^2$.

These quantum numbers have physical significance. The principal quantum number n indicates the general energy state or a major energy division. The second number, l, indicates a subdivision and is a measure of angular momentum. The third number, m_l, indicates a further division and is a measure of the component of angular momentum in a particular direction. The fourth number, m_s, indicates the spin.

There is a distinction between energy state and energy level. An energy *level* is the energy of an electron actually occupying a given energy or quantum state. An energy *state* is a possible energy configuration and can exist even if there is no electron occupying it. The Pauli principle applies to energy states, not to energy levels.

The order of occupation of states by electrons depends on both n and l. It may be further complicated by screening, i.e., partial shielding of the nucleus from the outermost electrons by the electrical charge density of electrons nearer the nucleus. Table 3-2 has been constructed from the rules stated above concerning the values l, m_l, and m_s may take for a given value of n. The reader can readily extend this treatment to shells with higher values of

Table 3-2. The number of quantum states in the first three shells

Shell	n	l	m_l	m_s	Number of states	
First	1	0	0	$+\frac{1}{2}$ $-\frac{1}{2}$	2	
Second	2	0	0	$+\frac{1}{2}$ $-\frac{1}{2}$	2	8
		1	+1	$+\frac{1}{2}$ $-\frac{1}{2}$	6	
			0	$+\frac{1}{2}$ $-\frac{1}{2}$		
			−1	$+\frac{1}{2}$ $-\frac{1}{2}$		
Third	3	0	0	$+\frac{1}{2}$ $-\frac{1}{2}$	2	18
		1	+1	$+\frac{1}{2}$ $-\frac{1}{2}$	6	
			0	$+\frac{1}{2}$ $-\frac{1}{2}$		
			−1	$+\frac{1}{2}$ $-\frac{1}{2}$		
		2	+2	$+\frac{1}{2}$ $-\frac{1}{2}$	10	
			+1	$+\frac{1}{2}$ $-\frac{1}{2}$		
			0	$+\frac{1}{2}$ $-\frac{1}{2}$		
			−1	$+\frac{1}{2}$ $-\frac{1}{2}$		
			−2	$+\frac{1}{2}$ $-\frac{1}{2}$		

n and, for example, verify the fact that in the fourth shell the number of possible quantum states is 32.

When dealing with atomic structure, we find it convenient to use a conventional notation based on the classification of the electronic states of the atom. The one generally adopted is as follows: The value of the principal quantum number n is first stated, thus indicating the shell to which reference is made. This followed by a symbol which serves to specify the value of l. States for which $l = 0, 1, 2, 3,$ and 4 are called $s, p, d, f,$ and g states, respectively. Thus, in this notation, the states which have $n = 2, l = 1,$ are $2p$ states. The number of electrons occupying these states is indicated by means of an exponent. For example, the symbol $2p^4$ indicates that there are four electrons in those states where $n = 2$ and $l = 1$.

In the hydrogen atom, which contains one electron, the energy of the electron in a given state is determined solely by the value of n. This is not true of other atoms, however, for the electrostatic attraction of any one electron to the nucleus is modified by the presence of other electrons, and thus the energy of an electron in a particular quantum state depends upon the value of l as well as that of n. The general effects may be summarized as in the following list:

1. The greater the value of n for a particular shell, the greater is the energy of all the states in that shell. In this respect the behavior is similar to that of the one-electron atom.
2. The greater the value of l for a given value of n, the greater is the energy.
3. All those quantum states in a free atom having the same values of n and l are associated with the same energy.

From the Pauli principle, we see that only one electron can occupy a given quantum state, as characterized by the four quantum numbers, i.e., no two electrons can have precisely the same energy. Accordingly, in an atom of Z electrons (in the normal state), the Z quantum states of lowest energy are each occupied by one electron. When describing the arrangement of the electrons in the quantum states it is convenient to imagine a nucleus of increasing atomic number and then consider which states are successively occupied as electrons are added to balance the increasing nuclear charge. The sequence of occupation of states is given by the order of the energy levels associated with them. This rapidly becomes complicated, however, since the energy of a state depends on both n and l, and is not a simple function of either. As a result, when successive electrons are introduced into an atom, the order in which occupation takes place is as follows: $1s, 2s, 2p, 3s, 3p, (4s, 3d),$ $4p, (5s, 4d), 5p, (6s, 5d, 4f), 6p, (7s, 5f, 6d)$. The quantum states indicated in parentheses have almost the same energy when they are the outermost states of the atom, and thus irregularities occur in their occupation by electrons, as indicated in Fig. 3-1 and Table 3-3.

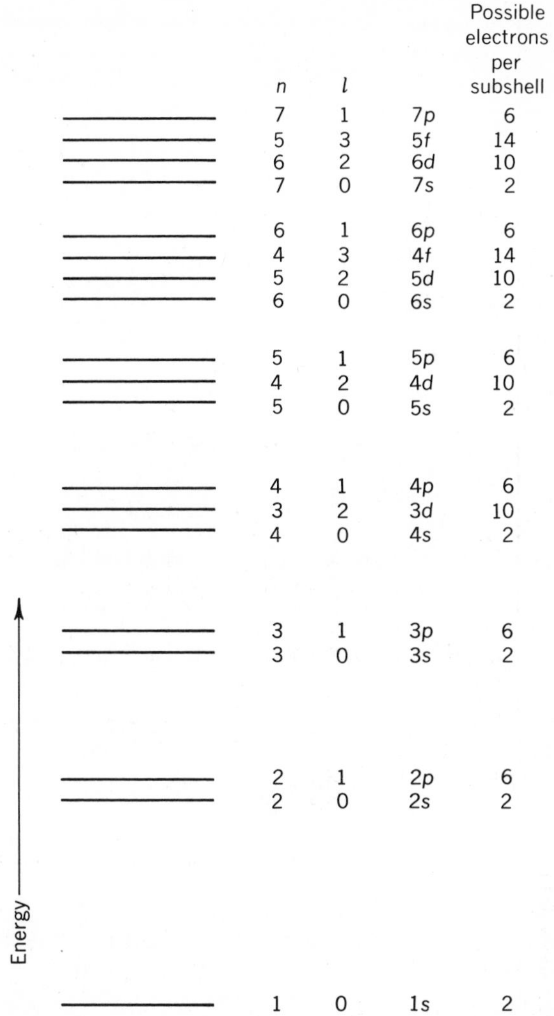

n	l	subshell	Possible electrons per subshell
7	1	7p	6
5	3	5f	14
6	2	6d	10
7	0	7s	2
6	1	6p	6
4	3	4f	14
5	2	5d	10
6	0	6s	2
5	1	5p	6
4	2	4d	10
5	0	5s	2
4	1	4p	6
3	2	3d	10
4	0	4s	2
3	1	3p	6
3	0	3s	2
2	1	2p	6
2	0	2s	2
1	0	1s	2

Energy →

Fig. 3-1. Schematic energy-level diagram for the ground state of the elements. Each level represents a possible energy state for an orbital electron. Divisions within a level (i.e., within a subshell) due to quantum numbers m_l and m_s are not shown.

The wave mechanics concept of the atom views the electron as a poorly defined cloud of electricity. Thus, the particle model with its orbits is modified so that the electron clouds are associated with orbitals, an orbital defining the most-probable volume in space where the electron may be. The orbital may be occupied or it may be empty. The size and shape of the orbital depend on which specific set of quantum numbers is under consideration.

Orbitals are normally represented by boundary surfaces containing most

26

Table 3-3. Periodic table of the elements with the outer electron configurations of neutral atoms in their ground states, atomic number, and atomic weight

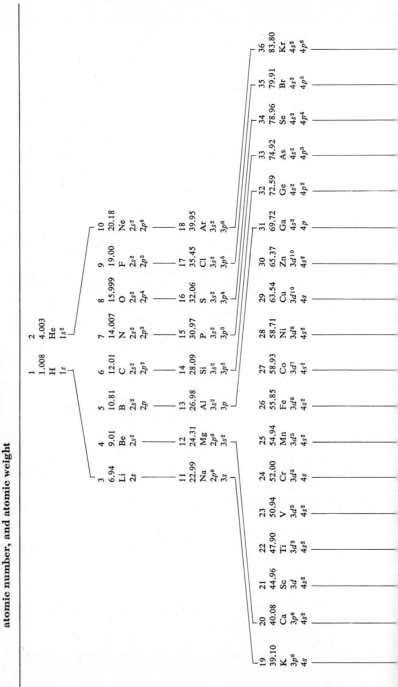

Z	Element	Atomic weight	Configuration
37	Rb	85.47	$4p^6\,5s$
38	Sr	87.62	$4p^6\,5s^2$
39	Y	88.91	$4d\,5s^2$
40	Zr	91.22	$4d^2\,5s^2$
41	Nb	92.91	$4d^4\,5s$
42 *	Mo	95.94	$4d^5\,5s$
43	Tc	99	$4d^6\,5s$
44	Ru	101.07	$4d^7\,5s$
45	Rh	102.91	$4d^8\,5s$
46	Pd	106.4	$4d^{10}$
47	Ag	107.87	$4d^{10}\,5s$
48	Cd	112.40	$4d^{10}\,5s^2$
49	In	114.82	$5s^2\,5p$
50	Sn	118.69	$5s^2\,5p^2$
51	Sb	121.75	$5s^2\,5p^3$
52	Te	127.60	$5s^2\,5p^4$
53	I	126.90	$5s^2\,5p^5$
54	Xe	131.30	$5s^2\,5p^6$
55	Cs	132.91	$5p^6\,6s$
56	Ba	137.34	$5p^6\,6s^2$
57	La	138.91	$5p^6\,5d\,6s^2$
58	Ce	140.12	$4f^2\,6s^2$
59	Pr	140.91	$4f^3\,6s^2$
60	Nd	144.24	$4f^4\,6s^2$
61	Pm	145	$4f^5\,6s^2$
62	Sm	150.35	$4f^6\,6s^2$
63	Eu	151.96	$4f^7\,6s^2$
64	Gd	157.25	$4f^7\,5d\,6s^2$
65	Tb	158.92	$4f^8\,5d\,6s^2$
66	Dy	162.50	$4f^{10}\,6s^2$
67	Ho	164.93	$4f^{11}\,6s^2$
68	Er	167.26	$4f^{12}\,6s^2$
69	Tm	168.93	$4f^{13}\,6s^2$
70	Yb	173.04	$4f^{14}\,6s^2$
71	Lu	174.97	$4f^{14}\,5d\,6s^2$
72	Hf	178.49	$5d^2\,6s^2$
73	Ta	180.95	$5d^3\,6s^2$
74	W	183.85	$5d^4\,6s^2$
75	Re	186.2	$5d^5\,6s^2$
76	Os	190.2	$5d^6\,6s^2$
77	Ir	192.2	$5d^9$
78	Pt	195.1	$5d^9\,6s$
79	Au	196.97	$5d^{10}\,6s$
80	Hg	200.59	$5d^{10}\,6s^2$
81	Tl	204.37	$6s^2\,6p$
82	Pb	207.19	$6s^2\,6p^2$
83	Bi	208.98	$6s^2\,6p^3$
84	Po	210	$6s^2\,6p^4$
85	At	210	$6s^2\,6p^5$
86	Rn	222	$6s^2\,6p^6$
87	Fr	223	$6p^6\,7s$
88	Ra	226	$6p^6\,7s^2$
89	Ac	227	$6d\,7s^2$
90	Th	232.04	$6d^2\,7s^2$
91	Pa	231	$5f^2\,6d\,7s^2$
92	U	238.03	$5f^3\,6d\,7s^2$
93	Np	237	$5f^5\,7s^2$
94	Pu	242	$5f^6\,7s^2$
95	Am	-243	$5f^7\,7s^2$
96	Cm	247	$5f^7\,6d\,7s^2$
97	Bk	249	$5f^8\,6d\,7s^2$
98	Cf	251	$5f^9\,6d\,7s^2$
99	Es	254	$5f^{11}\,7s^2$
100	Fm	253	$5f^{12}\,7s^2$
101	Md	256	$5f^{13}\,7s^2$
102	No	253	$5f^{14}\,7s^2$
103	Lw	257	$5f^{14}\,6d\,7s^2$

(Configuration assignments for the rare earths and the actinide elements are somewhat uncertain.)

(perhaps 95 %) of the electron density. The origin of the coordinate system is taken at the nucleus of the atom. All the *s* orbitals have spherical boundary surfaces with the bounding sphere being successively larger as *n* increases as indicated in Fig. 3-2. The electron density is not constant everywhere within these spheres. Figure 3-3 shows the electron density in thin spherical shells as a function of distance from the nucleus for a number of orbitals. This nonuniformity in electron density exhibits maxima that are equivalent to radii. Within the boundary surface of the l*s* orbital there is a single "fuzzy" spherical shell of high electron density. Within the 2*s* boundary surface, there are two concentric dense and fuzzy shells.

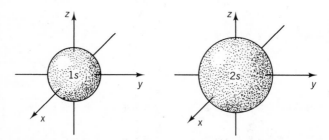

Fig. 3-2. Comparative boundary surfaces for l*s* and 2*s* orbitals.

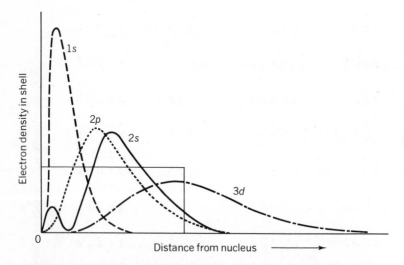

Fig. 3-3. Electron density distribution in l*s*, 2*s*, 2*p*, and 3*d* orbitals as a function of distance from the nucleus. The straight lines represent the distribution expected if the electron cloud were of constant density, dropping abruptly to zero at the edge of the boundary diagram.

The p orbitals all have the same shape with their boundary surfaces resembling distorted dumbbells. In contrast with the s orbitals, which are spherically symmetric, the p orbitals have directional properties as indicated in Fig. 3-4. An electron is somewhere in the dumbbell-shaped space with each lobe being equally probable. Both lobes constitute one p orbital. The electron density is not uniform but maximizes at some radius as indicated in Fig. 3-3. The p orbitals, unlike the s orbitals, have a plane of zero electron density (a nodal plane separating the two lobes) which is of significance in classifying the type of bonding between atoms.

The d orbitals are shown in Fig. 3-5. The d_{z^2} orbital is concentrated

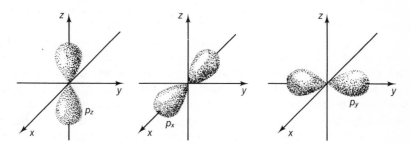

Fig. 3-4. The p orbitals.

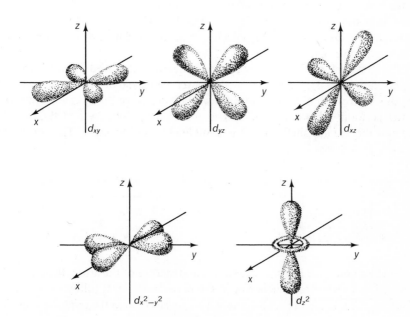

Fig. 3-5. The five d orbitals.

around the z-axis with the larger part of the volume being shaped somewhat like a p orbital while the rest is a doughnut-shaped cloud circling the middle. The other four orbitals are shaped like a clover leaf with four lobes and two nodal planes each. The shapes of the s, p, and d orbitals are important in bonding whereas the f, g, and higher orbitals are not.

3-3   THE PERIODIC TABLE

It is well-known that if the chemical elements are arranged in the order of increasing atomic number, there is a marked periodicity in properties. This is shown by listing the elements in a *periodic table*, such as Table 3-3, in such a way as to preserve the order of increasing atomic number when passing from left to right across the table, while connecting each group of chemically similar elements by vertical lines, as shown. Thus, the chemically similar alkali metals (Li, Na, K, etc.) all occur at the extreme left and are followed in the next column by the alkaline earth metals (Be, Mg, Ca, etc.). Likewise, the chemically inert gases (He, Ne, Ar, etc.) are found in the last column, preceded by the very characteristic group of halogen elements (F, Cl, Br, I).

Correlating this table with the electronic structures of the atoms gives the clue that allows interpretation of behavior in terms of atomic structure. The first element in the table is hydrogen, which has one electron. According to the rules, this electron normally occupies one of the two $1s$ states, since these have the lowest energy. The helium atom ($Z = 2$) has two electrons, both contained in the $1s$ state, which can accommodate two electrons provided they have opposite spins.

The first shell ($n = 1$) is now full, and therefore the third electron (forming lithium, $Z = 3$) must go into the lowest subshell of the second shell, i.e., the $2s$ level. Thus with lithium a new shell begins to form, and a new row of the periodic table commences. The elements in this row add their electrons to the second shell until it is completely filled at neon. The order of filling is as follows:

	Li	Be	B	C	N	O	F	Ne
Number of $1s$ electrons	2	2	2	2	2	2	2	2
Number of $2s$ electrons	1	2	2	2	2	2	2	2
Number of $2p$ electrons	0	0	1	2	3	4	5	6

When a new shell begins to form in the structure of the atom, the element in question commences a new row in the periodic table. It follows that these elements, the alkali metals, all contain a single electron in the outermost shell. In sodium, for example, following the element neon, there are 10 electrons filling the first and second shells and one in the lowest subshell of the third shell. The electronic structure of the third shell is as follows:

	Na	Mg	Al	Si	P	S	Cl	Ar
Number of 3s electrons	1	2	2	2	2	2	2	2
Number of 3p electrons	0	0	1	2	3	4	5	6

An important effect occurs in the elements immediately following argon. To complete the third shell, 10 more electrons are required to occupy the $3d$ states. For these elements, however, the $4s$ energy level is slightly lower than the $3d$, and so a fourth shell begins to form before the third is filled. Consequently, the order of occupation is:

	K	Ca	Sc	Ti	V	Cr	Mn	Fe	Co	Ni	Cu	Zn	Ga	Ge	As	Se	Br	Kr
No. of 3d electrons	0	0	1	2	3	5	5	6	7	8	10	10	10	10	10	10	10	10
No. of 4s electrons	1	2	2	2	2	1	2	2	2	2	1	2	2	2	2	2	2	2
No. of 4p electrons	0	0	0	0	0	0	0	0	0	0	0	0	1	2	3	4	5	6

Thus this new period commences with an alkali metal, potassium, and an alkaline earth metal, calcium. In the following series of elements, extending to copper, the $3d$ states become increasingly occupied. Hence there is a large group in this period with only one or two electrons in the outermost shell. All these elements are metals; those in which the third shell is expanding from 8 to 18 electrons are known as *transition* metals.

The same sort of behavior is found in the subsequent rows of the periodic table. The $5s$ states are occupied before the $4d$ and the $6s$ states before the $5d$; as a result, two additional groups of transition metals are obtained. In the final long period, the situation is complicated by the fact that during filling of the $5d$ states in the last group of transition metals, the $4f$ states begin to be occupied. This results in a group of elements known as the *rare earths*.

EXAMPLE 3-1

Iron has an atomic number of 26. What is its electronic structure? What is the electronic structure of uranium with an atomic number of 92?

Solution:

Ground state Fe	$1s^2\ 2s^2\ 2p^6\ 3s^2\ 3p^6\ 3d^6\ 4s^2$
Ferrous ion Fe^{2+}	$1s^2\ 2s^2\ 2p^6\ 3s^2\ 3p^6\ 3d^6$

Ferric ion Fe^{3+} $1s^2\ 2s^2\ 2p^6\ 3s^2\ 3p^6\ 3d^5$

Ground state U $1s^2\ 2s^2\ 2p^6\ 3s^2\ 3p^6\ 3d^{10}\ 4s^2\ 4p^6\ 4d^{10}\ 4f^{14}\ 5s^2\ 5p^6$
$5d^{10}\ 5f^3\ 6s^2\ 6p^6\ 6d^1\ 7s^2$

3-4   CLASSIFICATION OF THE ELEMENTS

In the above discussion such terms as *alkali metals, alkaline earth metals, inert gases*, and *halogens* were used. These terms apply to classes of elements which show similar behavior. Such groups may be considered as subgroups of the two much broader classes of metals and nonmetals. Ordinarily, we consider a material to be metallic if it has high thermal and electrical conductivity, if it is opaque and lustrous, and if it has considerable mechanical strength in combination with some ductility. All other materials would be nonmetals. This distinction, however, is not entirely satisfactory because there is no one obvious property of metals which is always absent in an otherwise nonmetallic material.

In the metallurgical sense, *metallic* applies to macroscopic objects such as copper wire, steel beams, aluminum aircraft, etc. In such objects there are large numbers of atoms coexisting in a special state, i.e., the metallic state with the physical and mechanical properties that typify metals.

From a chemical viewpoint, metals are elements which form strongly basic oxides and hydroxides and which combine with acids to form salts. Here the properties of individual atoms are of primary concern, whereas metallurgists are concerned with the properties of large numbers of atoms. Table 3-4 shows the elements classified as metals or nonmetals in the chemical

Table 3-4. Classification of the elements

Metals						Nonmetals	
Alkali metals	Alkaline earth metals	—	—	—	—	Halogens	Rare gases He
Li	Be			N	O	F	Ne
Na	Mg	Al		P	S	Cl	Ar
K	Ca	Ga	Ge	As	Se	Br	Kr
Rb	Sr	In	Sn	Sb	Te	I	Xe
Cs	Ba	Tl	Pb	Bi	Po	At	Rn
Fr	Ra						
Transition metals	Sc thru Zn			Either metals, nonmetals, or both			
	Y thru Cd			B	C		
	La thru Hg			Si			
	Ac						
Rare earths	Ce thru Lu						
	Th thru Mv						

sense. Comparison of this table with Table 3-3 shows that the rare earth metals can be considered as a transition series falling within another transition series.

3-5   ELECTRONIC STRUCTURE AND RESULTING BEHAVIOR

From Table 3-4 it is obvious that elements connected by vertical lines in the periodic table are chemically similar and that some structural similarity might be expected in the atoms of such elements. On examination, we see that the main factor in determining the chemical properties of an element is the number of electrons in the outermost shell. Thus, in general, the metallic elements with a small number of outer electrons appear at the left in the table, while the nonmetals, with a relatively large number of outer electrons, appear at the right. The elements in the middle of the periodic table exhibit both metallic and nonmetallic characteristics with one or the other predominating. The elements boron, carbon, and silicon can be considered either metals or nonmetals, or both.

The relation between electronic structure and chemical properties becomes clearer when the atoms of both chemically similar and dissimilar elements (Li, Na, K, F, Cl, and Br) are examined further. It is apparent that each alkali metal (Li, Na, K, etc.) has a single electron in its outer shell. Likewise, each halogen (F, Cl, Br, etc.) has seven outer electrons. The structure of the inner shells, conversely, provides no such basis for distinction between the two groups (and is frequently identical even in chemically dissimilar atoms). The conclusion is that the structure of the inner shells is of little importance in determining chemical characteristics and that the number of electrons in the outermost shell is of primary importance. Because of their special significance in the determination of chemical properties, these outer electrons are called *valence* electrons.

The electrons, especially the valence electrons, produce most of the properties of engineering interest: they determine the chemical properties; they establish the nature of interatomic bonding (and therefore strength and ductility); they control the size of the atom; they affect the electrical conductivity; and they influence optical characteristics.

The strong affinity which exists between certain chemical elements such as sodium and chlorine can be interpreted in terms of atomic structures. The rare gases (He, Ne, Ar, etc.) show little tendency to combine chemically, although combination of Xe with other elements has been established. The physical interpretation of the tendency for two or more atoms to join together in a molecule is based on the principle that a system is most stable when its total energy is at a minimum. Thus, in a collection of atoms, if the total energy is less when they are combined in some aggregate than when they are free, the group will be more stable in the aggregate. This is the same as saying that there is a force of attraction between the atoms and work must be done to separate atoms in an aggregate.

The facts that chemical properties depend on the configuration of the outer shell and that the rare gases are chemically inactive indicate the valence electrons in a rare gas must produce a very stable structure. The periodic table indicates eight valence electrons fill the outer shell and produce a "stable octet." This idea can be extended to explain the activity of other atoms if we assume their chemical behavior results from their tendency to attain the most stable arrangement of their electrons. The alkali metals, for instance, have only one valence electron. The easiest way for such atoms to obtain a stable octet is to transfer this electron to some other nearby atom. In sodium, for example, the transfer produces the electronic structure of neon, the rare gas which precedes sodium in the periodic table. This rare-gas configuration carries a net positive charge; i.e., it is a positive ion. Likewise, magnesium may give up two valence electrons to obtain this structure. Chlorine, on the other hand, with seven valence electrons, can achieve a rare-gas structure by taking up an extra electron to produce the electron configuration of argon. Chlorine, then, becomes a negative ion because it has a net negative charge.

Continuing, we readily see why there is a great chemical affinity between elements such as sodium and chlorine. If atoms of these elements closely approach each other, a valence electron of sodium is transferred to chlorine, producing a rare-gas structure in each. As a result, the sodium atom becomes a positive ion and the chlorine atom a negative ion. These ions are attracted to each other by the normal electrostatic attraction of unlike charges, and they associate as a stable compound. This mechanism is the basis of the formation of the compounds known as *ionic* or *heteropolar* compounds.

EXAMPLE 3-2

In metallic sodium (Na), the $3s$ electron is weakly bound to the atom with an ionization energy of 5.2 eV. Chlorine (Cl), on the other hand, has a large electron affinity with each Cl atom giving up 3.7 eV when an electron is added to its outer shell. Write the pertinent reactions and indicate the physical interpretation.

Solution:

$$Na + 5.2\,eV \longrightarrow Na^+ + e^-$$
$$Cl + e^- \longrightarrow Cl^- + 3.7\,eV$$
$$Na + Cl + 1.5\,eV \longrightarrow Na^+ + Cl^-$$

Since a net decrease in energy arises from electrostatic attraction of the ions, Na and Cl are capable of forming ionic bonds.

From the above argument, the elements may be roughly divided into two general classes: those that gain electrons to form stable outer shells (electronegative elements) and those that lose electrons (electropositive elements). Because their atoms possess few valence electrons (frequently only one), metals are electropositive elements, and it is easy for them to lose electrons in chemical combination. This is a fundamental way of (chemically) distinguishing metallic elements.

The most characteristic metallic elements (from the chemical viewpoint) are those with only one valence electron in their atoms, such as the alkali metals and the copper, silver, and gold group. The transition elements are also markedly metallic in their properties, but metallic properties become less obvious in elements with a greater number of valence electrons in their atoms. When the atoms of an element have several valence electrons [for example, tin (4) and bismuth (5)], there is uncertainty as to whether the atom will lose or receive electrons in chemical combination, and therefore both metallic and nonmetallic properties are often found. With six or seven valence electrons in the atom, the electronegative characteristics are strongly marked. Such atoms are typical nonmetals, i.e., chlorine, oxygen, and sulfur.

The properties of the metallic state (from the metallurgical viewpoint) do not depend directly on the structure of individual atoms. However, the features of atomic structure which produce metallic behavior in the chemical sense are also responsible for allowing atoms to exist in the kinds of aggregates typical of the metallic state.

3-6 BONDING BETWEEN ATOMS

A collection of atoms may exist in any of three possible states, i.e., gas, liquid, and solid. The gaseous or vapor state is distinguished by the fact that atoms in a gas are usually well separated from one another; thus any particular atom can travel a considerable distance (relative to its own size) before it collides with another. Metals can exist as vapors. For instance, the boiling point of mercury is low enough so that its vapor can be used instead of steam for power generation in a turbine. Zinc has a lower vapor pressure than mercury, but it is still sufficiently high to make recovery of zinc by distillation commercially feasible. In general, however, boiling points of metals are too high to be used in common metallurgical practices. Metals at sufficiently high temperatures behave as monatomic gases and obey the normal gas laws in the same way as nonmetallic gases.

In metallic vapors, as in other gases, by increasing the pressure, or lowering the temperature, or both, the atoms may be brought closer together and condensed to liquid. The random to-and-fro motion of atoms in the vapor is replaced by more restricted movements. However, the atoms still have considerable mobility in the liquid, and are restrained in their motions only by their neighbors and the walls of the container.

As the liquid cools, there is a continued decrease in mobility of the atoms until, at a certain temperature (known as the freezing point), the random motion of the atoms "stops" and they take relatively fixed positions with regard to one another, forming a solid. These positions might more properly be called centers of oscillation, for the atoms have by no means stopped moving.

Atoms are held in the solid state by relatively strong interatomic forces which generally are functions of temperature and pressure. These interatomic forces can be attractive or repulsive, and the equilibrium spacing of atoms in a solid is obtained when the opposing forces are balanced at a particular temperature and pressure.

If the potential energy of a pair of atoms is taken as zero at infinite spacing between them (in practice, a few hundred angstroms), then, as the atoms are brought together, the attractive force becomes operative and potential energy decreases (lower dashed curve in Fig. 3-6) since work is being done by the atoms. As spacing decreases, the repulsive force begins to be effective and creates a positive potential energy (upper dashed curve in Fig. 3-6) since work is being done on the atoms. The sum of these potential energies varies with distance between atoms as shown by the solid curve in Fig. 3-6. Increasing the pressure has the effect of decreasing the interatomic distance. Increas-

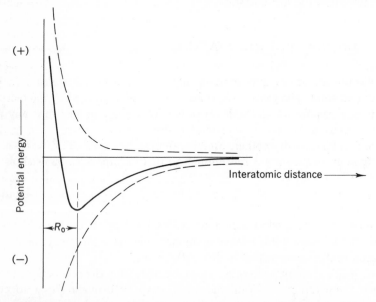

Fig. 3-6. The potential energy of a pair of atoms as a function of their distance of separation.

ing the temperature increases the kinetic energy of the atoms and thereby increases the interatomic distance.

The interatomic bonds are of various types: ionic, covalent, van der Waals, and metallic. Melting and boiling temperatures give crude indications of the strength of interatomic bonds.

3-6-1 IONIC BOND

The ionic (or heteropolar) bond is the type commonly found in inorganic chemistry and is probably the simplest example of interatomic bonding. This bond exists between certain unlike atoms or groupings of atoms. In the simplest case, an electron is transferred from a metallic atom to a non-metallic atom and the two resulting ions are held together by electrostatic attraction. In the case of potassium chloride, for example, both ions (K^+ and Cl^-) are spherically symmetric, having the rare-gas, closed-shell type of electronic structure. The combination of the energy to form the ions and the energy of electrostatic attraction is similar to that shown by the lower dashed curve in Fig. 3-6, assuming no overlapping of electron shells.

As the shells begin to overlap appreciably, another energy term becomes important. This is a repulsive interaction arising principally from the Pauli exclusion principle. Since all available $3s$ and $3p$ states are filled in both ions, if the approach is so close that the $n = 3$ shells start to overlap, then some of the electrons must go to higher energy states such as the $3d$ or $4s$. While it is difficult to calculate the precise shape of the repulsive energy curve, it will be of the form of the upper dashed curve in Fig. 3-6. It is clear that the sum of the attractive and repulsive energies must have a minimum which is shown in Fig. 3-6. The stable position of this system is an ion-pair bound together with the atomic nuclei at a (most probable) distance R_0 (Fig. 3-6) or d (Fig. 3-7). If two ion pairs cluster together to form an ion square (Fig. 3-8) the

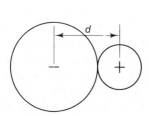

Fig. 3-7. Ion-pair formation.

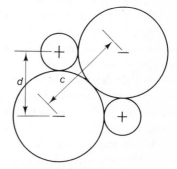

Fig. 3-8. Ion-square formation. $c = \sqrt{2d}$.

configuration becomes more stable. If a "large" number of ions are brought together in a single cluster by lining up ion squares in all directions, the result is not a random packing but an orderly patterned array, i.e., a macroscopic three-dimensional crystalline solid, e.g., KCl.

3-6-2 COVALENT BONDS

The covalent bond is formed by sharing of electrons between atoms rather than by electron transfer. An excellent example of covalent bonding is found in the chlorine molecule. Here, the outer shell of each atom possesses seven electrons. Each chlorine atom would like to gain an electron, and thus form a stable octet. This can be done by the sharing of two electrons between pairs of chlorine atoms, thereby producing stable diatomic molecules. In other words, each atom "contributes" one electron for the sharing process.

The nature of the covalent bonding can be demonstrated in terms of the hydrogen molecule. Consider the case of the hydrogen molecule ion (H_2^+) since the nature of bonding is the same and it provides a ready transition to understanding the hydrogen molecule. In this case, there are two nuclei, both positively charged, with one electron. It can be shown from quantum mechanics that there is a high probability of finding the electron near one nucleus or the other in the region between them. This situation gives rise to an attractive force between the nuclei. However, as the two nuclei approach each other, there is a repulsive force set up because both nuclei are positively charged. As in Fig. 3-6, the sum of attractive and repulsive forces has a minimum which determines the most probable distance between the two nuclei.

The electron in the H_2^+ ion "resonates" back and forth between the state with a high probability of finding the electron near nucleus A and the state with a high probability of finding the electron near nucleus B. Thus, on the average, the electron moves back and forth between the two nuclei, and it is this "resonance" which leads to the binding.

The nature of the covalent bonding can be demonstrated in terms of the simplest molecule, i.e., H_2. If two isolated H atoms [Fig. 3-9 (a)], each with its electron in the ground state $1s$ orbital, approach each other, the $1s$ clouds begin to overlap [Fig. 3-9 (b)]. Each electron is attracted to the other nucleus and the overlap increases (provided the electrons have opposite spins). The two atomic orbitals merge into one larger cloud called a molecular orbital. Within the molecular orbital, the two electrons are attracted to both nuclei. When the repulsive forces have balanced the attractive forces (Fig. 3-6), a molecule results with a stability that is significantly greater than that of the two isolated atoms. This covalent bonding is also known as *homopolar* or *electron-pair* binding. It is the common type of bonding in organic molecules and in many inorganic molecules.

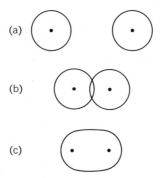

Fig. 3-9. Molecular orbital formation from s orbitals. (a) Two isolated atomic orbitals; (b) overlap; (c) the molecular orbital boundary diagram.

In the molecular orbital of H_2, the two electrons are equally shared between the two nuclei and can not be specifically identified with either of the nuclei. The two shared electrons are more likely to be found between the two nuclei and thus the electron density there is relatively large. The contour diagram of the molecular orbital is contracted somewhat parallel to the internuclear axis and is expanded at the center of the molecule. It should be obvious that the molecular orbital is *not* a simple superposition of the atomic orbitals.

Thus, whether in gas, liquid, or solid, any two atoms held together by a covalent bond should be regarded as linked together by the particular pair of electrons they share. The diamond form of carbon is an excellent example. Each carbon atom has four valence electrons and is able to form electron pairs with four neighboring atoms, thus achieving the stable configuration of eight electrons in the outer shell. This is reflected in the atomic arrangement of diamond in which any given atom is surrounded by four others symmetrically disposed at the corners of a tetrahedron at the center of which is the given atom as illustrated in Fig. 3-10.

In the covalent bond, the valence electrons are shared as electron pairs, which help to join each pair of neighboring atoms in the structure. In this bond, the number of geometrically closest neighbors (known as the coordination number) which a given atom may have is established by the number of valence electrons. A closed valence shell is a stable octet. If N represents the number of electrons present in the valence shell of an electrically neutral atom, then $8 - N$ is the number of electrons which are required to obtain a stable octet. This is also the number of covalent bonds which an atom must form to fill its valence shell and is therefore the number of adjacent atoms to which it must be bonded to satisfy its own valence requirements. Thus $8 - N$ is the coordination number in covalent bonding, but only when bonding is entirely normal covalent.

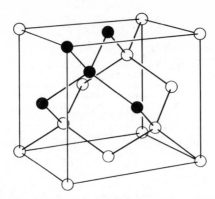

Fig. 3-10. Schematic representation of electron sharing between carbon atoms in the diamond structure. The heavy circles and lines represent a a single tetrahedron.

Diatomic molecules with unlike nuclei (heteronuclear) can also bond covalently with electron pairing schemes. For example, consider the HF molecule. The ground state of the F atom is

$$F(Z = 9): 1s^2 2s^2 2p_x^2 2p_y^2 2p_z^1$$

The half-empty $2p_z$ atomic orbital, when "aimed" at the $1s$ atomic orbital of the hydrogen atom could overlap to form a molecular orbital with a shape somewhat like that shown in Fig. 3-11 containing the $1s$ electron from the H and the $2p_z$ electron from the F (with spins paired) to form a bond. All the other F electrons are considered nonbonding.

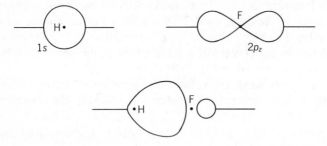

Fig. 3-11. Formation of the HF molecule. Orbitals have been omitted which are not involved in the bonding.

Next consider a larger and somewhat more complex molecule, H_2O. The electronic structure of O is

$$O(Z = 8): 1s^2 2s^2 2p_x^1 2p_y^1 2p_z^2$$

If two hydrogen atoms having electrons with proper spin approach the p_x and p_y orbitals head on, two bonds, each with a shape like that in HF (Fig.

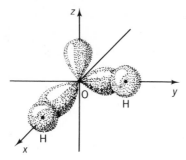

Fig. 3-12. Model of the water molecule. Orbitals not involved in the bonding have been omitted for clarity.

3-11), will form. We would thus expect the water molecule (Fig. 3-12) to be nonlinear with a bond angle of about 90°, the angle between the p_x and p_y orbitals. The actual angle is 105°. The 15° difference may be due to the mutual repulsion of the two hydrogen nuclei.

In homonuclear molecules like H_2, N_2, and O_2, the electron pair in the bonding molecular orbital is shared equally by the two nuclei. For heteronuclear molecules, however, this is not true. The electronegative elements tend to take more than their share of the electron pair in the bonding. Electronegativity decreases across the periodic table (Table 3-3) from *right* to *left* and down the table in a given column. As the difference in electronegativity between two atoms increases, the molecular orbital cloud of the bonding pair is distorted toward the more electronegative atom and the character of the bond departs more and more from pure covalency. In the extreme case of little (or no) sharing, there is an ionic bond.

Orbitals in the same atom which are nearly alike in energy have the ability to combine in an additive manner to produce hybrid orbitals. An *s* orbital can combine with a *p* orbital in the same atom to form two new and equivalent orbitals as shown in Fig. 3-13. The unsymmetrical hybrid orbitals have properties of both the *s* and *p* orbitals from which they were formed. They are "fatter" than the *p* orbitals but have directional characteristics like *p* orbitals. These specific orbitals are known as digonal or *sp* hybrids. This phenomenon of hybridization operates in the case of carbon in forming the diamond structure (Fig. 3-10). It is also of great interest in the structure of hydrocarbons and polymers in general.

3-6-3 THE VAN DER WAALS BOND

Fluorine, the first member of the halogen family and the remaining halogens, chlorine, bromine, and iodine, all form stable diatomic molecules. The melting points and boiling points of the condensed states of the halogens

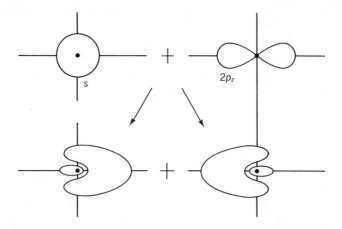

Fig. 3-13. Formation of *sp* digonal hybrid bonds. Orbitals not involved in the bonding have been omitted for clarity.

decrease with increasing atomic number. The diatomic molecules are formed with covalent bonds. The additional forces which hold the molecules together in the liquid or solid are of the van der Waals type, with strength of bond increasing rapidly as atomic number increases. At room temperature, fluorine and chlorine are gases while bromine is liquid and iodine is solid. The atomic arrangement in iodine (Fig. 3-14) consists of diatomic iodine molecules held

Fig. 3-14. The structure of solid iodine showing diatomic I_2 molecules bound together with van der Waals forces.

together by van der Waals bonds. The diatomic molecules of liquid bromine are less tightly bound than those of iodine.

Returning to our earlier model, i.e., an atom consisting of a positively charged nucleus surrounded by a cloud of negatively charged electrons, consider what happens when two of these atoms approach each other. If we assume that the electron clouds are spherical, static, and undeformable, no force results between the two atoms. Actually, the electron clouds result from motions of the individual electrons which are constantly moving around their respective atomic nuclei. If, at a given instant, atom A has a preponderance of electrons on the side closest to atom B, these electrons will tend to repel the electrons on the A side of atom B. Thus, the motions of the electrons in one atom affect motions of the electrons in other atoms, and a lowering of energy, hence an attraction, occurs if the electrons in one atom move in harmony with those in the other atom as the two atoms approach each other. This phenomenon produces the van der Waals force. This force is sometimes called a polarization force. It produces a potential energy proportional to $-1/r^6$ if potential energy is zero when the atoms are an "infinite" distance apart.

When two atoms approach one another from a large distance, attraction of the van der Waals type is always produced, and we can regard it as drawing the atoms together until the electron clouds overlap slightly. This overlapping means that the outermost electrons of each atom are attracted to the nucleus of the other atom, thus producing an attraction between the two atoms. At the same time, the electron clouds repel each other. As the atoms are drawn together, the repulsion increases until a point is reached where attraction and repulsion balance and equilibrium results.

The van der Waals attraction is the only long-range force which exists between neutral atoms and is the common interatomic force operating in the rare gases. However, the van der Waals attraction is weak in the rare gases, because the octets of electrons in the outer shells are so stable that electronic motion in any one atom is only slightly disturbed by motions of electrons in adjacent atoms. Nevertheless, at sufficiently low temperatures, where thermal motion is reduced, the van der Waals attraction is sufficiently large to account for liquefaction and solidification of the rare gases.

3-6-4 METALLIC BOND

In some respects, the metallic bond is similar to the covalent bond, and yet it is quite different. It is possible to consider, in an elementary manner, the metallic bond as a multiplicity of resonating covalent bonds. In the covalent bond, two atoms are linked together by particular pairs of electrons shared between them. In metallic bonding the valence electrons, which hold the atoms together, are not bound to individual atoms or pairs of atoms but move freely throughout the whole metal. Thus, the similarity to the covalent bond lies in a sharing of electrons. The difference is that these electrons belong to the metal as a whole, rather than to any particular atom.

In metallic atoms, the valence electrons lie much farther from the nucleus than do those in nonmetals. Thus the wide orbits of the valence electrons allow them to pass to regions remote from their parent atoms. Likewise, they never associate for any appreciable period of time with any one atom in the metal but drift through the whole assembly as a kind of electron gas. Only the valence electrons behave in this manner, for electrons in the inner shells lie much closer to the nucleus and are unable to move outside its field of influence. Consequently, as far as the inner electrons are concerned, an atom of a metal is much the same as an isolated atom, because these electrons are held firmly within the atom and remain in their characteristic quantum shells.

Thus, the internal structure of a metal is composed of two parts:

1. A collection of positive ions, each consisting of the core of an atom, i.e., the nucleus and the nonvalence electrons, and
2. A "gas" of free valence electrons swarming between the ions throughout the whole of the metal. The binding force holding the metal together is provided by the attraction of the positive ions to the negative valence electrons continually passing between them.

This picture of the metallic state is essentially that proposed in the classical free electron theory of Drude and Lorentz, formulated before the advent of quantum mechanics. More recent work has substantially modified the details of the above model but the main assumption, i.e., the idea of "free" electrons in the metal, is still useful.

3-6-5 MIXED BONDING

The above discussion of bonding has been in terms of ideal cases which are shown schematically in Fig. 3-15 and compared in Table 3-5. Mixed bonding, however, is possible. In fact, bonding between atoms in many materials can not be classified as one of the four ideal types but rather as a mixture of those types. This is indicated by a tetrahedral diagram which is viewed from the top in Fig. 3-16. The apices represent van der Waals, ionic, metallic, and covalent bonding. Various types of mixed bonding are represented on the edges. The metals gradually change from pure metallic bonding as in sodium to the less perfect metals such as tellurium and arsenic, finally reaching the pure covalent bonding of carbon in the form of diamond. Starting

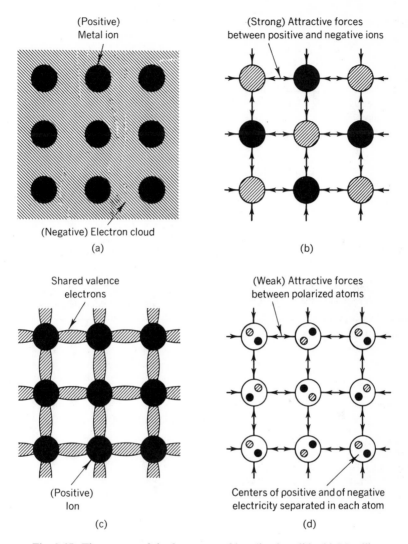

Fig. 3-15. The natures of the four types of bonding in solids. (a) Metallic bonding (b) ionic bonding (c) covalent bonding (d) Van der Waals bonding.

at diamond, we find graphite, benzene rings, and high polymers. Ultimately, we reach the rare gases and pure van der Waals bonding. Layer lattices develop between ionic and van der Waals bonding with various defect structures formed between ionic and metallic bonding. In general, there are no mixtures between metallic and van der Waals bonding. Figure 3-16 illustrates a continuity of the solid state of matter.

Table 3-5. Summary of atomic bonding

Type of bond	Number of electrons "shared"	Kinds of atoms involved	Remarks
van der Waals (molecular)	0	same	Weak electrostatic attraction due to unsymmetrical electrical charges in electrically neutral (as a whole) atoms or molecules
Ionic	1 (or more) transferred	different	Strong electrostatic attraction
Covalent normal	2	same or different	Electron pair "revolves" in common orbit about both nuclei, one atom supplying one electron
coordinate	2	different	Electron pair "revolves" in common orbit about both nuclei, one atom supplying both electrons, the other atom supplying none (quite rare)
Metallic	∞	same	General attraction of a very large number of positive (metallic) ions for a dispersed cloud of electrons

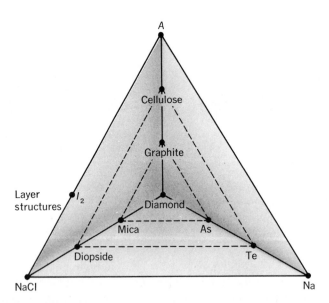

Fig. 3-16. A tetrahedron showing graphically the four principal types of bonding at the corners; mixed bondings are shown along the edges joining the corners.

EXAMPLE 3-3

In the ionic bond, the net potential energy of two ions is

$$U = \frac{Q_1 Q_2}{4\pi\epsilon_0 r} + \frac{b}{r^n}$$

The first term is an attraction where Q_1 and Q_2 are point charges, ϵ_0 is the permittivity constant, r is the distance between the ions. The second term is a repulsion where b and n are positive constants characteristic of the two ions. Calculate the net potential energy for the equilibrium distance, R_0, between ionic centers.

Solution:

$$\frac{dU}{dr} = 0 \quad \text{for} \quad r = R_0$$

$$\frac{dU}{dr} = -\frac{1}{r}\left(\frac{-e^2}{4\pi\epsilon_0 r} + \frac{nb}{r^n}\right)$$

Setting equal to zero, with $r = R_0$

$$\frac{b}{R_0^n} = \frac{1}{n}\left(\frac{e^2}{4\pi\epsilon_0 R_0}\right)$$

$$U = \frac{-e^2}{4\pi\epsilon_0 R_0^2}\left(1 - \frac{1}{n}\right)$$

Some typical examples of binding energies are:

Type of bond	Substance	Bond energy		Melting temp. (°C)
		Kcal/mole	*eV/atom*	
Ionic	NaCl	183	7.95	801
	Al_2O_3	3620	157	3500
Covalent	Ge	75	3.26	958
	C (diamond)	170	7.40	3550
Metallic	Na	26	1.13	98
	Al	74	3.22	660
	Fe	97	4.21	1535
	W	201	8.75	3370
van der Waals	CH_4	2.4	0.104	−184
	Cl_2	7.4	0.322	−103
Hydrogen	H_2O	12	0.522	0

3-7 ZONE THEORY

We have shown that the structure of matter can be visualized as consisting of positive ions and valence electrons, with cohesive energy involving the following:

1. The potential energy of electrostatic interaction between ions, between valence electrons, and between ions and electrons, where the ions are considered as point charges.
2. The potential energy of the van der Waals attraction.
3. The kinetic energy of the valence electrons.

Cohesive energy in matter is determined mainly, and in the case of monovalent metals almost entirely, by the sum of the kinetic energies of the valence electrons and their potential energies in the field of their related ions. If the total energy of the valence electrons could be calculated for various possible structures, the stability of observed structures could be explained.

The possible energy states of the electrons in a free atom have been discussed previously. If a large number of these free atoms are brought close together, an interaction occurs among them, and the energy states of the valence electrons are altered. A simple schematic illustration is given in Fig. 3-17. The total aggregate of atoms must be considered as a single system, and each quantum state can be occupied by only one electron according to the Pauli principle. In aggregates of atoms, this principle effectively applies only to valence electrons since the inner shells are essentially the same as in free atoms. As the distance between atoms is decreased, the quantum states are so modified that the energy states come sufficiently close together to be regarded as a continuum of energy, for all practical purposes.

These regions of essentially continuous energy states are separated by gaps of forbidden energy states (see Fig. 3-17). The gaps depend on the direction of motion of electrons in the solid and will occur at different energy levels for electrons traveling in different directions. In some aggregates of atoms, the gaps for differently directed electrons happen to be coincident, and thus there is a range of energy which is forbidden to electrons regardless of their direction of motion. These gaps divide the energy spectrum into bands or zones. The regions of high density of permitted energy states are often called *Brillouin zones*.

The energy states can be represented mathematically in terms of wave or quantum mechanics. It is sufficient for us to regard the zone theory as a sort of unified theory of matter which supplies a coordinated and physically acceptable explanation of many characteristics of materials. The calculation of distribution of energy states in the Brillouin zones, the extent to which zones overlap or have gaps between them, and the extent to which they are

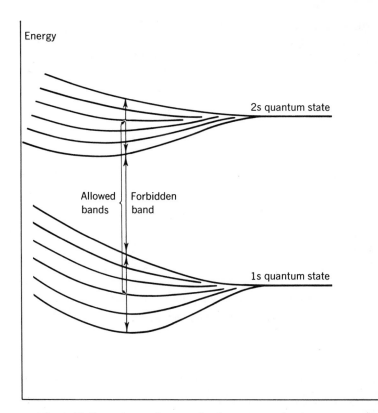

Fig. 3-17. Dependence of energy levels upon spacing between atoms; line of six hydrogen atoms, showing the incipient formation of allowed and forbidden energy bands. As the atoms are moved closer together, the coupling between atoms increases, splitting the energy levels as shown here. This is similar to a line of coupled electrical or mechanical oscillators.

filled with electrons are of prime importance in predicting the physical characteristics of solids.

One phenomenon which is readily explained in terms of the zone theory is the difference between conductors, semiconductors, and insulators. Conduction of electricity is possible only if electrons can increase their energy under the influence of an external field. Such an increase in energy becomes possible only if there are empty energy states in the Brillouin zone or if a filled zone overlaps an empty one. If the valence electrons just fill a Brillouin zone, and if there is a wide gap between this zone and the next empty zone, the electrons cannot acquire sufficient energy from the externally applied field to "jump" over the gap, and the material is thus a nonconductor or an insulator.

Electronic conductors such as metals are characterized by the presence (and insulators by the absence) of partly filled zones. If impurities are present to furnish intermediate energy states or if the energy gaps between the zones are small, there may be weak conductivity. Such substances are called semi-conductors. This phenomenon will be discussed further in Chapter 5.

QUESTIONS AND PROBLEMS

1. What are some of the consequences that would occur if Planck's constant were 6.63×10^{-27} Kcal-sec instead of 6.63×10^{-27} erg-sec? (1 cal $= 4.10 \times 10^7$ ergs).

2. Write the electron configuration for the following elements using their atomic numbers and the Pauli exclusion principle: Li $(Z = 3)$, Al $(Z = 13)$, Ti $(Z = 22)$, Fe $(Z = 26)$.

3. The periodic table of the elements is given in Table 3-4. Is this periodic in atomic weight or atomic number? What is the difference between atomic weight and atomic number? What is the significance of this difference?

4. What is the difference between the alkali metals and the halogens? How is the difference explained in terms of atomic and electronic structure?

5. What is the difference between the alkali metals and the alkaline earth metals? What is the source of this difference?

6. What are the rare earths? What are the transition elements? What is the reason for classifying these elements in this fashion?

7. What are the characteristics of metallic materials in comparison with other materials?

8. Discuss the similarities and differences of the homopolar and heteropolar bonds.

9. Discuss the similarities and differences of the covalent and metallic bonds.

10. Discuss the similarities and differences among van der Waals bonding, heteropolar, homopolar, and metallic bonding.

11. Does the zone theory of solids conflict with the theories of the types of bonding discussed in questions 8, 9, and 10?

12. What is the exclusion principle? Of what significance is it? Is its application limited to a single atom?

13. How can an "inert gas" like xenon (Xe) react with fluorine (F) to form stable compounds?

14. Why do carbon atoms in diamond bond covalently while lead atoms bond metallically when both have four valence electrons? What effects would you predict this difference in bonding to have on strength, ductility, electrical conductivity, etc.?

15. Could the hydrogen bond, generally classed as a dipole bond, be conceivably classed as: (a) an ionic bond, (b) a covalent bond, (c) a metallic bond, (d) some combination of these? If so, how?

16. Why do inert gas atoms form in condensed states only at low temperatures but not at room temperature?

17. Sterling silver is an alloy of 92.5 w/o Ag and 7.5 w/o Cu. Determine atomic percentages. Determine volume percentages, assuming atomic radii for both elements are the same in the alloy as in pure metal.

18. Copper has an atomic weight of 63.54 g/mole and a valence of 2. How many valence electrons per gram?

19. What is the weight percent of Al in alumina (Al_2O_3)?

4

AGGREGATES OF ATOMS

4-1 INTRODUCTION

In the preceding chapter we discussed types of bonding between atoms, but left much to be said about condensed states of matter. The liquid state is, in some respects, more complex than either the gaseous or solid states for there is no simple idealized model of a liquid from which a comprehensive theory can be built. It is sufficient to regard a liquid as a closely packed array of atoms or molecules in violent agitation in which the atomic arrangement is almost completely random. At the same time, however, if an instantaneous photograph of a liquid could be taken showing individual atoms, it would disclose many "clusters" in which the atoms have the close packing characteristic of solids, although the bulk of the material would show no regular arrangement of the atoms. As a consequence of close packing, liquids are much less compressible than gases. They differ from solids in that they can undergo viscous flow and change shape upon application of vanishingly small forces, whereas solids require finite applied forces before undergoing permanent changes of shape.

In liquids, atoms are relatively free to move about. If we decrease the temperature, however, mobility decreases accompanied by a decrease in specific volume as indicated in Fig. 4-1. On continued cooling, two different things can happen when the melting point is reached. If nuclei are available (see Chapter 8), crystallization will occur with a discontinuous change in volume and with a marked change in properties. If there are very few nuclei or if mobility is very low, crystallization will not occur readily and the volume will decrease at about the same rate as above the melting point, giving a supercooled liquid. When the transformation range is reached, the expansion coefficient decreases upon formation of a glassy material. This change is

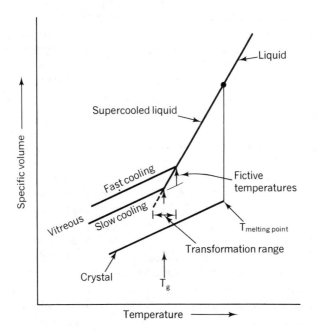

Fig. 4-1. Relationship among liquid, crystalline, and vitreous (glassy) states.

known as vitrification and produces a material with an amorphous structure much like the liquid but with such a high viscosity that it behaves much like a solid. The difference between a normal liquid and an amorphous or vitrified substance is primarily one of degree rather than of kind. A large number of engineering materials are solids at ordinary temperatures; thus the solid state is of primary interest. At one time it was the practice to classify solids into two groups:

1. Crystalline solids in which the atoms or molecules are arranged in a definite pattern in space, and
2. Amorphous solids in which the atomic or molecular arrangement is irregular. These are better classified as supercooled liquids.

4-2   CRYSTAL STRUCTURE

From the earlier discussion, it should be apparent that, when strong inter-atomic forces exist, atoms tend to pack closely together, the closeness of packing being particularly pronounced in the solid state. In this case, atoms can be regarded as hard spheres and the problem of close packing can be treated as one in which the whole assembly has a tendency toward efficient packing. A little thought or a few simple experiments with ping-pong balls quickly convince us that regular arrangements of the spheres generally lead to more compact assemblies than irregular arrangements. The same principle applies to arrangement of atoms in the solid state. We find that, where strong attractive forces are exerted, the atoms or molecules concerned arrange themselves in a regular three-dimensional pattern. It is this regularity which is the basis of crystallinity in materials; i.e., a crystal structure is nothing more than an orderly array of atoms or molecules. This definition of a crystal is distinct from the popular concept based on observation of external symmetry of crystals often seen during the study of elementary chemistry in which some crystals appear cubic, others needle-shaped and so on. This regular external shape is an outward form rather than a fundamental property of a crystal. It is obtained only when the conditions of crystallization are favorable to development of flat, geometric faces. In most instances, particularly with metals, these conditions are absent, and the crystals have irregular surfaces even though the internal arrangement is perfectly geometric.

Atomic arrays in crystals are conveniently described with respect to a three-dimensional net of straight lines. Consider a lattice of lines as in Fig. 4-2 dividing space into equal sized prisms which stand side-by-side with all faces in contact thereby filling all space and leaving no voids. The intersections of these lines are points of a *space lattice*, i.e., *a geometrical abstraction* which is useful as a reference in describing and correlating symmetry of

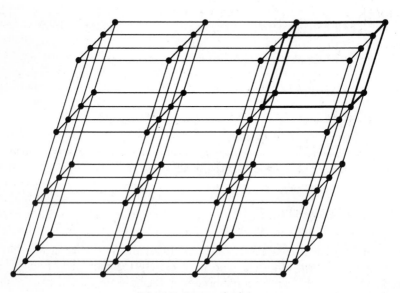

Fig. 4-2. A space lattice.

actual crystals. These lattice points are of fundamental importance in describing crystals, for they may be the positions occupied by individual atoms in crystals or they may be points about which several atoms are clustered. Since prisms of many different shapes can be drawn through the points of a space lattice to partition it into cells, the manner in which the network of reference lines is drawn is arbitrary. It is not necessary that the lines be drawn so that atoms lie only at corners of unit prisms. In fact, it is more convenient to describe some crystals with respect to prisms in which atoms lie at prism centers or at the centers of prism faces as well as at prism corners.

An important characteristic of a space lattice is that every point has identical surroundings. The grouping of lattice points about any given point is identical to the grouping about any other point in the lattice. In other words, if we could move about in the lattice, we would not be able to distinguish one point from another, for rows and planes near each point would be identical. If we were to wander among the atoms of a solid metal or chemical compound, we would find the view from any lattice point exactly the same as that from any other.

There are fourteen space lattices (Fig. 4-3); that is, no more than fourteen ways can be found in which points can be arranged in space so that each point has identical surroundings. There are, of course, many more than fourteen ways in which atoms can be arranged in actual crystals, thus there are a great number of crystal structures. Too often, the term "lattice" is loosely used as a synonym for "structure," an incorrect practice which is frequently confusing. The distinction can be clearly seen if we remember that a space lattice is an array of points in space. It is a geometrical abstraction which is useful only as a reference in describing and correlating symmetry of actual crystals. A crystal structure, however, is the arrangement of atoms or molecules which actually exists in a crystal. It is a dynamic rather than a static arrangement and is subject to many imperfections. Although any crystal structure has an inherent symmetry which corresponds to one of the fourteen space lattices, one, two, or several atoms or molecules in the crystal structure may be associated with each point of the space lattice. This symmetry can be maintained with an infinite number of different actual arrangements of atoms, making possible an endless number of crystal structures.

To specify a given arrangement of points in a space lattice, it is customary to identify a unit cell with a set of coordinate axes, chosen to have an origin at one of the lattice points. In a cubic lattice, for example, we choose three axes of equal length that are mutually perpendicular and form three edges of a cube. Each space lattice has some convenient set of axes, but they are not necessarily equal in length or orthogonal. Seven different systems of axes are used in crystallography, each possessing certain characteristics as to equality of angles and equality of lengths. These seven crystal systems are tabulated in Table 4-1 in conjunction with Fig. 4-4.

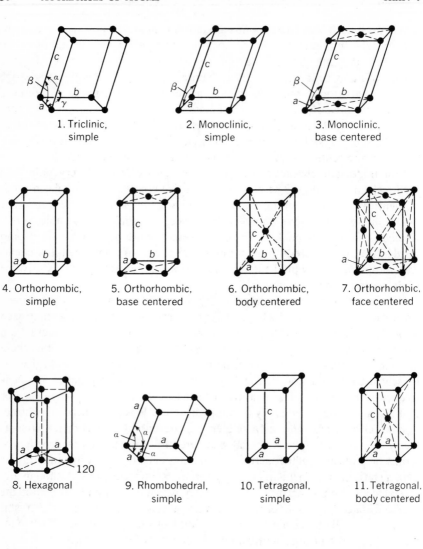

1. Triclinic, simple

2. Monoclinic, simple

3. Monoclinic. base centered

4. Orthorhombic, simple

5. Orthorhombic, base centered

6. Orthorhombic, body centered

7. Orthorhombic. face centered

8. Hexagonal

9. Rhombohedral, simple

10. Tetragonal, simple

11. Tetragonal, body centered

12. Cubic, simple

13. Cubic, body centered

14. Cubic, face centered

Fig. 4-3. The fourteen space (or Bravais) lattices as shown by their unit cells.

Table 4-1. The crystal systems

System	Parameters	Interaxial angles
Triclinic	$a \neq b \neq c$	$\alpha \neq \beta \neq \gamma$
Monoclinic	$a \neq b \neq c$	$\alpha = \gamma = 90° \neq \beta$
Orthorhombic	$a \neq b \neq c$	$\alpha = \beta = \gamma = 90°$
Tetragonal	$a = b \neq c$	$\alpha = \beta = \gamma = 90°$
Cubic	$a = b = c$	$\alpha = \beta = \gamma = 90°$
Hexagonal	$a = b \neq c$	$\alpha = \beta = 90°, \gamma = 120°$
Rhombohedral	$a = b = c$	$\alpha = \beta = \gamma \neq 90°$

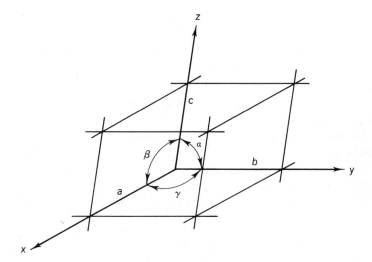

Fig. 4-4. Crystal axes.

The network of lines through the points of a space lattice (Fig. 4-2) divides it into unit cells. Each unit cell in a space lattice is identical in size, shape, and orientation to every other unit cell. It is the building block from which the crystal is constructed by repetition in three dimensions. The unit cells of the fourteen space lattices are shown in Fig. 4-3. All crystal structures are based on these fourteen arrangements. We might ask, for example, why a base-centered tetragonal lattice cannot be considered as a separate and discrete arrangement. It can be, provided we are willing to delete the simple tetragon from the fourteen. This is illustrated in Fig. 4-5 in which a base-centered tetragonal lattice cell is shown in solid lines. It becomes apparent that this can also be represented by the simple tetragonal lattice shown by dashed lines in Fig. 4-5. The c parameter (or dimension) is the same for the two cells, but the a parameter of the simple tetragonal cell is equal to the a

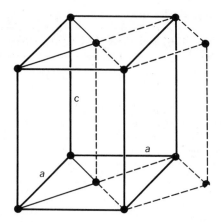

Figure 4-5. Relationship of base-centered tetragonal lattice (solid lines) to simple tetragonal lattice (dashed lines).

parameter of the base-centered tetragon divided by the square root of two. Since the two are interchangeable, only one can be considered as a basic type. The simpler lattice is normally taken to be basic.

The body-centered cubic, face-centered cubic, and hexagonal lattices (Fig. 4-6) are common and of prime importance in metals. The hexagonal close-packed structure is not a space lattice, since the surroundings of the interior atoms are not identical to those of the corner atoms. The actual crystal structure is based on the simple hexagonal lattice (Fig. 4-3, no. 8); however, a pair of atoms is associated with each lattice point as will be discussed in further detail shortly. Some of the metals associated with nuclear applications, such as uranium and plutonium, have crystal structures which are more complicated than these three relatively simple types. Crystalline ceramics, in general, also are more complex.

4-2-1 UNIT CELLS VS. PRIMITIVE CELLS

In the literature, we often find reference to unit cells and to primitive cells. The primitive cell may be defined as a geometrical shape which, when repeated indefinitely in three dimensions, will fill all space and is the equivalent of one atom. The unit cell differs from the primitive cell in that it is not restricted to being the equivalent of one atom. In some cases, the two coincide. For instance, in Fig. 4-3, all fourteen space lattices are shown by their unit cells. Of these fourteen, only nos. 1, 2, 4, 9, 10, and 12 are also primitive cells.

Primitive cells are drawn with lattice points at all corners, and each primitive cell contains the equivalent of one atom. For instance, considering a simple cubic unit cell (Fig. 4-3, no. 12), we see this unit cell portrayed as a

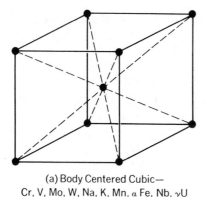

(a) Body Centered Cubic—
Cr, V, Mo, W, Na, K, Mn, α Fe, Nb, γU

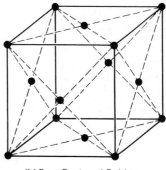

(b) Face Centered Cubic—
Al, Cu, Ag, Au, Pb, βNi, γFe

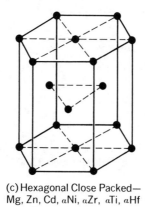

(c) Hexagonal Close Packed—
Mg, Zn, Cd, αNi, αZr, αTi, αHf

Fig. 4-6. Unit structure cells of the more common metals. Positions of centers of atoms (or atomic positions) given by the solid circles.

cube having an atom at each corner. However, at a given corner, this atom must be shared with seven other identical cubes which fill the volume surrounding this point. Thus, there is effectively only $\frac{1}{8}$ of the atom which can be assigned to that particular unit cell. Since there are eight corners in a cube, there is the equivalent of one atom, and thus the primitive cell and unit cell coincide.

Continuing with the cubic system, consider the body-centered cube shown in Fig. 4-3, no. 13. In this case, there is one atom at the center of the cube and one atom contributed by the eight corners. This cell, then, has two atoms and, to avoid confusion, should be termed a unit cell. For the face-centered cubic lattice, the unit cell is shown in Fig. 4-3, no. 14. There are six face atoms, but each face atom is shared by two cells. Consequently, each face contributes $\frac{1}{2}$ an atom. The faces thus contribute three atoms and the corners one for a total of four atoms in the unit cell. The face-centered cubic structure (FCC) can also be considered as four interpenetrating simple cubic cells.

In the study of crystals, the primitive cell has limited use, because the unit cell more clearly demonstrates the symmetrical features of a lattice. In other words, the unit cell can usually be visualized readily whereas the primitive cell can not. For example, the cubic nature of the face-centered cube is immediately apparent in the unit structure cell in Fig. 4-7 (solid lines), but it is not nearly so obvious in the rhombohedral primitive cell (dashed lines).

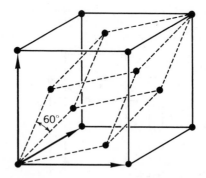

Fig. 4-7. Face-centered cubic lattice showing the unit cell (solid lines) · and the primitive rhombohedral cell (dashed lines).

Another instructive comparison of primitive and unit cells is given by the hexagonal structure. The hexagonal primitive cell has a base which is a parallelogram having equal sides and included angles of 60° and 120°. The third dimension is different from the base dimension and is perpendicular to the basal plane. At first it may be a little difficult to visualize this structure as a hexagon. The difficulty is reduced if three of these primitive cells are put together as shown in Fig. 4-3, no. 8, since this structure is easily seen to be a hexagonal prism. In the case of the hexagonal close-packed (HCP) structure, which is found in many metals, the unit cell can be pictured in at least three different ways as shown in Fig. 4-8.

EXAMPLE 4-1

Determine the number of atoms in the unit cell of the HCP structure.

Solution:

(a) Each position at the "corner" of the base hexagon is shared by six hexagons. Thus each such position in a unit cell is the equivalent of

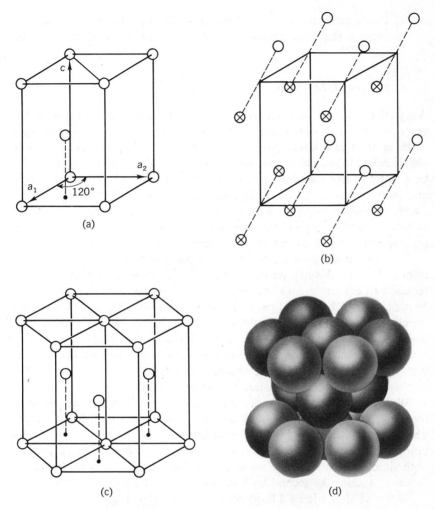

Fig. 4-8. The hexagonal close-packed structure as visualized in different but interchangeable ways: (a) As a parallelepiped unit cell with two atoms; (b) as a parallelepiped unit cell with two atoms associated with each lattice point; (c) as a hexagonal prism with six atoms, showing the symmetrical arrangement; (d) as a representation closer to fact than the open arrangement used for pictorial convenience in (c).

1/6 atom. Since there are twelve such positions, these are the equivalent of two atoms.

(b) Each position at the geometric center of the base hexagon is shared by two hexagons. Thus each such position in a unit cell is the equivalent of 1/2 atom. Since there are two such positions, these are the equivalent of one atom.

(c) There are three atoms completely within the hexagonal prism.
(d) Adding the three items above gives the equivalent of six atoms per unit cell.

4-2-2 PACKING OF ATOMS

A crystal structure is a regular array of atoms arranged on one of the fourteen space lattices. The least complicated crystal structures are those having a single atom at each lattice point. Polonium has the simplest structure, being simple cubic. In normal metals, the atoms (or positive ions) are held together by a cloud of free electrons so that each atom tends to be attracted equally and indiscriminately to all its geometrically nearest neighbors by the free electrons passing between them. This condition fosters the formation of closely packed structures of the types which can be demonstrated by efficiently packing uniformly sized spheres into a given volume.

The student can gain a better grasp of packing by conducting an experiment as follows. Assume we have a quantity of small spheres which we are required to efficiently pack into a box (a "two-dimensional" approach can be made using a handful of pennies). After some shuffling, it is obvious that the closest possible packing is obtained when the spheres are in contact and their centers occupy positions which correspond to the apices of equilateral triangles. It is also evident that there are two sets of triangles, one set with vertices pointing away from the observer (points up) and the other set with vertices pointing toward the observer (points down).

When a second layer is added, there is closest packing if the spheres in this new layer rest in the hollows formed by the spheres of the first layer. The centers of the spheres in the second layer will lie above the centers of the points-up triangles or above the centers of the points-down triangles but not both simultaneously. Which set is used is immaterial. For our discussion, however, assume the second layer is centered on the points-up triangles.

When we start adding a third layer, the spheres will again rest in the hollows formed by the spheres in the second layer. And again we have the option of placing the third layer on the points-up or on the points-down triangles. If we center the third layer on the points-down triangles, we find the third layer is directly above the first layer. If additional layers are added using an alternate stacking sequence, i.e., alternately centering the layers on the points-up and the points-down triangles, the sequence can be written as ABABABAB. . . . This arrangement of spheres, translated to an arrangement of atoms, is the hexagonal close-packed (HCP) structure.

We found earlier that many elements having covalent bonding form arrangements in which the coordination number is $8 - N$, where N is the number of valence electrons. What, then, is the coordination number of the HCP structure just demonstrated? The geometry of the structure shows that any one atom has twelve equidistant neighbors. It is apparent that, in any layer, a given sphere is tangent to six other spheres. When a sphere is placed in the adjoining layer it fits in the hollow formed by three spheres and thus is tangent to three spheres in the adjoining layer. Thus any given atom in the HCP structure is tangent to twelve other atoms, six in its own layer and three each in two adjoining layers.

When the third layer of spheres was added in the above discussion, we assumed this layer was centered on the points-down triangles of the second layer. What happens if the centers of the points-up triangles of the second layer are used instead? The distribution of spheres in the third layer is the same as in the first two layers but does not lie directly above either of these two layers. If a fourth layer is added, centered on the points-up triangles of the third layer, we find the fourth layer is directly above the first layer and duplicates it completely. The stacking sequence for this structure can then be written as ABCABCABC. . . . This arrangement has the same density of packing and the same coordination number as the HCP structure. However, it is the face-centered cubic (FCC) structure. This may not be readily apparent by comparison with Fig. 4-6(b) since the layers described above are actually octahedral or diagonal planes of the cube.

We can show that, in HCP crystals, the ideal axial ratio, c/a (i.e., the ratio of the unit cell height, c, to the distance, a, between neighboring sites in the basal plane), is 1.633 when calculated using hard, perfect spheres. In actual elements, however, this ratio varies from about 1.58 (Be) to 1.89 (Cd). Presumably this indicates that atoms are actually ellipsoidal rather than spherical since there is no reason to assume they are not in "contact."

Since the HCP and FCC crystal structures have the same density of packing and the same coordination number, we might expect the behavior of the two structures to be very much alike with regard to physical and mechanical properties. Actually, this is not the case as will be demonstrated later.

No other crystal structure can be represented by spheres packed together to occupy a total minimum volume. In other words, the maximum density of packing is found only in the HCP and FCC crystal structures. The BCC unit cell contains two atoms, and the coordination number is eight. There is partial compensation for this in the fact that there are six next nearest neighbors at distances only slightly greater than that of the eight nearest neighbors. Some characteristics of cubic structures are given in Table 4-2.

EXAMPLE 4-2

What is the coordination number of (a) iodine, (b) selenium, (c) arsenic, (d) copper?

Solution:

Parts (a), (b), and (c) are materials with covalent bonding and follow the $8 - N$ rule.

(a) $8 - N = 8 - 7 = 1$

(b) $8 - N = 8 - 6 = 2$

(c) $8 - N = 8 - 5 = 3$

(d) Copper is an FCC metal. In the (111) plane, any given atom is tangent to six other atoms in the same plane. It is tangent to three in an adjoining parallel plane. There are two such parallel adjoining planes. Thus the number of nearest neighbors, i.e., the coordination number, is twelve.

Table 4-2. Characteristics of cubic lattices

	Simple	Body-centered	Face-centered
Unit cell volume	a^3	a^3	a^3
Lattice points per cell	1	2	4
Nearest neighbor distance (diameter)	a	$a\sqrt{3}/2$	$a/\sqrt{2}$
Number of nearest neighbors	6	8	12
Second nearest neighbor distance	$a\sqrt{2}$	a	a
Number of second neighbors	12	6	6

4-3 LATTICE PLANES AND DIRECTIONS

It is desirable to have a system of notation for planes within a crystal or space lattice such that the system specifies orientation without giving position in space. Miller indices are used for this purpose. These indices are based on the intercepts of a plane with the three crystal axes, i.e., the three edges of the unit cell. The intercepts are measured in terms of the edge lengths or dimensions of the unit cell which are the unit distances from the origin along the three axes. For instance, the plane that cuts the x-axis at a distance from the origin equal to one half the x-dimension of the cell is said to have an x-intercept equal to $\frac{1}{2}$, and if it cuts the y-axis at $\frac{1}{2}$ the y-dimension of the cell, the y-intercept is $\frac{1}{2}$, regardless of the relative magnitudes of the x- and y-dimensions. If a plane is parallel to an axis, it intercepts the axis at infinity.

To determine Miller indices of a plane, we take the following steps:

1. Find the intercepts on the three axes in multiples or fractions of the edge lengths along each axis.
2. Determine the reciprocals of these numbers.
3. Reduce the reciprocals to the three smallest integers having the same ratio as the reciprocals.
4. Enclose these three integral numbers in parentheses, e.g., (hkl).

Thus, a plane shown intersecting the axes in Fig. 4-9(a) has intercepts 1, 1, 1, and therefore, has Miller indices (111). A plane having intercepts of

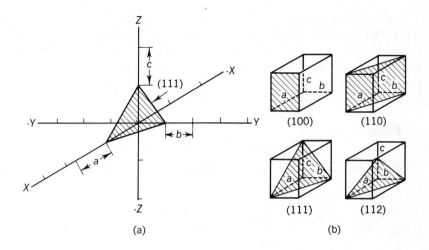

(a)

(b)

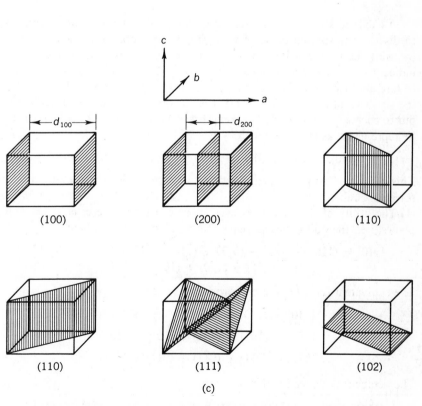

(c)

Fig. 4-9. Miller indices of lattice planes in two coordinate axis orientations.

1, 2 and infinity has reciprocal intercepts of 1, $\frac{1}{2}$, and 0 and Miller indices (210). Fig. 4-9 also shows some of the more important planes in the unit cell. It should be kept in mind that all planes parallel to the crosshatched ones have the same indices. If a plane cuts any axis, for example the y-axis, on the negative side of the origin, the corresponding index will be negative and is indicated by placing a minus sign above the index ($h\bar{k}l$).

EXAMPLE 4-3

A triclinic cell [Fig. 4-3(1)] has a plane passed through it with intercepts of $a/2$, $3b/2$, and c. Find the Miller indices of the plane.

Solution:

Step 1:	1/2	3/2	1
Step 2:	2	2/3	1
Step 3:	6	2	3
Step 4:	(623)		

A cube has six equivalent faces. If we have a definite orientation and wish to discuss one specific plane of these six, it is possible to specify this plane by using the proper Miller indices. Parentheses are used around the Miller indices to signify a specific plane. On the other hand, it is often advantageous to talk about planes of a "form," i.e., a family of equivalent planes such as the six faces of a cube. To do this it is customary to use the Miller indices but to enclose them in curly brackets (braces). Thus the set of cube faces can be represented as {100} in which

$$\{100\} = (100) + (010) + (001) + (\bar{1}00) + (0\bar{1}0) + (00\bar{1})$$

This notation thus provides a shorthand scheme to avoid writing the indices for all six cube faces.

The utility of the scheme is even more evident in the case of the (110) planes, i.e., the dodecahedral planes (in a cubic system), where

$$\{110\} = (110) + (101) + (011) + (\bar{1}10) + (\bar{1}01) + (0\bar{1}1) + (1\bar{1}0) \\ + (10\bar{1}) + (01\bar{1}) + (\bar{1}\bar{1}0) + (\bar{1}0\bar{1}) + (0\bar{1}\bar{1})$$

The equivalent form for the orthorhombic system is

$$\{110\} = (110) + (\bar{1}10) + (1\bar{1}0) + (\bar{1}\bar{1}0)$$
$$\{101\} = (101) + (\bar{1}01) + (10\bar{1}) + (\bar{1}0\bar{1})$$
$$\{011\} = (011) + (0\bar{1}1) + (01\bar{1}) + (0\bar{1}\bar{1})$$

The octahedral planes for the cube are

$$\{111\} = (111) + (\bar{1}11) + (1\bar{1}1) + (11\bar{1}) + (\bar{1}\bar{1}1) + (\bar{1}1\bar{1}) + (1\bar{1}\bar{1}) \\ + (\bar{1}\bar{1}\bar{1})$$

The stepwise procedure and notational scheme for expressing Miller indices are given in Table 4-3.

Table 4-3. Indices for lattice planes and directions

Procedure	Plane			Direction		
Intercept or projection on crystal axes (multiple or fraction of edge distance, i.e., lattice parameter)	x 1	y 2	z 3	x 1	y 2	z 3
Reciprocal of intercept	1	$\frac{1}{2}$	$\frac{1}{3}$	...		
Smallest integers with same ratio	6	3	2	1	2	3
Indices	(632) (Miller Indices)			[123] (Direction Indices)		
Generalized Indices	(hkl)			[uvw]		
Indices of family or set of equivalents	{632}			<123>		
Generalized indices	{hkl}			<uvw>		

Direction indices are defined in a different manner. A line is constructed through the origin of the crystal axes in the direction under consideration and the coordinates of a point on the line are determined in multiples of lattice parameters of the unit cell. The indices of the direction are taken as the smallest integers proportional to these coordinates and are closed in square brackets. For example, suppose the coordinates are $x = 3a$, $y = b$, and $z = c/2$, then the smallest integers proportional to these three numbers are 6, 2, and 1 and the line has a [621] direction. As further examples, the x-axis has direction indices [100], the y-axis [010], and the z-axis [001]. A face diagonal of the xy face of the unit cell has direction indices [110], and a body diagonal of the cell has direction indices [111]. Negative indices occur if any of the coordinates are negative. For example, the $-y$-axis has indices [0$\bar{1}$0]. A full set of equivalent directions, i.e., directions of a form, are indicated by carets: $<uvw>$. The procedure and notational schemes are given in Table 4-3.

Reciprocals are used in computing indices of a plane but are not used in computing indices of a direction. A frequent error assumes a direction is always perpendicular to a plane having the same indices. This is true for the cubic system but is not true, in general, for other systems.

EXAMPLE **4-4**

Consider the triclinic cell in the previous example with a line passing through the origin and having projections of $a/2$, $3b/2$, and c. Find the direction indices of the line.

Solution:

Step 1:   1/2   3/2   1
No reciprocals
Step 3:   1   3   2
Step 4:   [132]

The above plane and direction indexing scheme is applicable to all lattices with a sometime exception of the hexagonal. Here again, the basic method is applicable, but indices are often written in terms of four numbers rather than three. The unit cell [Fig. 4-8(a)] is defined by two equal and coplanar vectors at 120° to each other and by another vector at right angles to this plane. Because of symmetry, a third axis in the basal plane of the hexagonal prism [Fig. 4-8(c)], making an angle of 120° with each of the other two basal plane vectors, is often used with them. The indices in the hexagonal system, called Miller-Bravais indices, refer to these four axes and use the notation $(hkil)$, where i refers to the third basal plane index. The three indices, h, k, and i are not independent but are related through the equation $h + k + i = 0$. The side or prismatic planes are all similar and are symmetrically located. This is illustrated in Fig. 4-10 and by the standard notation for a family of

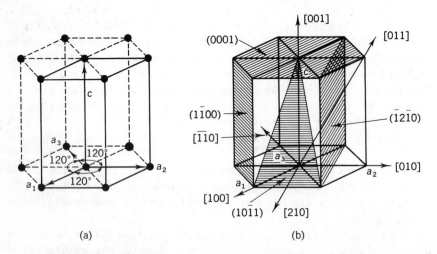

(a)   (b)

Fig. 4-10. Simple hexagonal structure cell showing indices of some planes and directions.

planes:

$$\{10\bar{1}0\} = (10\bar{1}0) + (01\bar{1}0) + (\bar{1}100) + (\bar{1}010) + (0\bar{1}10) + (1\bar{1}00)$$

If directions in a hexagonal lattice are expressed in terms of four vectors [$uvtw$] then the notation for a family of directions is typified by

$$\langle 2\bar{1}\bar{1}0 \rangle = [2\bar{1}\bar{1}0] + [\bar{1}2\bar{1}0] + [\bar{1}\bar{1}20]$$

for the three axes in the basal plane. The equation $u + v + t = 0$ applies.

4-4   LATTICE GEOMETRY AND ATOMIC RADII

In determining indices of planes and directions, the intercepts or projections are measured in terms of multiples of the edge lengths or lattice parameters of the unit cell.

A knowledge of lattice parameters and the unit cell enables us to determine a dimension which can be regarded, within limitations, as the radius of the atom. If we assume atoms are spheres, then, knowing the structure, it is a simple geometrical problem to calculate the radius of the sphere in terms of the lattice parameters. This is the closest approach we can make to defining the size of an atom for there is no fixed boundary which can be used to determine atomic size in the way the radius of a baseball can be determined. As we might expect, the atomic radius of a particular element is not the same in all crystals but depends upon the characteristics of the interatomic forces which are acting. The radius of an ion of a metal in an ionic crystal (i.e., a salt) is smaller, frequently considerably so, than the radius of the same element in a metallic crystal. For example, the radius of the sodium ion in salts is 0.98 A, whereas its radius in the metallic state is 1.85 A. It is obvious that, when we state the atomic radius of an element, the type of crystal from which the value is obtained must be specified. If comparison is limited to consideration of crystals with the same type of binding force, then the atomic radius of an element is fairly constant. Variations do occur, however, and atomic radius increases with increasing coordination number. The interatomic distances of some elements are given in Table 4-4. These values represent the distances between the centers of nearest neighbors in the crystal and may be regarded as the atomic diameter. We see there can be differences in atomic size of the same element in different crystal structures.

The distance between parallel planes with a given set of Miller indices is a useful bit of information which can be determined from geometry and from lattice parameters. Planes with large interplanar spacing have low indices and a high density of lattice points, whereas the reverse is true for planes of close spacing. If we consider the simple cube shown in Fig. 4-11, it is obvious

Table 4-4. Crystal structures of some elements

Element	Crystal structure	Closest interatomic distance (Angstroms)	Lattice parameters a (Angstroms)	Axial ratio c/a	Temperature of measurement, °C
Aluminum	FCC	2.862	4.049	—	20
Barium	BCC	4.35	5.025	—	20
Beryllium	HCP	2.225	2.285	1.57	20
Cadmium	HCP	2.979	2.979	1.88	20
Calcium	FCC	3.94	5.57	—	20
Chromium	BCC	2.498	2.884	—	20
Cobalt	HCP	2.506	2.507	1.62	20
	FCC	2.511	3.552	—	room*
Copper	FCC	2.556	3.615	—	20
Gold	FCC	2.884	4.078	—	20
Iridium	FCC	2.714	3.839	—	20
Iron	BCC	2.481	2.866	—	20
	FCC	2.585	3.656	—	950
Lead	FCC	3.499	4.950	—	20
Lithium	BCC	3.039	3.509	—	20
Magnesium	HCP	3.196	3.209	1.62	20
Molybdenum	BCC	2.725	3.147	—	20
Nickel	FCC	2.491	3.524	—	20
Platinum	FCC	2.775	3.924	—	20
Potassium	BCC	4.627	5.344	—	20
Rhodium	FCC	2.689	3.803	—	20
Rubidium	BCC	4.88	5.63	—	−173
Silver	FCC	2.888	4.086	—	20
Sodium	BCC	3.715	4.291	—	20
Strontium	FCC	4.31	6.087	—	20
Tantalum	BCC	2.860	3.303	—	20
Thorium	FCC	3.60	5.088	—	20
Titanium	HCP	2.89	2.950	1.60	25
	BCC	2.89	3.33	—	900
Tungsten	BCC	2.739	3.165	—	20
Uranium	BCO	2.762	a = 2.854 b = 5.869 c = 4.956	—	25
	Tet		a = b = 10.763 c = 5.652	—	720
	BCC	3.02	3.524	—	805
Vanadium	BCC	2.632	3.039	—	20
Zinc	HCP	2.664	2.664	1.86	—
Zirconium	HCP	3.17	3.230	1.59	—
	BCC	3.13	3.62	—	867

*FCC cobalt is metastable at room temperature.

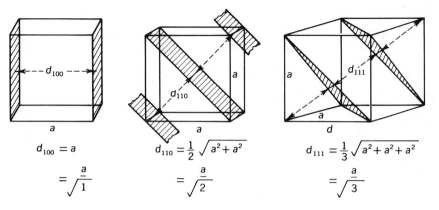

Fig. 4-11. Some plane spacings in a simple cubic lattice.

that, in this case, the most widely spaced atomic planes are those spaced at intervals equal to the lattice parameters, i.e., the (100) planes. The spacing is

$$d_{100} = a \tag{4-1}$$

The dodecahedral planes (110) bisect the face diagonal for a spacing of

$$d_{110} = \tfrac{1}{2}\sqrt{a^2 + a^2} = a/\sqrt{2} \tag{4-2}$$

The octahedral planes (111) trisect the body diagonal for a spacing of

$$d_{111} = \tfrac{1}{3}\sqrt{a^2 + a^2 + a^2} = a/\sqrt{3} \tag{4-3}$$

General equations for interplanar spacing for various unit cells can be written in terms of the Miller indices; namely:

$$\text{cubic:} \quad \frac{1}{d_{hkl}^2} = \frac{h^2 + k^2 + l^2}{a^2} \tag{4-4}$$

$$\text{tetragonal:} \quad \frac{1}{d_{hkl}^2} = \frac{h^2 + k^2}{a^2} + \frac{l^2}{c^2} \tag{4-5}$$

$$\text{orthorhombic:} \quad \frac{1}{d_{hkl}^2} = \frac{h^2}{a^2} + \frac{k^2}{b^2} + \frac{l^2}{c^2} \tag{4-6}$$

$$\text{hexagonal:} \quad \frac{1}{d_{hkl}^2} = \frac{4}{3}\left(\frac{h^2 + hk + k^2}{a^2}\right) + \frac{l^2}{c^2} \tag{4-7}$$

EXAMPLE 4-5

(a) Which planes are farthest apart in the simple cubic structure?
(b) Which planes are farthest apart in the BCC structure?

Solution:

(a) Applying Eq. (4-4), the spacing between (100) planes is the cube dimension *a*. For higher indices, the perpendicular distance between adjacent parallel planes will be less than *a*.

(b) Applying Eq. (4-4), it would appear that the (100) planes are the farthest apart. Examination of the structure, however, shows that planes of the (100) type are actually located at $a/2$ and a, thus having an effective spacing of $a/2$. The spacing between (110) planes [from Eq. (4-4)] is $a/\sqrt{2}$. This is the largest perpendicular spacing between planes in the BCC structure.

4-5   METALLIC STRUCTURES

The concept of the metallic bond arises from the attraction between positive ions and "electron gas." It is spherical in nature and does not put directional, spatial, or numerical limitations on the bond. The bond from any one atom is, therefore, capable of acting on as many neighbors as can be placed around that atom. We thus expect crystal structure to be determined mainly by geometrical considerations in which the basic "building block" is the atom. For a given metal, all the atoms are the same size, although the atomic size is not the same for all metals. It is true that many metals have either FCC or HCP structures, both of which have the densest possible packing. A number of others have the BCC structure. Essentially all of the common metals have one of these three structures as indicated in Fig. 4-6. The reason for one type of arrangement in preference to another depends on the average potential energy of the structure. In all cases, the most stable state is one of minimum free energy. Since nonnearest neighbors contribute to the potential energy of an atom, as well as the nearest neighbors, slight changes in the long-range order can alter the average potential energy of the whole structure. On this basis, the BCC structure with a coordination number of 8 but with six next-nearest neighbors may have a lower free energy than either the FCC or HCP structure with a coordination number of 12.

One of the exceptions, uranium, is a heavy metal with many unique aspects. The room temperature structure is orthorhombic as shown in Fig. 4-12. This structure is sometimes visualized as a stack of "corrugated" sheets of atoms parallel to the *a-c* plane with about 2.8 A between the atoms in the sheets and about 3.3 A between the sheets. Even though uranium has a number of clearly metallic characteristics, the bonding between the atoms is largely covalent, being somewhat analogous to arsenic, antimony, and bismuth.

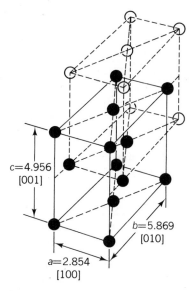

Fig. 4-12. Crystal structure of alpha uranium. The structure cell is shown by solid circles. Additional lattice points are indicated by open circles to better illustrate the structure.

4-6 CERAMIC STRUCTURES

From the definition in Sec. 1-2, it is obvious that ceramics is somewhat of a miscellaneous or catchall classification and thus includes a great variety of materials with a broad spectrum of structures and characteristics. It would be impossible to discuss all of them within the scope of this text. Therefore, we shall discuss only a few typical ones.

4-6-1 IONIC STRUCTURES

In those crystal structures having a large degree of ionic bonding (oxides, halides, silicates, for example), the structure is substantially determined on the basis of the way in which positive and negative ions can be stacked together on one of the Bravais lattices so that the electrostatic attractive forces are maximized and the electrostatic repulsive forces are minimized. Cations are generally smaller in size than anions. Many ionic structures can be constructed by having the smaller, positively-charged cations located in interstices between the larger, negatively-charged anions. The coordination number of anions around a cation is determined by the geometry required for the cation to stay in contact with each anion. This geometry is fixed by the ratio of the radius of the cation relative to the radius of the anion. The range of radius ratios for each coordination number is given in Table 4-5. These critical radius ratios are not always exact since the atom does not always have spherical symmetry as it may be distorted or polarized by neighboring ions. This is especially true for high atomic weight, large, easily distorted anions adjacent to small, highly charged cations. In addition, the presence of directional covalent bonding can have a definite effect.

Table 4-5.   Radius ratio and coordination number

Rr	CN	Symmetry
< 0.155	1 or 2	Linear or bent molecules
0.155–0.225	3	Trigonal planar environment of A
0.225–0.414	4	Tetrahedral environment of A
0.414–0.732	4	Square planar environment of A
0.414–0.732	6	Octahedral environment of A (example, NaCl)
0.732–1.0	8	Body-centered-cubic, or twisted-cubic environment of A
> 1.0	12	Close-packed structure of metals where all atoms are alike

Many oxides and halides crystallize in the rock salt structure illustrated in Fig. 4-13(a). In this arrangement, the large anions are in FCC packing with all octahedral interstitial positions filled with small cations. Thus the structure is two interpenetrating FCC lattices with the corner of one located at point $\frac{1}{2}$, 0, 0 of the other. Typical oxides which have this structure are MgO, CaO, SrO, BaO, CdO, MnO, FeO, CoO and NiO. All the alkali halides except CsCl, CsBr, and CsI also have this structure. The latter have a BCC type of structure as shown in Fig. 4-13(b).

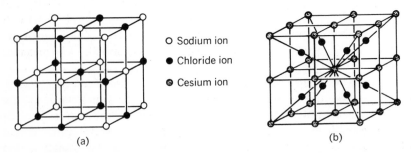

O Sodium ion

● Chloride ion

⊕ Cesium ion

(a)   (b)

Fig. 4-13. Crystal structures of typical ionic solids: (a) Sodium chloride; (b) cesium chloride.

A number of oxides, fluorides, and some intermetallic compounds have the fluorite (CaF_2) structure shown in Fig. 4-14. This is FCC with Ca at the cube corners and face centers and F at all quarter points along the cube diagonals. ThO_2, TeO_2, and UO_2 have this structure. The large number of vacant interstitial sites is advantageous when UO_2 is used as a nuclear fuel, as many of the fission products can be accommodated in the vacant lattice positions.

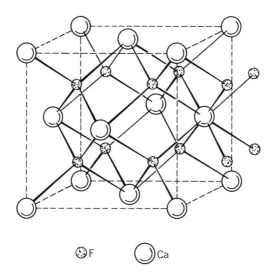

⊙ F   ◯ Ca

Fig. 4-14. Fluorite structure.

Beryllia (BeO) has a radius ratio of 0.25, requiring tetrahedral coordination of four oxygen ions about each beryllium ion. But each oxygen anion must be coordinated with four beryllium cations. This can be achieved by cubic close packing of the large oxygen anions with half the tetrahedral interstices with beryllium cations. This structure is shown in Fig. 4-15. It also occurs for ZnS and is generally known as the zinc blende structure. Note its similarity to diamond (Fig. 3-10). This structure can also be regarded as interpenetrating FCC lattices with the corner of one located at the position $\frac{1}{4}, \frac{1}{4}, \frac{1}{4}$ of the other. The so-called adamantine compounds having this structure have covalent bonding and are formed between elements which are symmetrically located in the periodic table (Table 3-3) with respect to the column containing C, Si, Ge, and Sn. An alternative structure for some of these compounds is the wurtzite structure (ZnS, ZnO) shown in Fig. 4-16 in which one kind of atom is located on HCP positions and atoms of the other kind are located at intermediate points. Each atom of one kind is symmetri-

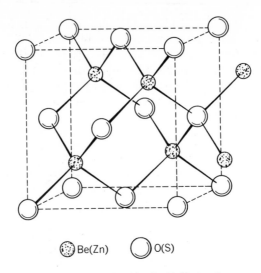

Be(Zn)   O(S)

Fig. 4-15. BeO (zinc blende, ZnS) structure.

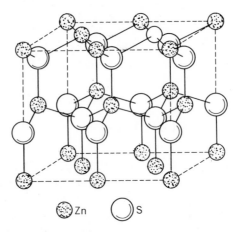

Zn   S

Fig. 4-16. Wurtzite (ZnS) structure (also oxygen positions in H_2O).

cally surrounded by four atoms of the other kind. Ice forms with the oxygen atoms arranged in a slightly distorted wurtzite structure.

Alumina (Al_2O_3) requires four aluminum cations adjacent to each oxygen anion. This can be achieved by approximately HCP packing of the oxygen ions with two-thirds of the octahedral interstitial sites filled with aluminum ions. Successive similar layers are built up so that maximum spacing of the aluminum ions is obtained.

A variation of the cubic structure is perovskite ($BaTiO_3$), which is FCC, having Ba at the cube corners, O at the face centers, and Ti at the body center (Fig. 5-21). The Ti is actually in an octahedral interstice in the FCC structure formed by Ba and O. This structure is found in several compounds.

Alloying of many of the transition metals with metalloids often results in the nickel arsenide (NiAs) structure, which has alternating layers of metal and metalloid atoms. Each layer is essentially a basal plane of the HCP structure.

Graphite (Fig. 4-17) forms a layer structure somewhat similar to that of NiAs, having strong covalent bonding between atoms in the hexagonal layer planes with weak (van der Waals) bonding between the planes. This gives a structure with very strong directional characteristics. This can be an advantage when graphite is used as a lubricant. It is a disadvantage when graphite is used as a moderator in nuclear reactors. Boron nitride (BN) also forms a layer structure with the B and N atoms alternating in position within the layer planes.

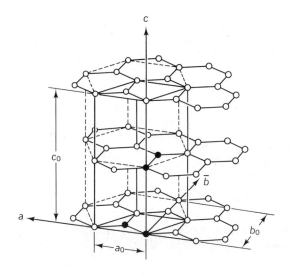

Fig. 4-17. Classical crystal structure of graphite. The ortho-hexagonal axes are indicated by a, b, and c. The four black atoms indicate a hexagonal structure cell. $a_0 = 2.46$ A, $b_0 = 4.28$ A, $c_0 = 6.71$ A.

4-6–2 COVALENT STRUCTURES

Covalent crystals form when a repeated structure can be built which is consistent with the highly directional nature of the covalent bond. Carbon, for example, forms four tetrahedral bonds (with sp^3 hybridization). In methane, CH_4, all four are used in forming the molecule so that no crystal can be built. In contrast, however, carbon provides the "classical" example of the covalent crystal in diamond which follows the $8 - N$ coordination number rule. This structure (shown in Fig. 4-18: see also Fig. 3-10) is found in carbon, germanium, silicon, tin at low temperatures, and in certain compounds. Each atom has four nearest neighbors giving a configuration that is variously called diamond cubic, body-centered tetrahedral, or tetrahedral cubic.

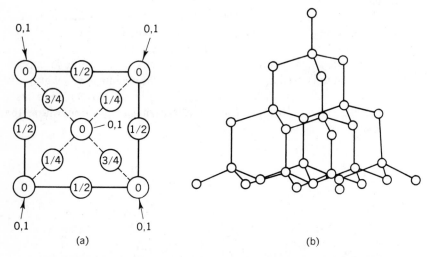

(a)   (b)

Fig. 4-18. The structure of diamond: (a) Projection on the basal plane (numbers in circle indicate height of atom above basal plane as a multiple of the cell dimension); (b) perspective drawing showing each atom with four nearest neighbors.

From the discussions on metallic structures, we might expect to find that a structure of only one kind of atom, such as carbon, would be relatively closely packed with a coordination number of 8 or 12. The fact that the coordination number is 4 is a direct consequence of the sp^3 hybridization shown in Fig. 4-19.

Covalent crystals, such as diamond and SiC, have high hardness, high melting points, and (in high purity specimens) low electrical conductivity at low temperature. In addition to purely covalent crystals, a number of other crystals have a significant contribution of covalent bonding. This has an influence on the crystal structure.

Many oxide and silicate structures can be built by combining regular tetrahedral (SiO_4) or octahedral (AlO_6, MgO_6) polyhedra. It has long been recognized that the silicates have a peculiarly complex chemistry. The silicates have the characteristics of having an enormous variety of compositions, being difficult to assign significant formulae, and a close morphological association of seemingly unrelated compounds.

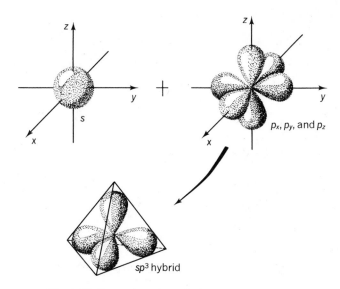

Fig. 4-19. Formation of tetrahedral sp^3 hybrids.

The key to all silicate structures is the tetrahedral coordination of silicon by oxygen with four oxygen ions invariably arrayed around a central silicon ion as shown for SiO_4^{4-} in Fig. 4-20. The SiO_4 tetrahedra can be arranged in various structures by sharing a corner, by sharing an edge, or by sharing a face in common. The shared corner is a more stable arrangement than the shared edge while shared faces very seldom occur. The SiO_4 tetrahedra can be arranged in structures such that corners are shared in a variety of ways as indicated in Fig. 4-20. There are four general types. In the orthosilicates (SiO_4^{4-}), the tetrahedra are independent. In pyrosilicates, two tetrahedra share one corner to form ions ($Si_2O_7^{6-}$). In metasilicates, two corners are shared to form a variety of chain or ring structures ($SiO_3^{2-} - (SiO_3)_n^{2n-}$). Layer structures ($(Si_2O_5)_n^{2n-}$) are made from tetrahedra which share three corners. In the various forms of silica (SiO_2) four corners are shared.

4-6-3 NONCRYSTALLINE SOLIDS

All of the above structures, including the silicates, are crystalline in the sense of a repetitive three-dimensional network. There are important ceramic materials, however, which do not have the periodicity characteristic of the crystalline state. These noncrystalline solids include the glasses, inorganic cements, and adhesives.

Noncrystalline solids are materials which lack long-range order based on a lattice. This also occurs in liquid structures in which each atom is sur-

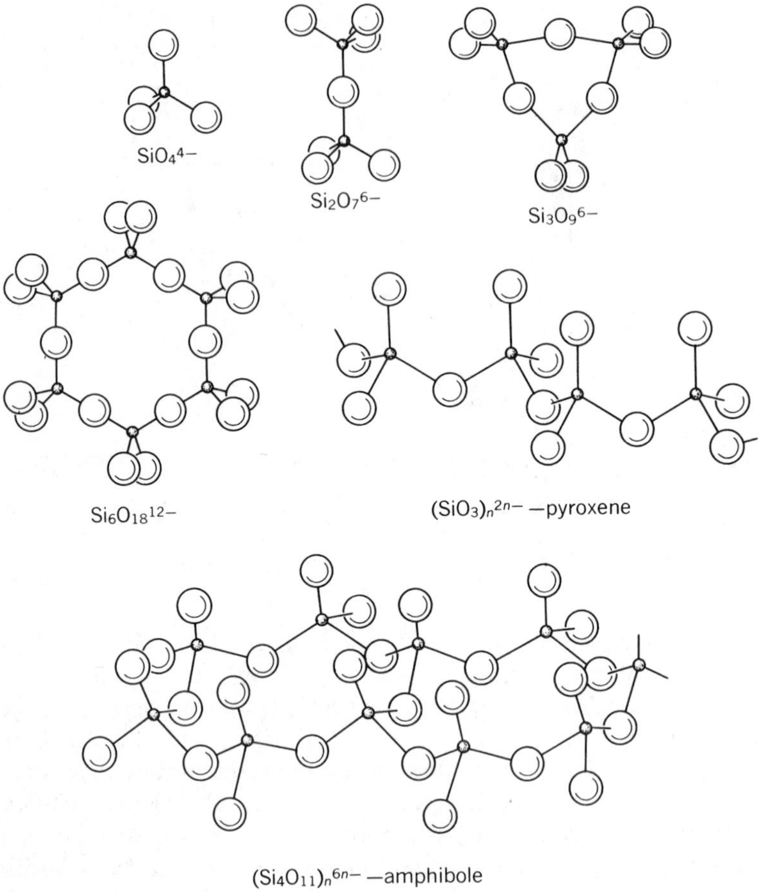

Fig. 4-20. Some silicate ions and chain structures.

rounded by about the same number of other atoms (nearest neighbors) but the location of the second coordination ring is not precisely fixed while the location of the third coordination ring is highly indefinite. The distinction between a liquid and a noncrystalline solid having this structure is the time required for each to react to some external force, e.g., mechanical stresses. The reaction time in the liquid is small compared with the reaction time in the noncrystalline solid.

Glasses such as fused silica have extended three-dimensional structures built from coordinated groups which have definite first order coordination but lack long-range periodicity. Comparable structures of a crystal and a noncrystalline network are shown in Fig. 4-21. Such a random network structure is the major one for noncrystalline solids and is found for oxide glasses,

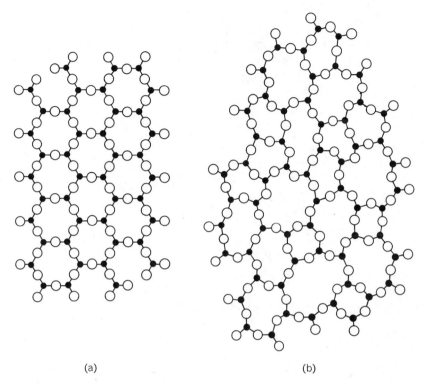

(a)   (b)

Fig. 4-21. (a) Regular crystalline lattice and; (b) corresponding irregular glassy network.

hydrogen-bonded inorganic solids, and organic compounds containing hydroxyl groups.

In addition to SiO_2, glasses can be formed from B_2O_3, GeO_2, P_2O_5, As_2O_5 and from combinations of these and other oxides such as silicates, borates, germinates, phosphates, and arsenates. BeF_2 readily forms a glass. Sulfur, Se, and Te, among the elements, form glasses. A number of organic compounds, e.g., alcohols and long-chain compounds, can be made to form glasses.

Formation of polycomponent glasses does not require that all constituents be glass forming cations. A glass can be formed so long as the composition has a high percentage of cations surrounded by oxygen in triangular or tetrahedral coordination, the polyhedra share corners, and some oxygen atoms are bonded to two cations but do not share bonds with other cations. A sodium silicate, for example, can form a glass in which the sodium atoms fit in holes in the network without contributing to formation of the network as shown in Fig. 4-22. In all cases, the network structure is a function of

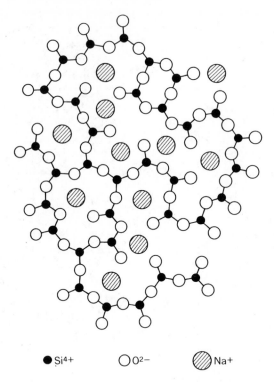

$\bullet$ Si^{4+} $\bigcirc$ O^{2-} $\oslash$ Na^{+}

Fig. 4-22. Schematic representation of a sodium silicate glass.

the composition ratio of oxygen to silicon and the size of the modifying ion and its ability to fit into the interstices in the random network.

4-7 POLYMERIC STRUCTURES

Polymers are generally organic amorphous or crystalline materials in which large chainlike macromolecules are the basic structural units. Organic structures can not be as systematically classified as can metallic, ionic, and covalent structures. In metals we are concerned with packing of "spherical" units (atoms) under the influence of (relatively) nondirectional forces, and thus the major concern is with geometry. In ionic structures, geometry is a major factor but it is modified by the requirement to balance the electrostatic forces to give a minimum free energy. In covalent structures the bonds are limited in number and strongly oriented in direction but still function between individual atoms. In polymers, however, the basic structural unit is not an atom but a complete molecule which may have an irregular and often complex shape. The interatomic bonding within the molecules is strongly covalent. The bonding between the molecules may be relatively weak van der Waals bonds, significantly stronger hydrogen bonds, or crosslinks between atoms of adjacent chains. Each substance has its own distinctive structure, and we can only discuss some structural features of common occurrence in very general terms.

Polymeric structure is formed by combining molecules of ordinary and repetitive size and composition (monomers) through processes such as polymerization and polycondensation. Crystallinity can exist in some polymers but they are never completely crystalline. It is convenient to classify structures based on the shape of the monomer. We assign the molecules to one of three groups: (1) small or symmetrically shaped; (2) long; or (3) more or less flat. This division is rather arbitrary and somewhat artificial since some molecules can be assigned to more than one group. There are also some molecules of complex compounds which cannot be properly assigned to any. In many structures, the molecules tend to assume as closely-packed an arrangement as is consistent with their shapes if intermolecular bonding is van der Waals. When other kinds of intermolecular bonding predominate, the molecular shape may have little effect in determining the structural arrangement. In general, however, the physical (macroscopic) arrangement depends on the chemical structure of the molecules, i.e., the type of monomer, the arrangement in the chains or networks, the extent of branching, and the stereoregularity of the macromolecules. Polymers can be classified according to: (1) the method of synthesis; (2) the kind of chain structure; (3) use or application; (4) properties.

4-7-1 ORGANIC MOLECULES

Organic molecules are formed when carbon atoms bond to other units. The saturated* hydrocarbons, for example, consist of repeating units of carbon with each carbon atom bonded to four other atoms with the internuclear axes directed toward the corners of a tetrahedron. The simplest example is methane, CH_4, whose bonding orbitals would look like those in Fig. 4-19. With the carbon atom at the center, the hydrogen atoms are positioned to define a regular tetrahedron.

Carbon atoms can also form other kinds of bonding units. This means different combinations of the $2s$ and $2p$ states in the carbon atom. The basic tetrahedron, discussed above, is shown in Fig. 4-23(a). The "carbonyl" bonding [Fig. 4-23(b)] has two single bonds and one double bond. The allene

---

*A molecule is *saturated* when there are only single bonds between carbon atoms. *Unsaturated* molecules are formed when more than one bond exists between two atoms.

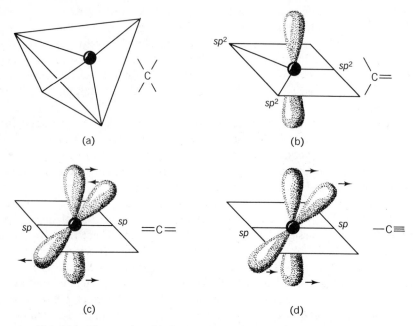

Fig. 4-23. The atomic orbitals of carbon can be combined in a number of different ways to give a variety of hybrid bonding orbitals: (a) Tetrahedral carbon atom; (b) carbonyl atom; (c) allene atom; (d) acetylene atom.

bonding unit [Fig. 4-23(c)] has two double bonds at 180° to each while the acetylene bonding unit [Fig. 4-23(d)] has a triple bond and a single bond at 180° to each other.

More complex molecules are formed when one or more of the hydrogen atoms is replaced by another atom or group. If one hydrogen is replaced by a CH_3 group, an ethane molecule, CH_3—CH_3, is formed. This molecule has one C—C bond and six C—H bonds. The formula, C_2H_6, gives no sense of the structure. The line formula [Fig. 4-24(a)] implies the molecule is planar, but the molecule actually has a form like that indicated in Fig. 4-24(b). The two C atoms are at the centers of two tetrahedra which share a corner (one apex from each tetrahedron) at the midpoint of the line connecting the two C nuclei. The six H atoms are on the other six apices of the two tetrahedra. In order to minimize the repulsion between the H nuclei, the tetrahedra are twisted so that they are separated by 60° [Fig. 4-24(c)].

Ethylene, C_2H_2, is a molecule which bears some similarity to ethane but also some definite differences. The atoms are coplanar. The total strength of the bond between the two carbon atoms is almost, but not quite, double that between the carbon atoms in ethane. In addition, the two CH_2 groups can not rotate relative to each other. The two latter facts indicate an unsaturated bond (double) between the C atoms as indicated in Fig. 4-25.

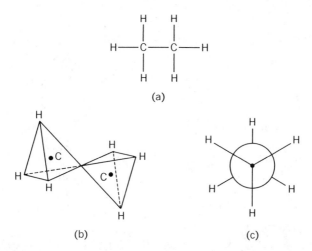

Fig. 4-24. Structure of the ethane molecule, C_2H_6: (a) Line formula; (b) tetrahedral CH_3 groups connected by a bond; (c) side view showing the relative orientation of the two CH_3 groups.

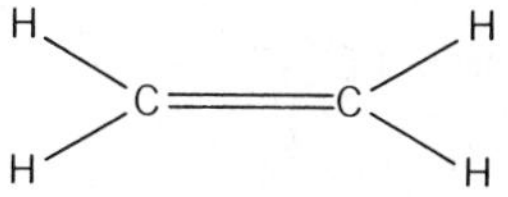

Fig. 4-25. Line formula of ethylene.

Benzene, C_6H_6, is another planar molecule in which the six C atoms are at the corners of a regular hexagon with each carbon bonded to a single H atom located outside the hexagon as shown in Fig. 4-26(a). This arrangement does not account for all the electrons present and thus there are three single bonds and three double bonds in the hexagon as indicated in Fig. 4-26(b). Since all C—C bonds in benzene are equivalent, the double bonds resonate between the positions shown in Fig. 4-26(b) and the alternate positions shown as single bonds. This benzene ring (or phenyl side group) appears in a number of compounds.

A large number of organic molecules are of interest in polymeric materials. For example, the unsaturated molecules, $X_2C{=}CX_2$ (vinyls) and $CX_2{=}CX$ $-CH{=}CX_2$ (dienes), where the X's represent any elements or organic groups which can bond with C, are the basic monomers for a number of commercial polymers.

Tetrahedral C atoms can "string" together to form a "backbone" for a polymer. When this chain grows in two directions only, a "linear" polymer molecule is formed as shown in Fig, 4-27(a). Each C atom is joined to two others by covalent bonds that are formed by overlapping of sp^3 hybrid orbit-

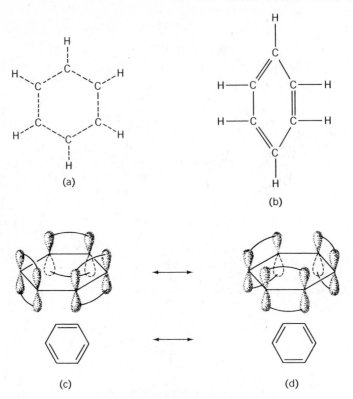

Fig. 4-26. Structure of benzene, C_6H_6: (a) Position of nuclei; (b) line formula; (c, d) the two resonance structures.

als. In the linear hydrocarbon series, C_xH_{2x+2}, pure compounds have been isolated with x as large as 94. Polymethylene, in comparison, is a mixture with x varying from very low values to as great as 10^5 or 10^6. In like manner, other H atoms can be replaced to form an infinite variety of organic compounds. Branched or nonsymmetric hydrocarbons form by further addition of C to the linear chain [Fig. 4-27(b)]. Covalent bonds may also be formed between C and other elements.

4-7-2 SYNTHESIS OF POLYMERS

Polymer molecules are formed from monomers through two principal mechanisms: (1) condensation polymerization (step-growth), and (2) addition polymerization (chain-growth).

Condensation polymerization is similar to the low molecular weight condensation reaction between two reactive groups such as an acid and an

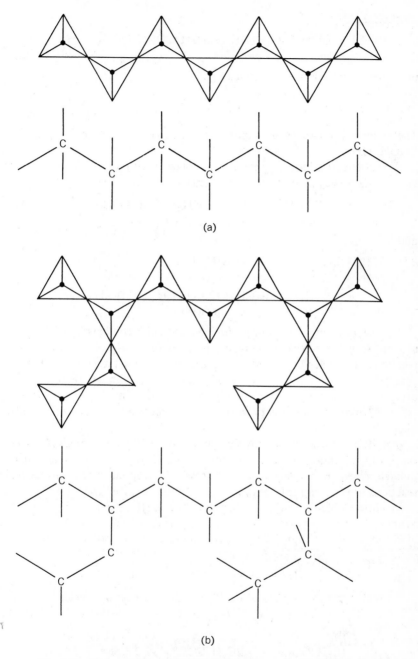

Fig. 4-27. Carbon atoms can be joined through covalent bonding to form the backbone for polymer molecules: (a) Backbone for a linear polymer; (b) branching in linear carbon chains.

alcohol, e.g.,

$$CH_3COOH + C_2H_5OH \longrightarrow CH_3COOCH_2CH_3 + H_2O$$
$$\text{(acid)} \qquad \text{(alcohol)} \qquad\qquad \text{(ester)} \qquad\qquad \text{(water)}$$

or between an acid and an amine, e.g.,

$$R\ COOH + R'NH_2 \longrightarrow R\ CONH\ R' + H_2O$$
$$\text{(acid)} \qquad \text{(amine)} \qquad\quad \text{(amide)} \qquad\quad \text{(water)}$$

The condensation reaction requires molecules with functional groups, e.g., COOH, OH, etc.

For polymers to form, molecules are needed with at least two functional groups. Examples of condensation of macromolecules are:

$$n\ H_2N(CH_2)_{10}COOH \longrightarrow H-[NH(CH_2)_{10}CO]_n-OH + (n-1)H_2O$$
$$\text{(11-aminoundecanoic acid)} \qquad\qquad \text{(nylon-11)}$$

and

$$n\ HOOC(CH_2)_4COOH + n\ H_2N(CH_2)_6NH_2 \longrightarrow$$
$$\text{(adipic acid)} \qquad\qquad \text{(hexamethylene diamine)}$$

$$HO-[CO(CH_2)_4CONH(CH_2)_6NH]-H + (2n-2)\ H_2O$$
$$\text{(nylon-66)}$$

All reactions involve the *stepwise* intermolecular condensation of reactive groups. Any two species can react. The monomer disappears early in the reaction. The molecular weight obtained depends on the conversion and is generally limited to about 100 to 300 monomers although there are some exceptions (e.g., nylons). A relatively long reaction time is generally needed. All molecular species are present at the same time and there are often byproducts.

Addition polymerization involves monomers of the $C=C$ or $C=C-C=C$ types which contain unsaturated double bonds. It is a chain reaction and only electrons are displaced in the process. The reacting species in the chain reactions are free radicals or ions.

Addition polymerization consists of three distinct steps:

1. *Initiation* (opening of the double bond by a free radical or ion):

$$\text{Free radical:}\quad \text{cat} \longrightarrow 2R*$$
$$R* + M \longrightarrow RM*$$

(cat = catalyst or initiator; $R*$ = a free radical produced by decomposition of the catalyst; M = monomer, e.g., styrene, vinylacetate, etc.) e.g.,

$$(C_6H_5COO)_2 \longrightarrow 2C_6H_5COO* \longrightarrow 2C_6H_5* + 2\ CO_2$$

$$\text{Cation:}\quad R^\oplus + {}_{**}C-C \longrightarrow R-C-C^\oplus$$
$$\text{Anion:}\quad R^\ominus + C-C_{**} \longrightarrow R-C-C^\ominus$$

2. *Propagation:*

$$RM^* + (n-1)M \longrightarrow RM_n^*$$

(where n = number of monomer units in the chain)

3. *Termination:* (e.g. by combination or coupling)

$$RM_m^* + RM_n^* \longrightarrow RM_m + {}_nR$$

An example of addition polymerization for a linear polymer is

$$n\ CH_2 = CH-[CH_2-CH]-$$
$$\qquad\qquad | \qquad\qquad |$$
$$\qquad\qquad X \qquad\qquad X$$

where X = Cl, Ph, $\overset{\displaystyle O}{\overset{\|}{C}}OCH_3$, etc.

Addition polymerization for a network or crosslinked polymer is typified by

In all three cases, the chain reaction is initiated when a reactive radical or ion (R^*, $R^\oplus$, or $R^\ominus$) approaches a double bond in the monomer and polarizes one of the electron pairs. Cationic polymerization works best with monomers which have an "electron rich" double bond. Anionic polymerization works best with monomers having an electron deficient double bond such as methacrylonitrile, vinylidene cyanide, etc. Macromolecular size is very large, e.g., 10,000 to 100,000 monomers in a single chain.

Coordination (stereoregular) polymerization is a form of addition polymerization in which, by using special catalysts, the orientation of a monomer can be directed in approaching a growing chain end, resulting in a definite order of chain configuration as shown in Fig. 4-28.

Copolymerization (somewhat similar to solid solution in metals: see Chapter 7) involves two or more kinds of monomers in the chain. If two monomers are used, rather than one, there are four propagation steps instead

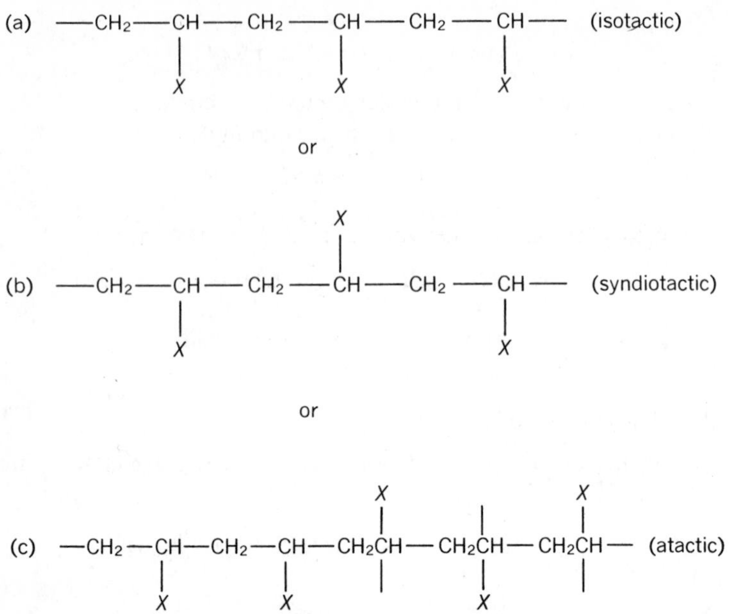

(a) —CH$_2$—CH——CH$_2$——CH——CH$_2$——CH——   (isotactic)
              | | |
              X   X   X

or

                              X
                              |
(b) —CH$_2$——CH——CH$_2$——CH——CH$_2$——CH——   (syndiotactic)
              | |
              X   X

or

                    X   X
                    | |
(c) —CH$_2$—CH—CH$_2$—CH—CH$_2$CH—CH$_2$CH—CH$_2$CH—   (atactic)
              | | | |
              X   X   X

Fig. 4-28. Sterisomers from coordination polymerization: (a) Isotatic:
all side groups on the same side; (b) syndiotatic: side groups alternate
from one side to the other; (c) atactic: side group arrangement is random.

of one. The polymerization reaction can be either condensation or addition.
The various types of copolymers which result depend on monomer reactivity
ratios.

In general, depending on the specific polymerization mechanism, the
resulting polymer can have: (1) linear, branched, or crosslinked chains;
(2) regular or random placements of the monomer in the chain; (3) monomers
of the same kind (homopolymer) or of two or more kinds (copolymers) of
different types; (4) short or long chains; i.e., low or high macromolecular
weight. There are many more ways of varying the constitution of a polymer
than that of any other material (i.e., metals and ceramics), leading to a very
broad spectrum of properties (characteristics) for the needs in design and
application. A number of commercially important organic polymer molecules are given in Table 4-6.

4-7-3 CHAIN STRUCTURES

Polymer chains join together to form three-dimensional structures. The physical properties developed in polymers depend, to a high degree, on the way
in which the chains are bonded together. All atoms within a polymer chain

Table 4-6.   Some commercially important organic polymer molecules

Name	Structure
Polyamide (nylon)	$(-(-CH_2-)_n NHC(-CH_2-)_m C-NH-)_x$ (with two C=O groups)
Polybutadiene	$(-CH_2-CH{=}CH-CH_2-)_x$
Cellulose nitrate	see structure below
Polychloroprene (Neoprene)	$(-CH_2-CH{=}CCl-CH_2-)_x$
Polyethylene	$(-CH_2-CH_2-)_x$
Polyformaldehyde (acetal resin)	$(-CH_2-O-)_x$
Polypropylene	$(-CH_2-CH-)_x$ with CH_3
Polystyrene	$(-CH_2-CH-)_x$ with C_6H_5
Polytetrafluoroethylene (Teflon)	$(-CF_2-CF_2-)_x$
Polyvinylchloride	$(-CH_2-CH-)_x$ with Cl

Cellulose nitrate structure:

$$
\begin{array}{c}
NO_3 \quad NO_3 \\
| \qquad | \\
CH-CH \\
\diagup \qquad \diagdown \\
(-CH \qquad\qquad CH-)_x \\
\diagdown \qquad \diagup \\
CH-O \\
| \\
CH_2 \\
| \\
NO_3
\end{array}
$$

have a full complement of primary bonds, thus the chains are held together by secondary bonds only. The long chain molecules are very flexible and have a tendency to bend and take a variety of paths through the solid rather than remain straight. When chains touch or cross each other, a secondary bond is formed. When there are a small number of secondary bonds, the material is readily deformed under applied forces. As the number of secondary bonds increases, the material becomes stiffer and stronger. The degree of order can range from almost complete randomness [Fig. 4-29(a)] to almost complete perfection, i.e., a crystalline material.

The simplest structure is the chain or linear polymer molecule, in which there is length but no appreciable thickness relative to the length. There is no implication of a straight or rodlike structure. If a large number of such chains are put together to form a three-dimensional solid, one result may be a "completely" random (amorphous) structure as in Fig. 4-29(a). An analogy to this might be a dish of cold spaghetti that has become very sticky. A number of chain polymers assume a helical configuration rather than linear. In such

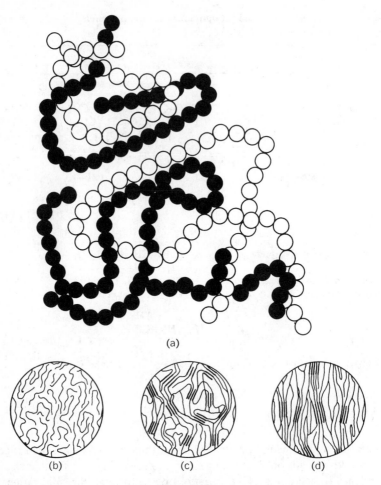

Fig. 4-29. Schematic representation of polymeric structure: (a) Five chains in a polymer with each circle representing a mer. Crystallinity is present where the chains align; (b) amorphous polymer; (c) crystalline polymer and; (d) oriented crystalline polymer.

cases, the amount of potential "entanglement" between chains is greater than for a linear chain.

If the chains are straight and parallel, the maximum number of secondary bonds are formed. In this case, the parallel polymer chains (either zigzag, e.g., polyethylenes, or helical, e.g., polypropylene) are packed together in a regular pattern which essentially fulfills the requirements for a crystal. A crystallographic unit cell can be designated which adequately describes this local structure. Such crystallites are shown schematically in Fig. 4-29(b) in an amorphous matrix. Under certain circumstances, it is possible to produce well-defined single crystals of polymers in which the individual polymer chain is folded back and forth many times within the crystal.

Some polymers can crystallize while others cannot. The side groups can be very important in determining this. Polymers such as polyethylene and polytetrafluoroethylene have small atoms which are symmetrically arranged on the carbon "backbone." These chains can readily align with each other and produce a material with a rather high degree of crystallinity although complete crystallinity has yet to be achieved. Chains with bulky or large side groups do not crystallize readily. Polystyrene and polyvinyl acetate, for example, have large side groups on one side of the chain and are somewhat difficult to crystallize. Polymethyl methacrylate has side groups on both sides of the chain with crystallization being very unlikely.

Bulky side groups reduce the number of secondary bonds but they also make it difficult for one chain to slide past another (sort of like mechanical interference). This tends to strengthen the polymer. It is difficult to crystallize copolymers as different portions of the chain have different structures and thus do not align readily.

Random branching or crosslinking can have a significant effect on structure and crystallization. Branching is typified by polyethylene which was first produced commercially by free radical polymerization of ethylene at high temperature and high pressure. In this process, the growing radical (R*) removes an H atom from a polyethylene molecule (P), terminating the growth of (R*) but starting a new chain (branch) growing from the original chain (P). In addition to these long branches, short branches (a few carbon atoms long) are formed when the growing free radical "bends" around and "bites" itself, removing an H atom from a nearby methylene group. Both kinds of branching are shown in Fig. 4-30.

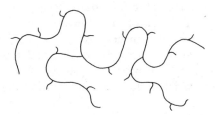

Fig. 4-30. A polymer chain with both long and short branches.

Crosslinking occurs when two chains are joined together by a third chain, an atom, or a group of atoms. The bonds formed in crosslinking are primary rather than secondary. Crosslinking thus strengthens the polymer but inhibits crystallization. Vulcanization of natural rubber is a good example of crosslinking, as in Fig. 4-31 which shows two polyisoprene chains. During vulcanization, double bonds between C atoms are broken and S, which has

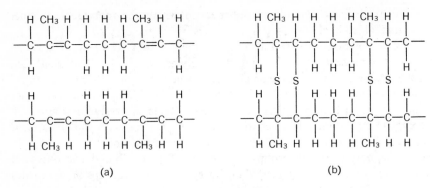

Fig. 4-31. Schematic vulcanization of rubber: (a) Two polyisoprene chains; (b) broken carbon double bonds provide sites for sulfur to bond the chains together through crosslinking.

been added, bonds two chains together by providing a crosslink. In this process, the rubber structure is converted to a network in three dimensions or a "space" polymer. Some other examples of space polymers are given in Table 4-7.

Coordination polymerization allows production of polymers which have the same chemical composition but different chain structures and thus different properties. For example, the vinyl polymers have a chain structure similar to polyethylene except that one hydrogen is replaced by a side group. The side groups can be attached in different ways as indicated in Fig. 4-28. Isotactic polystyrene, for example, has been crystallized but atactic polystyrene has not.

The physical properties of polymers are influenced by a number of structural parameters, the most important being: chemical composition; chain regularity; type and bulk of substituents; type and frequency of polar groups; type and number of branches, crosslinks or hydrogen-bonds; chain flexibility; stereoregularity; molecular weight; and molecular weight distribution. Extensive ranges in properties can be developed by control of branching, molecular weight, and crystallinity as indicated for polyethylene in Fig. 4-32.

4-7-4 CLASSIFICATION BY PROPERTIES

As control is exercised over the structural parameters, the properties of polymers can be drastically varied. Nonetheless, it is possible to classify them in three groups, based on their reactions to stress and temperature. These groups are:

1. *Thermoplasts* or *thermoplastic materials* that retain an ability to be repeatedly formed by heat and pressure.

Table 4-7. Some organic space-polymer molecules

Polymer   Schematic representation of structure

Diamond

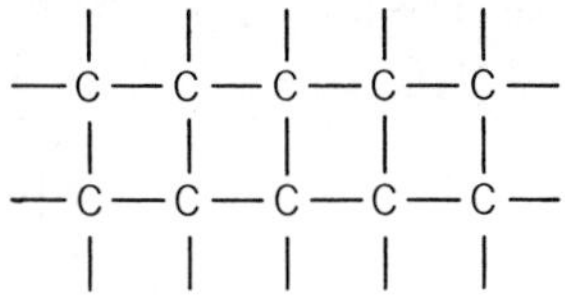

Vulcanized
rubber   (one possible representation)

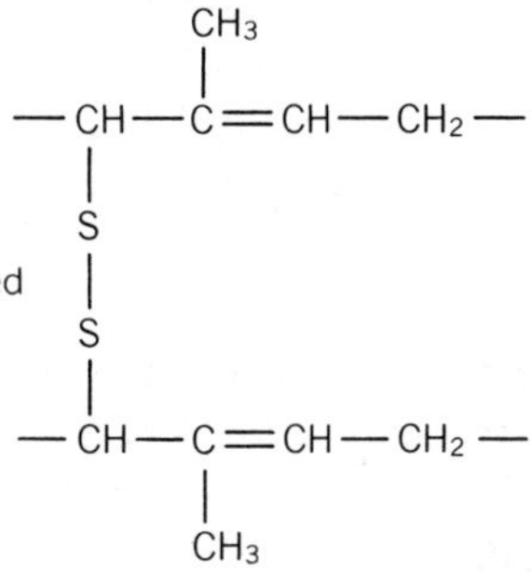

Phenolic

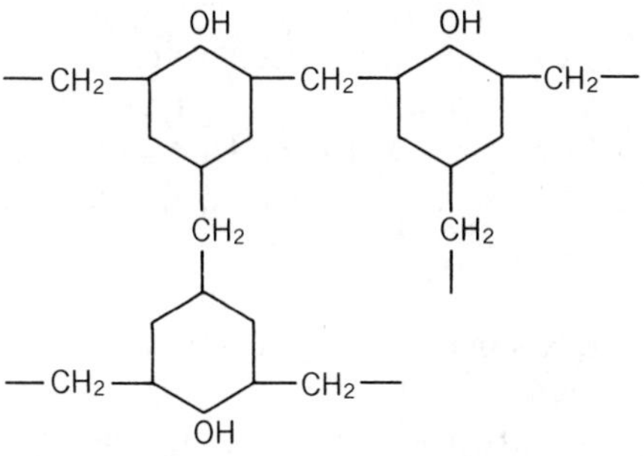

Epoxy

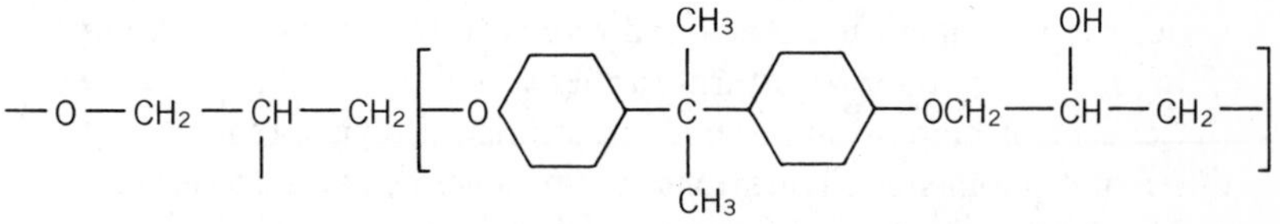

(one possible structure)

2. *Thermosets* or *thermosetting materials* that lose the above property
   by changing into a hard, rigid substance when subjected (once only)
   to the heat and pressure required for forming.
3. *Elastomers* that are thermoplastic during forming but which are "vulcanized" or "cured" to a rubberlike state (after forming) that is
   normally highly elastic.

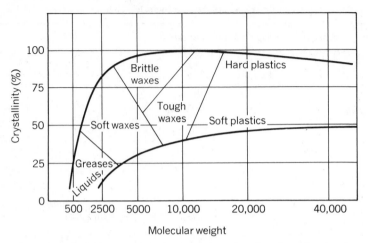

Fig. 4-32. Physical properties of polyethylene as a function of molecular weight and crystallinity. [R. B. Richards, *J. App. Chem. 1, 370* (1951).]

A thermoplast softens before decomposing upon heating. As the temperature increases, the random motion of the atoms about their equilibrium positions increases and bonds break as a result. The weakest bonds, i.e., generally secondary ones, break first and allow the chains to slide past each other. The material thus softens and flows. Upon cooling, new secondary bonds are formed and the material reverts to its original structure. This is very desirable for fabrication purposes but such polymers cannot be used to support loads at elevated temperatures.

Thermosets decompose before softening upon heating. In many thermosets, strength and hardness increase somewhat as temperature is increased, since polymerization is accelerated by heat and any monomers not previously polymerized tend to do so. Network structures are generally thermosets.

Elastomers can undergo large elongations (up to several hundred percent) under room temperature loads and return to the original shape upon release of load. The behavior of natural rubber (polyisoprene) is typical and involves the effects of stereoisomers and crosslinks. Polyisoprene has at least two different monomer configurations, having a very small structural difference but rather drastic property differences. Cis-polyisoprene has bonding positions on the same side of the chain while trans-polyisoprene has bonding positions on opposite sides of the chain. Cis-polyisoprene has a "kinked" chain in comparison with trans-polyisoprene with chains which are flexible and not straight when unstrained. In natural rubber, these chains tend to coil. As load is applied, the chains uncoil, straighten, and thus produce the extensive elongation. Vulcanization (see Sec. 4-7-3) is required to make rubber and the synthetic elastomers, such as neoprene and polybutadiene.

4-8 POLYMORPHISM

A change in temperature or pressure, if not accompanied by melting or vaporization, may cause a solid to change its internal arrangement of atoms. The ability of a material to have more than one *crystal structure* is called *allotropy* or *polymorphism*. The transition from one crystal structure to another at a specific temperature and/or pressure is known as a *polymorphic change*. As an example, consider iron which exists at room temperature as body-centered cubic (α-Fe). When heated above 910°C, the crystal structure changes to face-centered cubic (γ-Fe). On further heating, to 1400°C, it reverts to body-centered cubic (δ-Fe). At 1540°C, it melts. The changes are reversed upon cooling. The atoms are the same in the different crystal structures, but there are differences in physical and mechanical properties. For example, γ-Fe will dissolve up to 2.0 w/o carbon but α-Fe will dissolve only 0.025 w/o carbon.

A pressure-induced transformation results in a denser form of crystal structure. For example, cerium transforms under pressure to a crystal structure which is 16% denser than at atmospheric pressure, although both crystal structures are face-centered cubic. In most practical applications, however, we are concerned with temperature-induced transformations. Polymorphic transformations are confined neither to a specific limited portion of the periodic table nor to one or two crystal structures. These transformations take place in many elements and in many compounds, and involve many different crystal structures. Uranium is an excellent example of transformation between relatively complex crystal structures.

A polymorphic transformation from FCC to HCP (or vice versa) is not too startling after a little reflection. The arrangement of atoms in the (111) octahedral planes in FCC is identical to that in the (0001) basal planes in HCP. In HCP, alternate planes are directly above each other, while in FCC, there are two intermediate planes between identical ones. Assuming that an atom has the same radius in both structures, a given atom in either structure has 12 nearest neighbors and 6 second nearest neighbors. The distance between the atom and its neighbors is the same in both structures. There is a difference in the distance from the given atom only with the third nearest neighbor (and more remote ones). Polymorphic changes from HCP to FCC occur in calcium, scandium, and lanthanum on cooling, and in cobalt on heating.

Iron, titanium, zirconium, thallium, and lithium exhibit polymorphic transformations from BCC to a close-packed structure (either FCC or HCP depending on the element) on cooling. Iron changes again, back to BCC, on further cooling, as noted above.

In addition to these changes in the type of unit cell, order-disorder transformations and rotations of unsymmetrical groups of atoms are considered polymorphic transformations. In all cases, total displacement of atoms or groups of atoms is very small, although the inference is often made that these transformations are violent rearrangements. For example, consider iron which is BCC at room temperature and FCC above 910°C. The BCC unit cell contains 2 atoms/cell while the FCC unit cell has 4 atoms/cell. While these appear quite different, the transformation occurs with little strain. While the atoms in both structures are about the same size, the lattice parameters of the two cells are different. In addition, the cube axes are not parallel. The (110) planes in BCC iron become (111) planes in FCC iron and [111] directions in the BCC become [110] directions in the FCC. The change from BCC to FCC is accompanied by a decrease in volume of a little less than 2%.

EXAMPLE 4-6

Using the hard sphere model for atoms and assuming the atomic radius stays constant, calculate the volume change in a polymorphic transformation from FCC to BCC.

Solution:

For FCC, the atoms touch along a face diagonal, i.e.,

$$4r = (a^2 + a^2)^{1/2} \longrightarrow a = 4r/\sqrt{2}$$

For BCC, the atoms touch along a body diagonal, i.e.,

$$4r = (a^2 + a^2 + a^2)^{1/2} \longrightarrow a = 4r/\sqrt{3}$$

The volume (fraction of cube)/atom is

$$V_{FCC} = \frac{a^3}{4} = \frac{(4r/\sqrt{2})^3}{4}$$

$$V_{BCC} = \frac{a^3}{2} = \frac{(4r/\sqrt{3})^3}{2}$$

Volume change is

$$\frac{\Delta V}{V} = \frac{\dfrac{(4r/\sqrt{3})^3}{2} - \dfrac{(4r/\sqrt{2})^3}{4}}{\dfrac{(4r/\sqrt{2})^3}{4}} = +9.3\%$$

If the volume change is calculated as going from BCC to FCC, the change is -8.1%.

Polymorphism is not limited to metals and alloys. It is found in a number of ceramics. Zirconia, ZrO_2, has a monoclinic structure at room temperature which changes to a tetragonal structure at about 1000°C. In this case, there is a large change in volume which results in disruption in ceramic bodies made from zirconia. Al_2O_3 normally has a hexagonal structure which is stable at all temperatures but a cubic form can be formed under certain circumstances. Carbon exists at room temperature as graphite or diamond as indicated earlier. Materials such as BN, TiO_2, As_2O_3, ZnS, FeS_2, $CaTiO_3$, Al_2SiO_5, and many others exhibit polymorphism. Crystalline silica, SiO_2, is found in a number of different polymorphic forms which correspond to the various ways in which the tetrahedral groups can share corners.

Crystalline polymers can also have polymorphs. Polytetrafluoroethylene, (PTFE), for example, has a twist in the $CF_2A—CF_2X$ fluorocarbon chain rather than the planar zigzag favored in a paraffin chain. Below 19°C, the helix has 26 CF_2 groups for 360° of twist. Above 19°C, the helix has 30 CF_2 groups for 360° of twist. A third crystalline phase occurs at high pressure. PTFE is not the only polymer having polymorphism as polytrimethylene oxide has three known crystal structures and polymorphism is known to exist in some polyolefins in isotactic form.

4-9 POLYCRYSTALLINE MATERIALS

Most crystalline materials are seldom found as single crystals but as polycrystalline aggregates in which the whole body is made up of a large number of small interlocking crystals or grains. The term "grain" refers to individual crystals in a macroscopic piece, but grain and crystal are commonly used interchangeably. Each grain in the aggregate is connected to its neighbors by a grain boundary which is generally of irregular shape and bears no relation to the internal symmetry of the crystal. Orientation of crystal axes in different grains is usually random, although in certain cases there is a preferred distribution of orientations. While it is possible to prepare materials as single crystals and to carry out experimental studies on these, nearly all commercial materials are polycrystalline.

The process of melting, casting, and freezing in a mold, by which nearly all metals and alloys are prepared, favors formation of a polycrystalline aggregate. To examine this, let us consider the cooling curve shown in Fig. 4-33. This is a schematic illustration of the temperature of metal in a container in which the metal is initially molten (i.e., entirely liquid) when heat is removed at a constant rate. As heat is removed, the temperature of the melt decreases with time until some temperature just below the melting (or freezing) point is reached. At that time, there is a change of curvature, the temperature rises to the melting point, remains at this temperature for some time, and then decreases rapidly.

So far as can be determined, there is no change in the melt until its temperature has decreased somewhat below its melting point, i.e., it has undergone a slight supercooling. Nuclei of the crystal then appear spontaneously at many points throughout the liquid. If the liquid were transparent and the

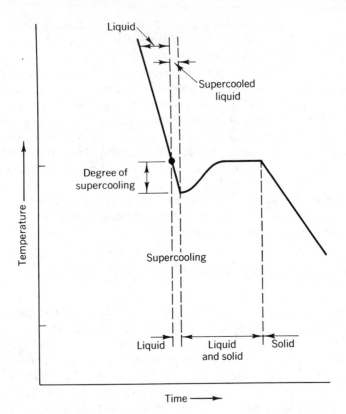

Fig. 4-33. Schematic cooling curve of a pure metal with heat being removed at a constant rate.

nuclei large enough to be visible, the appearance would be somewhat as shown in Fig. 4-34(a). (Nuclei here do not mean nuclei of atoms but rather collections of many atoms which form crystal nuclei. This latter term will be used to avoid possible confusion.) At each of these sites, atoms from the liquid attach themselves to the crystal nucleus, giving up their energy of motion as latent heat of fusion, and adopting relatively definite positions with respect to each other, thereby enlarging the crystal. These atoms are not motionless in the crystal at temperatures above absolute zero but oscillate about mean positions in the structure. If an observer were inside one of these growing crystals and were to look about him in any direction, he would find a periodic recurrence of atoms in whatever direction he looked. This atomic array is characteristic of solids as previously discussed.

As more heat is removed, crystals grow by orderly acquisition of additional atoms from the liquid. The appearance at an intermediate state during freezing would be somewhat as shown in Fig. 4-34(b) or (c). During growth,

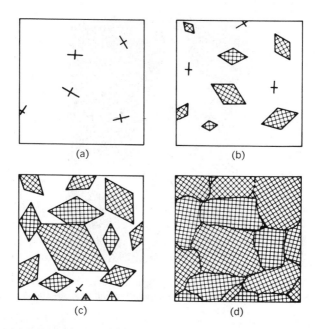

Fig. 4-34. Freezing of a uniformly cooled pure metal: (a) Nuclei form; (b) nuclei grow into crystals; some additional nucleation; (c) further crystal growth; mutual obstruction begins; (d) solidification completed as irregular grains.

additional crystal nuclei form continually. The number of nuclei forming per second in a unit volume of liquid is called the rate of crystal nucleation (N). The linear rate at which these crystal nuclei grow is called the rate of crystal growth (G). This rate of growth may be expressed in millimeters per second in a direction normal to the surface of the particle.

The actual size of the grains after complete solidification [Fig. 4-34(d)] is a function of both rate of crystal nucleation and rate of crystal growth. If grain size is termed S, then $S = f(G/N)$. It is difficult to evaluate this expression since both rates change with temperature or degree of supercooling. In general, crystalline materials show a finer grain size when cooled rapidly than when cooled slowly. Experiments on solidification also show that the rate at which crystal nuclei form below the melting point increases with decreasing temperature, i.e., increased supercooling. Nucleation and growth will be discussed in some detail in Chapter 8.

At normal rates of cooling and solidification, a large number of crystal nuclei are formed in the liquid, and since each one grows into a crystal, a great number of fine grains is found in the solid. In many instances, the crystals growing during the freezing process develop rapidly along certain crystallographic directions and slowly along others, resulting in the formation of

Fig. 4-35. Nickel crystals showing typical dendritic structure. Crystals were deposited by molten lithium upon attacking type 304 stainless steel.

long branchlike arms or dendrites, as shown in Fig. 4-35. As freezing continues, the outward growth of the dendrites is arrested by interference of dendrites from neighboring nuclei. The remaining liquid freezes in the spaces between the interlocking mesh of dendrites until the whole mass is solid.

A grain boundary is formed at every location where neighboring crystals make contact with one another, so that every interior grain is completely enveloped by a grain boundary. Considering their mode of formation, it is not surprising that grains in metallic materials have irregular outlines; each grain must take a shape forced on it by the configurations of neighboring grains.

The above discussion dealt with solidification. Formation of one crystal structure from another, as in polymorphic changes, also involves crystal nucleation and growth, and results in polycrystalline aggregates of new grains with grain boundaries.

4-9-1   GRAIN BOUNDARIES

The currently held view of a grain boundary is that it is a transition (a few atoms thick) between two adjacent lattices (Fig. 4-36). The arguments and evidence for this are substantially as follows:

1. Atoms in a grain boundary should be within range of interatomic forces of *both* crystals, and should thus take up positions which are a compromise between those required by the competing crystals. Since

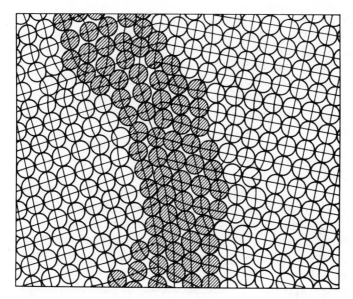

Fig. 4-36. Schematic representation of a boundary between two grains. Note the increasing irregularity in the lattices near the boundary. Shaded "atoms" constitute the grain boundary material.

interatomic forces act over a relatively short range, and are negligible beyond a few atomic diameters, the boundary layer ought to be only a few atoms thick.

2. Several properties of grain boundaries, e.g., energy, etching behavior, and ability to nucleate precipitates of second phases, are known to depend on the relative orientation of the adjoining grains. This is consistent with the idea that the atoms in the boundary find it increasingly difficult to fit simultaneously on both adjoining grains as the orientations become more dissimilar.

3. Measurements of energies of grain boundaries, in combination with theory, indicate that the thickness of the boundary should be about 2–4 atoms thick.

Whereas a grain boundary thickness of a few atoms has been confirmed by field ion micrography, the regions of irregularity in commercial metals are often wider, since impurities commonly concentrate at the grain boundaries during solidification.

4-9–2 CRYSTAL IMPERFECTIONS

The discussion so far has implied the existence of an ideal crystal, i.e., one with a perfectly regular structure throughout. In actual crystals, we find that there are irregularities in the structure and that the ideal structure is an over-simplification. For instance, atoms have been treated as being at rest in their lattice sites, whereas they are actually vibrating rapidly about their mean positions, the amplitude of vibration increasing with increasing temperature. Consequently, the arrangement of atoms in a crystal may be fairly irregular at any given instant, but over a finite time period (large in comparison with the period of vibration), the average positions of the atoms are regular, despite the thermal motion.

There are other sources of irregularities in crystals that produce departures from perfect periodicity which persist for long periods of time. It is customary to refer to such irregularities as imperfections, flaws, faults, or defects. One obvious source of such imperfections is the impurities normally found in materials. These impurities may be small particles (such as slag inclusions in metals) imbedded in the structure, or foreign atoms distributed throughout the lattice. Foreign atoms generally have atomic radii and electronic structures differing from those of the host atoms and therefore act as centers of distortion.

Experimental work has led to the conclusion that defects also exist quite independently of impurities. The properties of all materials are a function of the size, shape, and orientation of the structural units in the material. Certain properties, e.g., yield strength, tensile strength, hardness, impact strength, etc., are markedly influenced by these variables and are called structure-sensitive properties. Other properties, e.g., specific heat, density, lattice parameters, coefficient of thermal expansion, etc., are much less affected by these variables and are called structure-insensitive properties.

Structure-insensitive properties can usually be explained satisfactorily by deductions based on ideal crystals. In the case of structure-sensitive properties, however, the concept of imperfections in the lattice is necessary to account for the lack of agreement between experiment and theory. For instance, estimates of stress required to rupture crystals, based upon the assumption of a perfect lattice, give values 100 to 10,000 times greater than those observed. The most reasonable conclusion is that there are minute flaws in real crystals which act as centers of weakness.

X-ray diffraction studies indicate that a real crystal does not have perfectly uniform orientation but has a mosaic structure. This is a kind of substructure into which a "single" crystal is divided, as illustrated in greatly exaggerated fashion in Fig. 4-37. The lattice is composed of small blocks (or crystallites) such that the orientation is uniform within each block but varies slightly (from a few minutes of arc to a few degrees) between one crystallite and another. The approximate size of these crystallites is about 1000 atoms across. At their boundaries (called subboundaries) there are regions of misfit due to the slight differences in orientation.

It seems probable that the mosaic structure is formed as a result of

Fig. 4-37. Schematic representation of mosaic structure (greatly exaggerated).

conditions which exist during freezing. As the liquid freezes, it is likely that the various dendritic branches growing from any crystal nucleus will be slightly out of alignment with one another due to such factors as temperature gradients and impurity atoms which cause lattice distortion. The orientation throughout the crystal, therefore, will not be uniform but will deviate somewhat from theoretical. In addition, a volume change occurs on freezing which, for most materials, is a contraction. Thus, during the final stages of freezing the last traces of liquid trapped within the interlocking mesh of dendrites are prevented from contracting freely by the restraining action of the surrounding solid, and small regions are formed in which there are abnormal interatomic spacings.

Certain phenomena, diffusion in crystals in particular, can be reasonably explained only by assuming the presence of imperfections of atomic dimensions (point defects) in the lattice. The Frenkel defect [Fig. 4-38(a)] is formed by displacement of an atom (ion) from its normal position to an interstitial position. A crystal containing such defects has an equal number of vacancies and interstitialcies. The term interstitialcy serves to emphasize the concept that when the defect migrates through the crystal it probably does so by a sequence of jumps of the kind shown in Fig. 4-38(a) rather than by the migration of a single atom from one interstitial position to another. Schottky defects, Fig. 4-38(b), are vacancies, i.e., points in the space lattice which simply are not occupied by atoms (ions).

Thermodynamic reasoning indicates that a certain proportion of these point defects should be expected as a condition of equilibrium in crystals. In real crystals, however, the number of these defects is likely to be much

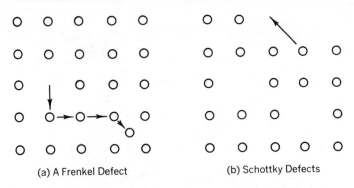

(a) A Frenkel Defect   (b) Schottky Defects

Fig. 4-38. Types of crystal defects on an atomic scale.

greater than that required for equilibrium in ideal crystals. The effect of point defects will be discussed in topics such as diffusion and effects of radiation on materials.

Certain phenomena cannot be explained in terms of point imperfections but can be explained in terms of line imperfections, e.g., dislocations. An *edge dislocation* can be visualized as an internal boundary of an extra plane of atoms as shown in Fig. 4-39. The edge dislocation is particularly useful in explaining slip in plastic flow during mechanical working and will be discussed in Chapter 6. A *screw dislocation* is a continuous helicoidal plane of atoms (Fig. 4-40) rather than a series of parallel planes. This type of dislocation is especially useful in explaining crystal growth as well as slip in plastic defor-mation.

Although these two simple line imperfections are sufficient for our pur-pose, more complex arrangements occur in real materials. There can be

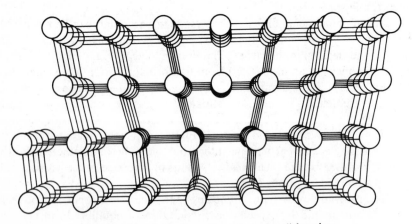

Fig. 4-39. Atomic configuration of an edge dislocation.

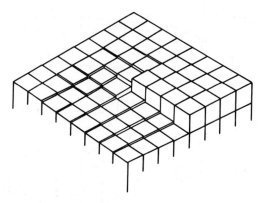

Fig. 4-40. A schematic screw dislocation.

curved-edge dislocations, bent dislocations, dislocation rings, and combined edge and screw dislocations.

Various laboratories have produced perfect, single crystals, called "whiskers," a few mm long and a few microns in diameter, which have phenomenally high resistance to plastic deformation. If sufficiently small crystals are grown, it is possible to produce a few which are relatively free from dislocations and have strength properties approaching the theoretical.

In the same context, the grain boundaries discussed above can be regarded as surface imperfections; i.e., we can think of grain boundaries and subboundaries as composed of various arrays of dislocations. Fig. 4-41 illustrates a simple dislocation boundary for two adjacent grains having a common axis normal to the paper and a rotation relative to each other through an angle θ. They are joined by localized distortion with edge dislocations shown by the symbol $\perp$. As the angle θ increases, the spacing between adjacent dislocations decreases until the boundary is composed entirely of dislocations. Experiment indicates that the model is valid for grain boundary angles up to about 26°.

4-9-3 EXTENSION OF CONCEPTS

One might infer from the foregoing discussion in this section that the information applies to metals. While this is true, the information also applies to nearly the same extent when applied to crystalline ceramics. The exceptions are such things as crystal formation from a melt which is not practical since most ceramics have relatively high melting points. Some differences in fabrication are discussed in Chapter 12.

In crystalline polymers, the morphology is complex and not nearly as well understood as for metals and ceramics. When a polymer is cooled from a

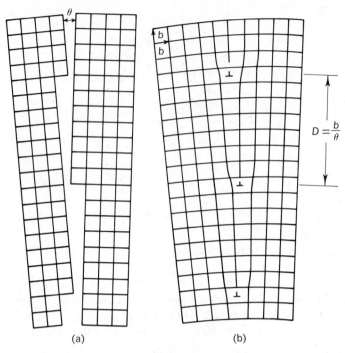

$$D = \frac{b}{\theta}$$

(a)   (b)

Fig. 4-41. A simple grain boundary. Constructed of edge dislocations.

melt and crystals are formed, the final density is always intermediate between that of the amorphous state and the theoretical density of a perfect crystal. The explanation is that the materials consist of two distinct regions—crystallites and amorphous regions. The crystallites are perhaps of the order of 100 A in size with an individual polymer chain passing through a number of crystallites and amorphous regions. Since the chains extend from one crystallite to another, growth of individual crystals is blocked before they can come into actual contact.

QUESTIONS AND PROBLEMS

1. Define a crystalline solid. Are there any substances which approximate or simulate the properties of solids but which are not considered to be solids? If so, what kind of substance are they? Of what practical significance are they?

2. What is a space lattice? How many space lattices are possible? How many crystal structures are possible? What is the distinction, if any, between a space lattice and a crystal structure?

3. Show that a face-centered tetragonal structure is equivalent to a body-centered tetragonal structure.

4. Discuss the similarities and differences of primitive cells and unit cells.

5. How many atoms in a gram of iron? How many primitive cells and how many unit cells in a gram of iron at room temperature? At 2000°F?

6. Assuming atoms to be hard spheres in contact, calculate for FCC, BCC, and simple cubic structures, the following:
   (a) cube volume in terms of atomic diameter D;
   (b) number of atoms per unit cell;
   (c) density in terms of number of atoms/D^3.

7. Show that the HCP structure, Fig. 4-6(c), has the equivalent of 6 atoms per unit cell. What fraction of the volume is occupied by atoms?

8. Show that the ideal c/a ratio for a hexagonal close-packed structure is 1.633.

9. The density of magnesium is 1.741 gm/cm³. Its atomic weight is 24.32. What are the three parameters of its hexagonal close-packed structure?

10. Calculate the fraction of volume occupied by atoms in FCC structure. How does this compare with the analogous calculation for HCP structure?

11. Calculate the size of the largest atom that can fit interstitially in a BCC structure (as a fraction of the radius of the BCC atoms). The positions with the largest interstitial holes have coordinates of the type $0, \frac{1}{2}, \frac{1}{4}$ in the BCC unit cell.

12. Calculate the diameter of the largest atom which could fit interstitially into an FCC structure without distorting it. The positions with the largest interstitial holes have coordinates of $0, 0, \frac{1}{2}$ or $\frac{1}{2}, \frac{1}{2}, \frac{1}{2}$.

13. Assume that a carbon atom (radius = 0.7 A) is forced into a position between two corner atoms of BCC iron (radius = 1.24 A). How much would the edge be expanded?

14. The FCC cell has the equivalent of four atoms. How many octahedral voids are there per unit cell?

15. Show that the numbers given in Table 4-2 are correct.

16. In a space lattice, every point has identical surroundings. Show that the HCP structure does not satisfy this requirement.

17. Manganese sulfide (MnS) has the NaCl structure with a density of about 4.0 g/cm³. Determine the size of the unit cell.

18. The crystal structure of solid GaAs is the "zincblende" structure; i.e., except for the two different kinds of atoms, the structure is essentially that of diamond. Discuss the type or types of binding between the atoms.

19. Polytetrafluoroethylene has a structure like polyethylene but fluorine replaces the four hydrogen atoms in each mer (C_2F_4). If it has an average molecular weight of 22,500 g/mole, how many mers per molecule?

20. Polyethylene as well as a number of other polymers can form amorphous or crystalline structure. Which is easier to crystallize: a polymer with a high or low degree of polymerization? Why?

21. Compare and discuss the similarities and dissimilarities of crystallinity in metals, ceramics, and polymers with particular reference to characteristics and properties.

22. What are Miller indices? Of what significance or use are they? In what matter are they determined? What form of notation is used?

23. Discuss the similarities and differences of Miller indices and direction indices.

24. In a cube, the six faces constitute a family of planes, or planes of a form. Do the six faces of a tetragonal cell also constitute planes of a form? If not, how many families are there?

25. Determine the Miller indices of the family of close-packed planes in FCC structure. Determine the direction indices of the family of close-packed directions in FCC structure.

26. Determine direction indices of the family of close-packed directions in BCC structure.

27. Demonstrate that Eqs. 4-4, 4-5, 4-6, and 4-7 are correct.

28. Which planes are farthest apart in a
    (a) simple cube,
    (b) body-centered cube,
    (c) face-centered cube?
    What is the distance between these planes in terms of the lattice parameter a?

29. Show that the [101] direction is perpendicular to the (101) plane and the [111] direction is perpendicular to the (111) plane in a cubic structure but not in others such as tetragonal or orthorhombic structures.

30. Show that a direction [uvw] lies in a plane (hkl) in cubic crystals if $uh + vk + wl = 0$

31. What is the meaning of coordination number? How is it used? Of what significance is it?

32. What is polymorphism or allotropy? Of what practical significance is it?

33. Discuss the differences and similarities of liquid and solid metal.

34. Describe how separate grains are formed when a metal solidifies.

35. What two characteristics determine the number of grains that will form during solidification? How do these characteristics operate?

36. What is meant by supercooling? Of what practical value is it?

37. Iron is said to have several allotropic forms, e.g., alpha, gamma, and delta iron. What takes place in the iron when it changes from one allotropic form to another? How is this allotropic change produced?

38. What volume change, if any, is found in iron when it changes from alpha to gamma and from gamma to delta?

39. Pure iron is BCC below 910°C and FCC above. (a) Assuming the same atomic radius in both structures, what is the percent change in volume in a polymorphic transformation from one structure to the other? (b) Measurements show a lattice parameter of 2.89 A for BCC and 3.66 A for FCC at 910°C. Recalculate the percent change in volume. (c) What is the atomic radius in each structure in part (b)?

40. How does a grain boundary differ from the body of crystalline material? What effects can the grain boundary have on the properties and behavior of materials?

41. What are crystal imperfections? What is the source of these defects? What are the effects of these imperfections on the properties and behavior of materials?

5

ELECTRICAL, MAGNETIC, AND THERMAL PROPERTIES

5-1 INTRODUCTION

In Chapter 3, the discussion ended with a brief introduction to band theory. We also indicated a further discussion after looking at aggregates of atoms in Chapter 4.

5-2 BAND THEORY

Consider bringing together two hydrogen atoms (each with a nucleus and a single electron) to form a molecule. We know the spins of the electrons can be the same, or they can be opposed. If the spins are parallel, the electrons will repel each other as a result of the Pauli exclusion principle, whereas they will attract each other if the spins are opposite. Thus, with parallel spins, the two electrons will tend to remain as far apart as possible. If the electrons are regarded as a cloud of negative charge surrounding the nuclei, then Fig. 5-1(a) shows how this cloud might look. If, however, the electrons have opposite spins, both of them can exist in the same orbit around either proton and would thus be expected to spend much time between the two protons [Fig. 5-1(b)]. Actually this electron cloud serves to join the two nuclei, partly by electrostatic attraction and partly by the sharing of both electrons. This attractive force joins the two atoms to form a molecule which is held together by a covalent bond (a special case of covalency). The repulsive force between the two atoms in Fig. 5-1(a) results in a so-called antibonding state.

If the electrons are closer, on the average, to the two nuclei than they would be when the two atoms are far apart, then the energy is lower. As the atoms come closer, the energy decreases further as shown in Fig. 5-2 where

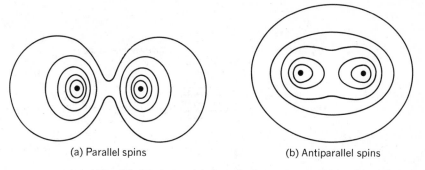

(a) Parallel spins   (b) Antiparallel spins

Fig. 5-1. Equipotentials for a hydrogen molecule.

energy is shown schematically as a function of interatomic distance. The increase in energy which occurs when two atoms having electrons with identical spins are brought together is also shown in Fig. 5-2. This separation or splitting of the energy level E_1 is designated by labeling the two levels parallel and antiparallel to indicate identical and opposite spins, respectively.

An obvious extension of the above discussion is to expect a splitting of all other levels, E_2, E_3, etc., in the hydrogen molecule (Fig. 5-3). Extending the discussion in a different direction, we would expect that bringing together a large number, N, of hydrogen atoms to form an aggregate (perhaps a crystal) would result in splitting of each energy level into N distinct energies close to the original value. These groups of discrete but closely spaced energy levels are known as energy bands as shown in Fig. 5-4.

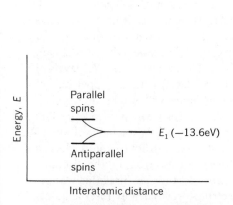

Fig. 5-2. Splitting of lowest energy level in the hydrogen molecule (schematic).

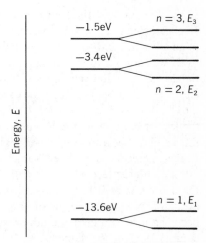

Fig. 5-3. Splitting of energy levels in the hydrogen molecule (schematic).

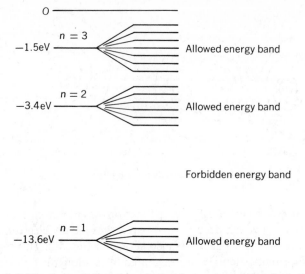

Fig. 5-4. Energy bands for a hypothetical hydrogen crystal (schematic).

5-2-1 ELECTRON ENERGY LEVELS IN AGGREGATES

Each band in Fig. 5-4 is a result of splitting of the original levels, E_n, into N new levels, one for each of the N atoms in the aggregate. Since interaction between electrons is small, the splitting in the levels is also small. This is in contrast to the large separation in energy between the E_1 (-13.6 eV) and E_2 (-3.4 eV) levels. The bands formed by the N split levels are known as *allowed bands*, and the spaces between them are known as *forbidden bands*. In effect, electrons, when present, will have a high probability of being found in an allowed band and a low probability of being found in a forbidden band.

If we cool our hydrogen aggregate to $0°$K, then all the electrons will have the lowest possible energy. According to the exclusion principle, the N electrons will drop, two at a time (one with spin up, the other with spin down), into the lowest group of N levels [Fig. 5-5(a)]. Thus we will have $N/2$ levels of the lowest band occupied by electrons (two per level), and the upper half of this band will be unoccupied. All higher energy bands will also be unoccupied. If the aggregate is warmed somewhat, the electrons will gain energy and move to higher levels [Fig. 5-5(b)]. Likewise, if we apply a small voltage to the aggregate at $0°$K (6 volts across an aggregate 1 m long would give an available potential energy of 6×10^{-10} eV), an electron in the topmost occupied level can gain enough energy to move to a higher (previously unoccupied) level in the same band. This electron (previously at rest) acquires kinetic energy and moves under the influence of the electric field. This hydrogen aggregate would then be an electrical conductor.

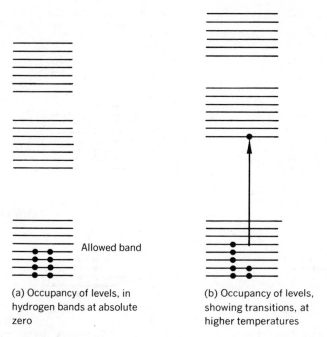

(a) Occupancy of levels, in
hydrogen bands at absolute
zero

(b) Occupancy of levels,
showing transitions, at
higher temperatures

Fig. 5-5. Occupancy of levels in energy bands in hypothetical hydrogen crystal (schematic).

If we could make a crystal of helium, we would find the energy bands are similar to those of hydrogen. The discrete levels in the helium atom will split into allowed bands in the crystal with forbidden bands between the allowed bands. There are 2 electrons per atom, rather than 1 as in hydrogen, and therefore each of the N levels in the lowest band is occupied (Fig. 5-6). If an electric field is applied across the crystal, the only way the electrons with the highest energy can accept more energy is to gain enough to cross a forbidden band and move into the bottom of the next allowed band. Any reasonable value of an electric field will provide much less energy than the energy equivalent to the width of the forbidden band. Thus the electron cannot accept any energy at all, and we would expect the hypothetical helium crystal to be an insulator.

Lithium is the next element in the periodic table (Table 3-3) and has three electrons, two in the $1s$ shell and one in the $2s$ shell (its valence electron). In keeping with the above discussion, we would expect an energy band diagram as shown in Fig. 5-7. The $1s$ band corresponds to a filled shell and is narrow, since the interaction among inner shell electrons is smaller than for the valence electrons. The next lowest band, the $2s$, is only half-filled, just as the hydrogen $1s$ was only half-filled. Thus we expect Li to be a good electrical conductor and to have metallic behavior. This is, in fact, what we observe.

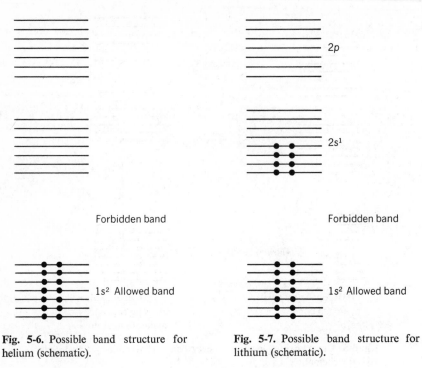

Forbidden band   Forbidden band

Fig. 5-6. Possible band structure for helium (schematic).

Fig. 5-7. Possible band structure for lithium (schematic).

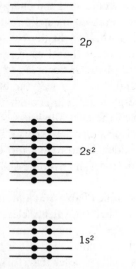

Fig. 5-8. Possible band structure for beryllium (schematic).

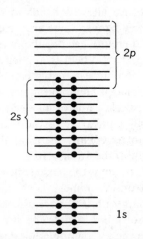

Fig. 5-9. Assumed band structure for beryllium showing overlapping bands (schematic).

In the case of beryllium, we would expect the two lowest bands, the $1s$ and $2s$, to be filled (Fig. 5-8), and thus Be should be an insulator. We know, however, that Be is a conductor. A possible explanation for this (Fig. 5-9) is overlapping bands. When the energy levels split into bands during formation of the aggregate, they may split (and widen) to such an extent that some of the bands merge; i.e., the expected forbidden band does not develop. In Fig. 5-9, the highest filled band and the lowest empty band overlap, with the result that we find a larger band which has both filled and empty levels.

In practical discussion, we need not consider any bands below the highest occupied one, since electrons cannot normally acquire enough energy to leave any of the lower completely filled bands. The highest occupied band represents the energy levels associated with the valence electrons and is called the valence band.

5-3   CONDUCTORS AND INSULATORS

An electron in a metal can contribute to conduction by moving under an applied electrical field of any intensity. An electron in an ionic or covalent bond, however, is too tightly bound to move under an electrical field of usual intensity. Intermolecular bonding is much weaker, but breaking an intermolecular bond does not contribute to conduction since molecules have no net electrical charge. Thus aggregates formed with only covalent, ionic (solid-state electrolytes are an exception), or molecular bonding cannot conduct electricity and are, by definition, insulators. Such materials would be expected to have essentially infinite resistance. Resistivity of real materials, while very high, is far from infinite, and most materials will conduct a limited amount of electricity under certain conditions.

Electrons can move through an ideal metallic crystal without resistance, but in actual crystals, electrons collide with phonons, dislocations, vacancies, impurity atoms, and other lattice imperfections. The residual resistivity is due to solute atoms, impurities, and dislocations and is usually independent of temperature. The total resistivity is the sum of the residual and thermal contributions. Increasing the temperature introduces thermal vibrations which impair the periodicity of the lattice and thus impede motion of the electrons.

The electrical resistivity, a characteristic property of a material, independent of dimensions, is

$$\rho = \frac{m}{e^2 N t_c} (\text{ohm} - \text{m}) \tag{5-1}$$

where m is the mass of the electron, e is its charge, N is the number of "free" valence electrons per unit volume and t_c is the mean free time between collisions (about 10^{-14} sec for most metals), i.e., a measure of mobility.

The reciprocal of electrical resistivity is called electrical conductivity, thus,

$$\sigma = \frac{1}{\rho} = \frac{e^2 N t_c}{m} \left(\frac{1}{\text{ohm} - \text{m}}\right) \tag{5-2}$$

Electrical resistivity, and all electrical properties, of oxides are very sensitive to both purity and perfection of the crystal. Resistivity is very high at low temperatures and decreases rapidly, usually exponentially, as temperature increases. This may be due to ionic conduction (rather rare) or electronic semiconduction. Since the manner of preparation of the oxides can have significant effect in the properties, handbook values for electrical properties of oxides (and for all insulators, for that matter) should be used with considerable reservations, especially for high temperature applications.

EXAMPLE 5-1

Compare the electrical resistivities of aluminum (Al) and alumina (Al_2O_3). Explain the difference.

Solution:

From Appendix B, the resistivity of aluminum is about 2.9 microhm-cm while that of alumina is greater than 10^{14} microhm-cm. Metallic aluminum has "free" electrons which can respond to an electrical potential and thus can transport electrical charge. In alumina, however, the valence electrons from the metal have been transferred to the oxygen ions. It might be noted that the Al^{3+} ions still have electrons. These cannot conduct electricity, however, since they cannot leave the parent atoms. Only valence electrons are involved in conducting electricity.

EXAMPLE 5-2

How long will it take for 10^{25} electrons to pass through a 75-watt light bulb which is operating at 30 volts dc?

Solution:

We recall that $E = IR$ and $W = EI$.

$$(10^{25} \text{ electrons})(1.6 \times 10^{-19} \text{ amp-sec/electron}) = 1.6 \times 10^6 \text{ amp-sec}$$

$$\text{current} = 75 \text{ watts}/30 \text{ volts} = 2.5 \text{ amp}$$

$$\text{time} = \frac{1.6 \times 10^6 \text{ amp-sec}}{2.5 \text{ amp}} = 6.4 \times 10^5 \text{ sec} = 178 \text{ hr}$$

5-4 SEMICONDUCTORS

Diamond, the classic example of a material having covalent bonding, has an electrical resistance of about 10^{12} ohm-m. As we go down the Group IV column in the periodic table (Table 3-3), the bond strength decreases markedly, i.e., in effect, the fraction of valence electrons available for conduction increases noticeably. This is reflected in the resistivity, which decreases. Diamond, Si, and Ge have the covalently bonded diamond cubic structure. Each atom is joined to each of its four nearest neighbors by sharing an electron pair (Fig. 3-10). Each atom thus has the equivalent of a closed valence shell, and the binding of the electrons to the atoms is very strong. This binding is responsible for the wide forbidden energy gap (5.6 eV) in diamond. Moving down the Group IV column in the periodic table, we find the same effect, although the bond strength decreases. Si has a gap of 1.1 eV, Ge about 0.75 eV, gray Sn about 0.1 eV, whereas white Sn and Pb are metals and have no energy gaps.

5-4-1 INTRINSIC SEMICONDUCTORS

Intrinsic semiconductivity can best be understood by considering the diamond cubic structure of Si and Ge. This structure (Fig. 3-10) has a tetrahedral configuration of four covalent bonds. This has been represented by a two-dimensional network in Fig. 5-10, where atoms are located at the nodes of the net, and the shared electron pair is indicated by a double link.

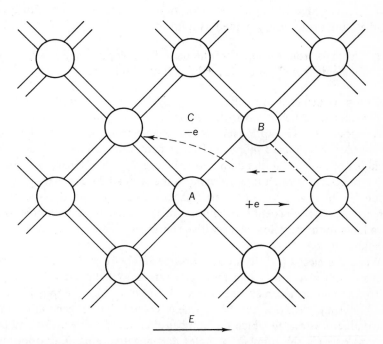

Fig. 5-10. Schematic representation of diamond crystal structure with a conducting electron at C and a hole at AB under the influence of an electric field.

All atoms are at rest at $0°K$, and all electrons are securely held by the covalent bonds, giving a perfect insulator. As the temperature increases, the thermal vibration of the atoms will occasionally develop enough energy to eject an electron from a bond. This produces a defect, a hole, in the bond, as shown by the single link at AB in Fig. 5-10. The ejected electron, C, is available for conduction and drifts away if an electric field is applied. At some later time, the missing link is restored by capture of an electron from a neighboring bond. The hole thus shifts to a neighboring bond, from there to another bond, and so on.

The hole, shifting from bond to bond, drifts in the general direction of the field while the electron drifts in the opposite direction, i.e., against the field. Thus both the electron and the hole contribute to conduction, and therefore both are considered to be imperfections of the covalent crystal structure. The structure becomes intrinsically imperfect through production of this defective bond. This is the source of the name intrinsic applied to the semiconductivity.

5-4-2 EXTRINSIC SEMICONDUCTORS

Extrinsic semiconductivity is due to a factor apart from the crystal structure itself; i.e., it is due to impurities incorporated in the structure. For Si and Ge, these impurities are provided as follows:

1. Elements from Group V of the periodic table, P, As, Sb.
2. Elements from Group III of the periodic table, Al, Ga, In.

Group V elements are called n (negative) impurities or *donors*. Group III elements are called p (positive) impurities or *acceptors*.

An atom from Group V has five valence electrons, one more than Si or Ge. It needs only four to share an electron pair with each of the four nearest neighbors of Si or Ge in the diamond structure. Thus the Group V atom has one extra electron which it can *donate* to the semiconductor. An atom from Group III, however, has only three valence electrons and thus needs one more to form four covalent bonds. It has one defective bond, or a hole, which is available for conduction as a positive charge by *accepting* an electron from a neighboring bond.

When the electron drifts away from the donor [Fig. 5-11(a)], the latter becomes a positive ion. When the hole drifts away from the acceptor [Fig. 5-11(b)], the latter becomes a negative ion. In both cases, the ions introduce additional imperfections in the covalent crystal. More generally, an n-semiconductor is one in which the number of conducting electrons and positive ions exceeds the number of holes and negative ions. The opposite is true for a p-semiconductor.

Extrinsic semiconductors are formed by the deliberate addition of a

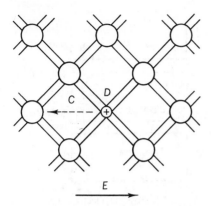

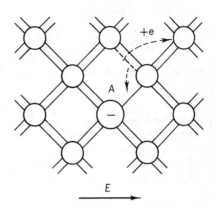

(a) n-type with donor positive ion D embedded in diamond with an excess electron available for conduction

(b) p-type with acceptor negative ion A embedded in diamond with a hole available for conduction

Fig. 5-11. Schematic representation of extrinsic semiconductors based on diamond

second element to an intrinsic semiconductor. The extrinsic conductivity effect is significantly greater than that in the basic intrinsic semiconductor. The amount of impurities (dopes) seldom exceeds 1 part in 10^5 and normally ranges between 1 part in 10^6 and 1 part in 10^7. Obviously, the semiconductor itself must be of still higher purity. Such high purity can be obtained, thanks to the development of the zone melting technique which has been used to purify Ge to an impurity content less than 1 part in 10^{10}.

Consider addition of P to Ge to form an n-semiconductor. The binding energy of the electron in Ge is about 0.05 eV if we assume its mass in the crystal is the same as its mass in free space. An electron moving in the periodic electric field of a crystal has its motion affected by the combined electric fields of all nuclei in the lattice. The electron no longer acts as if its mass were the free mass, but rather it acts as if it had an effective mass of about 20% of the free mass. Thus the effective binding energy of the excess electron is about 0.01 eV. This is the energy needed to place it in the conduction band from its position in the donor level in the energy gap or forbidden zone (Fig. 5-12).

Addition of B to Ge, however, to form a p-semiconductor, gives rise to acceptor levels in the energy gap about 0.01 eV above the top of the valence band (Fig. 5-12).

Consider a semiconductor having both donor and acceptor levels as shown in Fig. 5-13. At very low temperatures, the electrons will have the lowest possible energy and thus will all be located in the valence band and the donor level. There will be the usual concentrations of electrons since the donor and acceptor atoms are (physically) relatively far apart. As tempera-

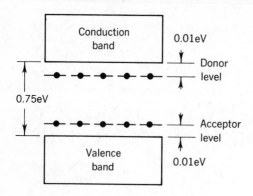

Fig. 5-12. Donors and acceptors for germanium.

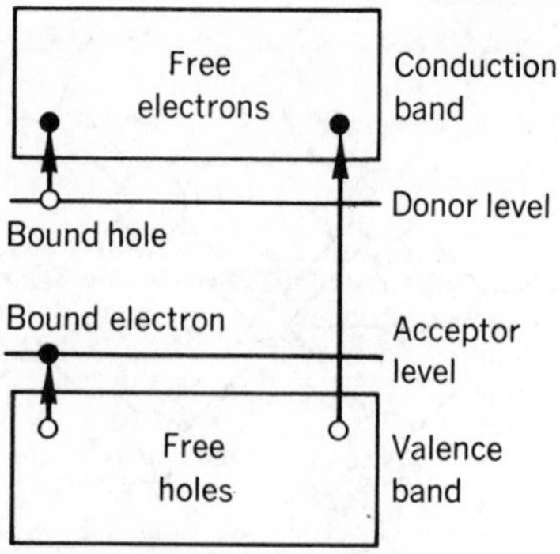

Fig. 5-13. Possible electron transitions in a doped semiconductor.

ture increases, electrons are excited from the valence band into the acceptor level and from the donor level into the conduction band. At still higher temperatures, electrons may be able to get directly from the valence band into the conduction band. Electrons which move from the valence band into the acceptor level leave free holes behind and introduce bound electrons into the acceptor level. Electrons leaving the donor level for the conduction band leave bound holes behind and introduce free electrons into the conduction band.

5-4-3 RESISTIVITY OF SEMICONDUCTORS

Considering conducting electrons and holes separately, the corresponding resistivities are:

$$\rho_e = \frac{m_e}{e^2 n_e t_e} \tag{5-3}$$

and

$$\rho_h = \frac{m_h}{e^2 n_h t_h} \tag{5-4}$$

where m_e and m_h are the "variable" electron mass and the "mass" of a hole, t_e and t_h are mean free times, and n_e and n_h are the number of conducting electrons and the number of holes per unit volume, respectively. The concepts of a "variable" electron mass and the "mass" of a hole can be made clear only in detailed terms of quantum mechanics.

The total resistivity is:

$$\frac{1}{\rho} = \frac{1}{\rho_e} + \frac{1}{\rho_h} \tag{5-5}$$

Since conductivity is the reciprocal of resistivity, we have

$$\sigma = \sigma_e + \sigma_h = e^2 \left(\frac{n_e t_e}{m_e} + \frac{n_h t_h}{m_h} \right) \tag{5-6}$$

The ratios t_e/m_e and t_h/m_h depend on details of crystal structure and only moderately on temperatures. On the other hand, n_e and n_h increase exponentially with increasing temperature (a property common to all semiconductors). This gives a corresponding *decrease* in resistivity. This contrasts sharply with normal conductors in which there is essentially a linear *increase* in resistivity with temperature.

Conductivity, the inverse of resistivity, can also be viewed in a somewhat different manner. Total conductivity

$$\sigma = n_n q \mu_n + n_p q \mu_p \tag{5-7}$$

where n_n is the number of negative carriers (electrons) and n_p is the number of positive carriers (holes) with μ_n and μ_p representing their respective mobilities. The term q is the charge per carrier, which in this case is 1.6×10^{-19} amp-sec. Properties of some semiconductor materials are given in Table 5-1.

Table 5-1 Properties of some semiconductors at 25°C

Material	Energy gap, eV	Intrinsic conductivity, $ohm^{-1}\text{-}cm^{-1}$	Electron mobility, n, $cm^2/volt\text{-}sec$	Hole mobility, p, $cm^2/volt\text{-}sec$
Diamond	6	10^{-18}	1800	1200
Silicon	1.1	5×10^{-6}	1400	480
Germanium	0.75	0.02	3900	1900
Tin (gray)	0.08	10^4	2000	1000
GaP	2.2	—	450	20
GaAs	1.4	10^{-8}	8500	4500
GaSb	0.7	—	5000	850
InSb	0.17	—	80 000	700
InAs	0.33	100	30 000	500

5-4-4 TEMPERATURE DEPENDENCE OF SEMICONDUCTIVITY

The exponential dependence of n_e and n_h on temperature can be understood by treating the conducting electron and hole as products of dissociation of a covalent electron. Under equilibrium, this process is balanced by a reverse process, i.e., regeneration of the covalent electron through capture of a conducting electron by a hole. We can write the following equation, similar to dissociation in a chemical compound:

$$\text{covalent electron} \rightleftharpoons \text{conducting electron} + \text{hole}$$

Like chemical dissociation, the process shifts more or less to the right, depending on temperature and the strength of the bond. At any given temperature, however, there is a constant ratio of charge carriers to covalent electrons. Dissociation is thus governed by the same law of mass action which governs normal chemical reactions. Thus $n_e n_h$ is constant, but this constant changes with temperature and strength of covalent bond.

The above applies to both intrinsic and extrinsic semiconductors. In an n-semiconductor, however, emphasis is primarily on n_e while in a p-semiconductor, emphasis is primarily on n_h.

EXAMPLE 5-3

The electrical resistivity of pure silicon (Si) is 3.0×10^3 ohm-m at room temperature (27°C). The conductivity is 2.67 mho/m at 250°C. Estimate the size of the energy gap.

Solution:

$$\sigma = \sigma_0 e^{-E_g/2kT}$$

$$\sigma_{RT} = \sigma_{300} = \frac{1}{3 \times 10^3} = \sigma_0 e^{-E_g/(2k)(300)}$$

$$\sigma_{250°C} = \sigma_{523} = 2.67 = \sigma_0 e^{-E_g/(2k)(523)}$$

$$\frac{\sigma_{300}}{\sigma_{523}} = \frac{1}{(3 \times 10^3)(2.67)} = \frac{\sigma_0 e^{-E_g/600k}}{\sigma_0 e^{-E_g/1046k}}$$

$$\ln\left(\frac{1}{3 \times 10^3 \times 2.67}\right) = \frac{-E_g(\frac{1}{300} - \frac{1}{523})}{2(1.38 \times 10^{-16})}$$

$$E_g = 1.76 \times 10^{-12} \text{ erg} = 1.1 \text{ eV}$$

5-4-5 SOME SEMICONDUCTOR MATERIALS

The above discussion was in terms of a single material from Group IV elements, specifically Si and Ge. In addition, there are at least 15 combinations of elements, III–V, II–VI, or IV–IV, which are of importance in commercial use. These are all intrinsic semiconductors which can be made extrinsic by addition of appropriate doping elements. SiC, for example, which is used for diodes, becomes p-type by adding Al or n-type by adding N. This group of semiconductors has specific uses for such things as diodes, photoconductors, infrared detectors, lasers, phosphors, etc.

Polymeric semiconductors are significantly harder to obtain than inorganic types. It is necessary to find substances which have: (1) electronic (rather than ionic) conduction; (2) a resistivity between 10^8 ohm-cm and 10^{-4} ohm-cm (midway between metals and insulators); and (3) a resistivity that decreases with increasing temperature. A very few materials have superficial analogy to inorganic semiconductors. These are perylene-iodine complexes (brittle, crystalline organic solids) with resistivities about 1 ohm-cm and some tetracyanquinodimethane complexes with resistivities about 10^{-2} ohm-cm.

Modification of the conductivity of organic materials is not very satisfactory. Most organics behave as p-type materials, most likely because of impurities or contact effects. Attempts to change the conductivity type have been unsuccessful with very minor exceptions. Current rectification has been observed at metal-organic interfaces, suggesting a p-n junction which is probably a surface rather than a bulk property. Conductivity cannot be altered by doping because the organics are not sufficiently pure initially.

In recent years, amorphous semiconductors have been developed for use in a variety of switching devices including threshold switches and memory switches. Raw materials are conventional semiconductor materials such as Si, Ge, Te, and As. After mixing, fusing in a furnace, and solidification, the material is a glassy and disordered substance. These materials lack the regular periodicity of crystals but there is local "short range order" in which each atom sees its neighboring atoms in much the same relative places as it would in a crystal. This short range order does have a direct bearing on semiconductor properties. The first switch (1958) based on amorphous material used an amorphous layer of tantalum oxide (about 90 namometers thick) on tantalum metal with an electrolyte solution forming part of the control electrode. Switching occurred when the film was sufficiently polarized. It is conceivable that development of amorphous semiconductors may lead to analogous developments in polymeric materials.

5-5 PHOTOCONDUCTIVITY

Photoconductivity is the ability of a material to conduct electric current when exposed to light. Because light is electromagnetic radiation, it interacts with matter in much the same way as X rays do, by setting electrons in vibration. The energy imparted by light is much less than that imparted by X rays, and thus only the valence electrons are affected. At the same time, the energy is much larger than that produced by thermal vibration. In many semiconductors, such as Se, Cu_2O, CdS, and ThS, light will dissociate some covalent bonds and thus substantially increase the number of conducting electrons and holes. If an electric circuit includes such materials, there will be an almost instantaneous increase in current when light shines upon them. This change in current is sufficient to activate relays in various light detectors such as the "photoelectric eye."

Visualize a valence electron in an insulator (e.g., Fig. 5-8), which light shines upon. The electron can obtain energy from the light (using Planck's formula) in the amount

$$E = h\nu \tag{5-8}$$

but we know that

$$v\lambda = c$$

and thus

$$E = \frac{hc}{\lambda} \tag{5-9}$$

where h is Planck's constant, v is the frequency, λ is the wavelength, and c is the speed of light. For light with a wavelength of about 7000 A, we would have an energy of about 2 eV. This is in the range of forbidden band width of many materials.

If we shine light on an insulator (like Fig. 5-8), some of the electrons near the top of the highest filled band will acquire enough energy to move into the lower levels in the next highest band, i.e., the conduction band. This is shown schematically in Fig. 5-14.

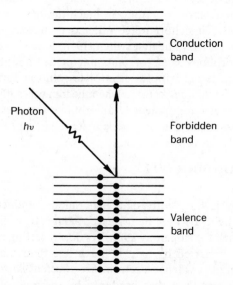

Fig. 5-14. Optical excitation in photoconductivity.

Photoconductivity can be developed in both ceramic and polymeric materials.

EXAMPLE 5-4

Assume a single crystal of germanium is doped with 10^{16} arsenic atoms per cm³. (a) Assuming that all the As atoms are ionized, what is the conductivity? (b) If the crystal is exposed to a pulse of light such that absorption produces 7×10^{15} electrons/cm³, what happens to the conductivity immediately after the exposure?

Solution:

(a) From Eq. (5-7)

$$\sigma = n_n q \mu_n$$
$$= (10^{16})(1.6 \times 10^{-19})(3900)$$
$$= 6.25 \text{ mho/cm}$$

(b) This increases the carrier electrons by

$$\frac{7 \times 10^{15}}{10^{16}}, \text{ or about } 70\%.$$

5-6 PHOTOEMISSION

Albert Einstein received his Nobel Prize for the theory of photoemission. This theory depends on the concept that the interchange of energy between radiation and matter takes place in quanta of energy $h\nu$. If a material (e.g., a metal) is placed in a vacuum and monochromatic light is shone on it, emission of electrons will occur if the frequency of the light is greater than a minimum value ν_0. The current developed is proportional to the intensity of the incident light.

We recall that the valence band of a metal is only partly filled with electrons (Fig. 5-7 or 5-9). There is a potential barrier which the electrons must overcome to escape into the vacuum. The height of this barrier is ϕ (Fig. 5-15), and an electron must receive a minimum of ϕ to escape. Thus, if the

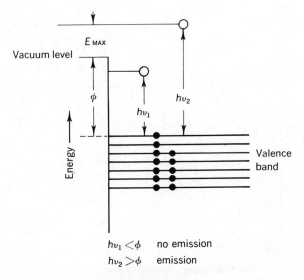

Fig. 5-15. Photoemission.

photon of incident light satisfies the condition $hv > \phi$, an electron will escape into the vacuum, and the number of electrons escaping will be proportional to the number of incident photons. If $hv < \phi$, no electrons escape. The maximum energy E_m of the emitted electron is $hv - \phi$.

5-7 MASERS AND LASERS

The term "maser" is an acronym formed from Microwave Amplification by Stimulated Emission of Radiation. Masers can operate at frequencies (or wavelengths) other than those in the microwave region. The optical maser, using visible light, is called a laser. There are also the iraser (infrared), uvaser (ultraviolet), and xraser (x ray), all based on the same general principles of quantum electronics.

If we shine light with a wavelength of 0.405μ on potassium, this corresponds to the transition from the $4s$ level to the $5p$ level (Fig. 5-16). A

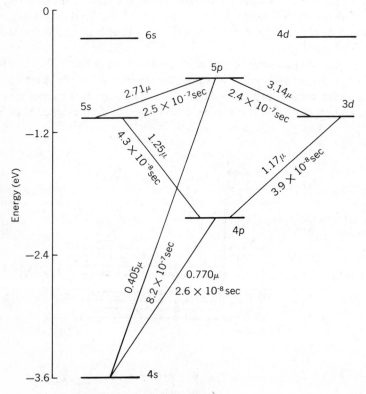

Fig. 5-16. Energy levels in a potassium laser.

large number of electrons will be given the proper quantity of energy to be raised to the $5p$ level. This is known as *optical pumping* and is analogous to the optical excitation of electrons from the valence band to the conduction band in photoconductivity. When the light is removed, the pumped electrons will give up their energy and return to the ground state, either by emission of light with wavelength 0.405μ or by two alternative three-step processes (Fig. 5-16). This emission is a random process.

Fig. 5-16 also shows the average lifetime for spontaneous emission. Thus electrons in the $5p$ state drop to the $3d$ state in about one-fourth of a microsecond, on the average, after reaching the $5p$ state. This is accompanied by emission of a 3.14μ line. Electrons which are pumped to the $5p$ level and drop to the $5s$ or $3d$ levels will return to the ground state, since the average lives in these states and in the $4p$ state are so much shorter than the average lifetime in the $5p$ state. Thus, after pumping, a large fraction of the pumped electrons will be in the $5p$ state, and a small fraction will be in the $5s$, $3d$, or $4p$ states. Before pumping, however, essentially all the electrons are in the ground state, and the effect of pumping is to shift the occupancy of the lowest and highest energy levels. This is known as a *population inversion* and is a necessary requirement for laser action.

Shining a weak beam with a wavelength of 2.71μ on the potassium would give $5s$ electrons enough energy to excite them into the $5p$ level. There are few $5s$ electrons to absorb this radiation. There are, however, many electrons in the $5p$ state, and they are stimulated by the incident radiation. If there were a level above the $5p$ at an energy corresponding to the wavelength of 2.71μ, electrons would make this transition, and an incident photon would be absorbed for each electron making the transition. It is not obvious that the same incident radiation can cause a downward transition with the electron giving up a photon of the same energy as the incident photon. In effect, the stimulated electron is pushed into a lower state, and the original incident photon is amplified to two photons of identical energies leaving the laser. Thus the creation of an inverted electron population and the possibility of stimulated emission lead to the amplifying properties of masers.

The spontaneous emission of light corresponding to the transition from $5p$ to $5s$ is random; i.e., there is no relation between the arrival of any one photon and all the others involved. For stimulated emission, however, the outgoing photon is triggered by the incoming one, so that the two are in phase and reinforce each other. Since this is true of all the incident and emitted waves, the laser is a *coherent* device.

The process just described applies to a *three-level* maser since two fast transitions, $5s \rightarrow 4p$ and $4p \rightarrow 4s$, act essentially as a single one. This is shown schematically in Fig. 5-17(a), in which the three levels are labeled E_1, E_2, and E_3. Population inversion can be obtained in other materials where there is an interchange in role of the two upper levels [Fig. 5-17(b)]. Some

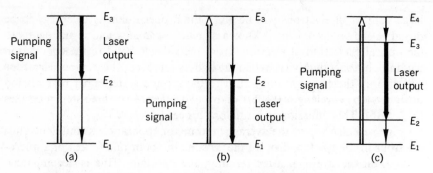

Fig. 5-17. Forms of lasers. Two forms of three-level laser are shown in (a) and (b). A four-level laser combining (a) and (b) is shown in (c).

materials combine the features of both three-level masers to give a *four-level* maser [Fig. 5-17(c)]. The $E_4 \rightarrow E_3$ and $E_2 \rightarrow E_1$ transitions are fast and the transition $E_3 \rightarrow E_2$ is slow. Thus, level E_3 supports the inverted population, and the $E_3 \rightarrow E_2$ transition is the lasing one.

Among the properties that make lasers unique is the extremely high-energy-density beam produced. With a high-energy laser, the beam can be focused to give an energy density of 10^8 watts/cm². Using special techniques, this can go as high as 10^{13} watts/cm². Lasers have applications in spectroscopy, micro-machining, optical ranging, nonlinear optics, optical information processing, and, no doubt, many others not yet visualized.

The pulsed ruby laser (1960) was the first optical maser and provides a good example from which other masers and lasers have been developed. Although the ruby laser operates on the general principles outlined above, there are some significant differences in detail. The energy band diagram (Fig. 5-18) shows three energy levels, but E_3 is a band (actually two bands)

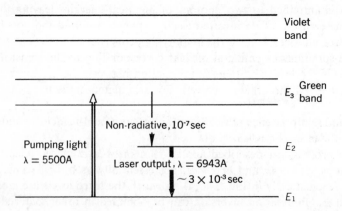

Fig. 5-18. Energy diagram for ruby laser.

rather than a single level and E_2 is a metastable level. This corresponds to a forbidden transition so there is no radiation in going from E_3 to E_2. The pumping frequency used for the lower band is 5500 A (yellow-green). Individual chromium ions, suspended in synthetic sapphire (Al_2O_3), absorb green light, convert part of it into heat, and emit the rest of the energy as red light. In normal fluorescence, each of the Cr ions does this independently with the radiation given off in all directions in short random bursts. The laser converts this emission into a cooperative phenomenon so that the ions emit in a cohesive and cooperative manner.

A ruby crystal is machined to a cylinder and the ends are ground parallel to optical tolerances. These ends are then silvered, one opaque and the other somewhat transparent. The silvering causes the light output to be reflected back and forth. This radiation causes other Cr ions to give up their energy in phase with one another and in a direction perpendicular to the mirrors. As a result, the laser emits a highly collimated beam of coherent red light with a wavelength of 6943 A. This type of laser is a pulsed laser in which the input signal has to be synchronized with a xenon flashtube which supplies the pumping.

A continuous ruby laser is possible by increasing the amount of energy from the pumping source. Pumping light from a very high intensity source is focused by means of mirrors onto the large end of a trumpet-shaped sapphire crystal whose small end contains the Cr_2O_3 which makes the ruby. Internal reflection channels the light down to the ruby end, permitting a large amount of energy to be supplied on a continuous basis. The laser output beam is transmitted out through the half-silvered extraction mirror. About 1 kw of power must be supplied to the lamp for continuous operation with an output power of a few milliwatts. Removal of this heat is a major problem in continuous operation. While the foregoing discussion was in terms of the ruby laser, a crystalline ceramic, there are a number of other materials, both ceramic and polymeric, which can be made to lase.

5-8 DIELECTRIC POLARIZATION

Dielectric polarization is a characteristic phenomenon of electrical insulators. It appears whenever electric charges are displaced relative to each other, and the displacements are reversible. Devices based on this phenomenon include condensers, rectifiers, resonators, amplifiers, transducers, memory devices for computers, and many others.

There are two classes of dielectrics: polar and nonpolar. A nonpolar dielectric is one which has no dipoles when the material is not in an electric field. (A dipole is a pair of equal, but opposite, electric charges which are arbitrarily close.) Consider a material composed of a large number of identical atoms. Since nuclear radius is extremely small compared with atomic radius, each atom can be visualized as a positive point charge at the center of a negative charge-cloud [Fig. 5-19(a)]. The nucleus has a charge of $+Ze$, where Z is the atomic number and e is 1.6×10^{-19} coulomb. The electron cloud has a charge of $-Ze$.

Application of an electric field will displace the nucleus slightly in the direction of the field, and the centers of the electron clouds will move slightly in the opposite direction, giving a separation of d [Fig. 5-19(b)]. Thus application of the field creates an equivalent dipole [Fig. 5-19(c)] at the site of each atom, thereby establishing induced polarization.

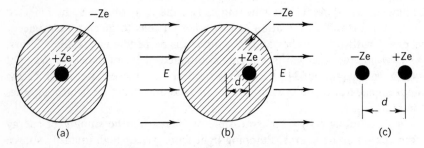

Fig. 5-19. Electronic polarization showing (a) nucleus and electron charge cloud, (b) effect of applied electric field, (c) the equivalent dipole.

Two types of induced polarization can exist. That described above is called electronic polarization. If the material is composed of positive and negative ions, these can undergo a relative displacement and develop ionic or atomic polarization.

Some molecules have dipoles even in the absence of an electric field. These are permanently polarized. The water molecule, for example, has the structure shown in Fig. 5-20(a). Since H has one valence electron and O has six, there is a tendency to form the stable octet through a covalent type of bonding. The atoms become charged because of this electron transfer, and two dipoles are established as in Fig. 5-20(b). The permanent dipole is the vector sum of the two OH dipoles. The structure of CO_2 is shown in Fig. 5-20(c). In this case, the CO dipoles cancel on addition, giving zero permanent dipole. In most materials, the dipoles will have random orientations (isotropy), and the macroscopic vector sum will be zero. When an electric field is applied, the dipoles tend to line up in the direction of the field, giving an *orientational polarization*.

Space charge polarization comes from mobile charges which are present because (1) they are impeded by interfaces, (2) they are not supplied by an electrode nor discharged by an electrode, and (3) they are trapped in the material.

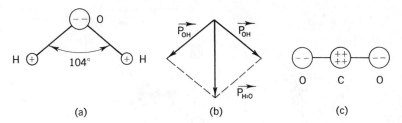

Fig. 5-20. Structure of molecules: (a) Water molecule with (b) equivalent dipoles; (c) carbon dioxide molecule.

In gases, the dipole field of any given molecule has essentially no effect on any other molecule. In liquids and solids, however, each dipole, permanent or induced, is subject to the field of all other dipoles. Thus the total field acting on a dipole, called the local field, is more than the applied field. Polarization can either designate the bound-charge density or the dipole moment per unit volume. These are essentially equivalent.

5-8-1 FERROELECTRICITY

In each case of induced polarization discussed above, the displacement of centers and the resulting polarization were proportional to the applied electric field and disappeared when the field was removed. There are other materials in which the centers of negative and positive charge, having once moved sufficiently from each other, remain permanently displaced. In this case (with materials such as $BaTiO_3$), the induced polarization builds up rapidly at first, passes through an inflection point, becomes linear, and the material seems to act like a normal dielectric. Extrapolation of this linear portion back to zero field is known as the *spontaneous polarization*. When the field is removed, the polarization decreases but does not retrace its buildup path. Instead, at zero applied field, we find a *remanent polarization*. Application of a reversed field will give a similar behavior in the opposite direction. Alternation of the direction of the field will give a typical hysteresis curve.

This phememon is known as ferroelectricity, by analogy to the spontaneous magnetism in ferrous materials. Development of spontaneous and remanent polarization is dependent on crystal structure. In the case of $BaTiO_3$ (Fig. 5-21), the unit cell is not quite a cube but is actually a tetragon with the c dimension about 2% greater than the a dimension, at and below room temperature. The difference decreases with increasing temperature until the structure is cubic at 120°C. Ferroelectricity does not develop in barium titanate at 120°C and above. The actual situation is quite complex, but it is known that, in the tetragon, the Ti ion has two possible positions of stable equilibrium, one just above and the other just below the geometric center

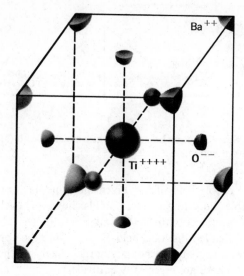

Fig. 5-21. Unit cell of barium titanate ($BaTiO_3$).

and that it oscillates between them. When an applied field acts upward, the Ti ion moves toward the upper position. Being off center, it creates an effective dipole. When the local field is strong enough to overcome thermal vibration, the Ti ion becomes locked in this position, and the unit cell acquires a permanent dipole parallel to the direction of the applied field. Ferroelectrics include $PbTiO_3$, KH_2PO_4, KH_2AsO_4, $NaCbO_3$, $NaTaO_3$, $LiTaO_3$, $LiCbO_3$, and WO_3.

5-8-2 DIELECTRIC STRENGTH

For normal dielectrics, the higher the field intensity, the greater the separation of charge centers (on the atomic level), and the greater the magnitude of restoring forces produced by the electrostatic attractions and repulsions. At some critical separation distance, however, the restoring forces reach a maximum. Beyond this point, the charges begin to move apart under the influence of the field, and the dielectric ceases to be an insulator. The corresponding field intensity is known as the *dielectric strength* or *breakdown strength*. Dielectric strengths of 10^9 and 10^{10} v/m are expected in solids on the basis of theoretical calculations. Strengths of nearly this magnitude are found in molecular solids such as impregnated papers and polymers. In ceramics, however, experimental values are about 100 times less. This is presumed to be due to imperfections such as vacancies, dislocations, cracks, etc.

Dielectrics are idealized on the basis of the concept of layers of materials.

If the layers are parallel to the capacitor plates, then the capacitive elements are in series and both the inverse capacitances and inverse conductivities are additive. If the layers are normal to the capacitor plates, however, the capacitances are additive.

The structure of most ceramic dielectrics consists of crystalline phases dispersed in a glass matrix or crystalline phases separated by a vitreous boundary layer. The resultant properties are intermediate between those of single crystals and those of glass. Both the measured dielectric constant and the loss factor increase with temperature increase, especially at lower frequencies. The chief contribution to dielectric losses comes from the glass portion of the structure. Careful control is necessary to obtain ceramics with low dielectric losses.

EXAMPLE 5-5

Two capacitor plates (2 cm $\times$ 4 cm) are parallel and 0.71 cm apart in a vacuum. (a) What voltage is required to produce a charge of 10^{-10} coulombs on the plates? (b) If a dielectric material with a relative dielectric constant of 2 is placed between these plates, what voltage is required to give the same density charge?

Solution:

(a) $$\mathfrak{D}_0 = \epsilon_0 E,$$

where $\mathfrak{D}_0$ is the charge density, ϵ_0 is the permittivity constant (8.85×10^{-14} coulomb/volt-cm) and E is the electric field or voltage gradient, V/d.

$$10^{-10} = (8.85 \times 10^{-14})(2 \times 4)\left(\frac{V}{0.71}\right)$$

$$V = 100 \text{ volt}$$

(b) $$K = \mathfrak{D}_m/\mathfrak{D}_0,$$

where K is the relative dielectric constant and $\mathfrak{D}_m$ is the charge density with the dielectric material present.

$$\mathfrak{D}_m = K\mathfrak{D}_0 = K\epsilon_0 E$$

$$10^{-10} = 2(8.85 \times 10^{-14})(2 \times 4)\left(\frac{V}{0.71}\right)$$

$$V = 50 \text{ volts}$$

5-9 MAGNETIZATION

Magnetization, like electric polarization, originates in phenomena of interaction between electric charges on the atomic scale. Magnetization, however, appears only when the electronic charges move relative to each other. The first recorded application of magnetic properties precedes application of dielectric properties by several centuries. Engineering applications, however, were not made until the first half of the 19th century. Insertion of a magnetized iron core in the rotor of an electric generator certainly has greatly increased the supply of electric energy. Recent applications of magnetism in communications engineering are at least as noteworthy. Magnetic amplifiers, modulators, and rectifiers in combination with transistors, compete successfully with similar devices based on vacuum tubes. Magnetic memory devices in digital computers are outstanding for efficiency and compactness.

The influence of materials on the magnetization current in an inductive circuit element is analogous to the effect of polarization on dielectric properties. In real materials, a magnetizing field, H, creates magnetic dipole chains similar to the electric dipole chains created in reaction to an applied electric field, E. The magnetic dipole moment per unit volume is a product of the number of elementary magnetic dipoles per unit volume and their magnetic moments. These moments, in turn, are proportional to the magnetizing field and rapidly follow changes in it. Magnetic properties can be measured as the ratio of the magnetization to the applied field.

In dielectrics, the electric flux density (D) and the polarization (P) have the same dimensions of dipole moment per unit volume and the electric field strength has the units of force per unit charge of torque per unit dipole moment. In magnetic materials, the magnetic field strength and magnetization have the units of magnetic dipole moment per unit volume while the magnetic flux density has units of torque per unit dipole moment.

5-9-1 DIAMAGNETISM

Magnetism is a manifestation of electric charge in motion. Thus the electrons orbiting around an atomic nucleus will produce magnetic effects. This is known as diamagnetism and is analogous to the induced polarization discussed above. A diamagnetic material is one in which there is no permanent polarization but one in which polarization can be induced. The induced polarization tends to reduce the total internal field. Thus there is a negative magnetic susceptibility. Diamagnetic materials have no permanent dipoles.

5-9-2 PARAMAGNETISM

In a material which is permanently polarized, i.e., one which has atomic dipoles even in the absence of an applied magnetic field, the dipoles interact weakly (or not at all). Thus they are oriented at random and have a low net magnetization. These materials are paramagnetic. When paramagnetic materials are placed in an applied magnetic field, the permanent dipoles are subjected to a torque which lines them up with the field. A diamagnetic effect is also developed by the orbiting electrons. This effect is very weak and is completely masked by the paramagnetism.

5-9-3 FERRO-, FERRI-, AND ANTIFERRO-MAGNETISM

Magnetization in solids is greatly affected by the stability of the structure. The transition elements, especially iron, are outstanding in this respect. The close interaction between neighboring atoms in solids restricts the freedom of orbital motions of their electrons, and, as a result, the electron spins are the only major contributors to the permanent magnetic dipoles. Permanent magnetic moments occur only in those systems in which unpaired electrons are present. Depending on the configuration of spins in the atom and on the orientation of magnetic dipoles in the unit cell, there may be total cooperation, partial cooperation, or total cancellation of individual magnetic dipoles. The properties which correspond to these conditions are known as ferromagnetism (characteristic of totally cooperating iron), ferrimagnetism, and antiferromagnetism, respectively. Both ferromagnetism and ferrimagnetism produce a net magnetization in the bulk of the material. The magnetization persists in the absence of an applied magnetic field.

5-9-4 DOMAINS

Ordinary iron is not magnetic and becomes so only when placed in a magnetic field, such as that of a solenoid. A current from a flashlight battery will produce this magnetism but is much too weak to originate the interaction forces which are responsible for ferromagnetism. Thus there must be some internal arrangement which helps line up the dipoles associated with electronic orbital and spin motion in such a way as to produce the large internal fields which are the essential difference between ferromagnetic and paramagnetic materials. The explanation is based on domain theory.

Consider a piece of iron which has never been magnetized. This will usually be a polycrystalline material with random orientation. Such a material is shown schematically in Fig. 5-22 where each grain is divided into several domains, each of which has a distinct direction of polarization. There are three mutually perpendicular orientations of the dipoles because iron is BCC and has easy directions of magnetization along the cube edges which are physically equivalent. Domain formation is based on the concept of minimization of energy. This, in turn, involves magnetostatic energy, exchange energy, anisotropy energy, and magnetostriction.

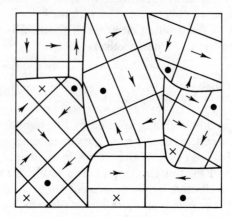

Fig. 5-22. Domains in ferromagnetic material (arrows indicate direction of polarization).

5-9-5   MAGNETOSTATIC ENERGY

Magnetostatic energy lowers the total energy. Consider two bar magnets having parallel magnetizations positioned side by side as shown in Fig. 5-23(a). The mutual force developed between them causes them to assume the antiparallel orientation [Fig. 5-23(b)], which is an arrangement of stable equilibrium and lower energy. The energies associated with a ferromagnetic material are quite high, as indicated in Fig. 5-24 for monocrystalline iron. (Energy density in a magnetic field is proportional to the square of the field.) In order to decrease this energy, domains form in the manner of the bar magnets (Fig. 5-23) so that adjacent regions of opposite magnetization are developed [Fig. 5-25(a)]. A large number of domains should be formed [Fig. 5-25(b)], since there is a greater decrease in energy as the number of domains

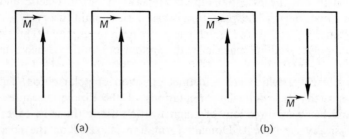

(a)   (b)

Fig. 5-23. Two bar magnets in (a) parallel orientation and (b) antiparallel orientation.

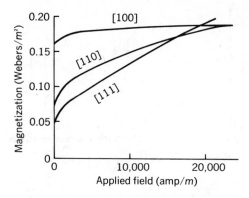

Fig. 5-24. Magnetization curves for monocrystalline iron along three principal directions.

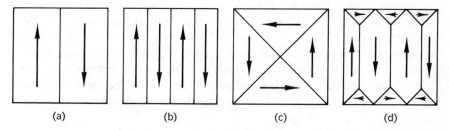

(a)   (b)   (c)   (d)

Fig. 5-25. Domain formation as a means for reducing dipole interaction: (a) Initial formation; (b) further reduction through formation of more domains; (c) domains of closure; (d) domains of closure observed in iron.

increases. A further decrease in energy is produced by domains at right angles since both directions represent easy magnetization [Fig. 5-25(c)]. These additional domains are called domains of closure and have been observed in the manner shown schematically in Fig. 5-25(d).

5-9-6 WALL ENERGY

The above phenomenon would tend to create an infinite number of domains of infinitesimal volume, a condition which is nonexistent. Thus there must be an additional physical effect which increases the energy and opposes domain formation. This effect is associated with the formation of the walls (interfaces) between the domains. The wall energy is composed of exchange energy and anisotropy energy. As shown in Fig. 5-26, the domain wall represents a transition over a small number of dipoles, and only a small amount of energy is required to move the wall.

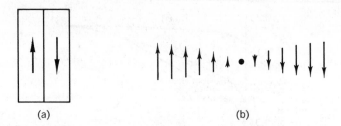

(a)   (b)

Fig. 5-26. (a) A domain wall; (b) transition region corresponding to a wall.

5-9-7 EXCHANGE ENERGY

As indicated above, the magnetic properties of iron result principally from the orbital and spin moments of the unfilled inner electronic shells. The antiparallel spin alignment (Figs. 5-1 and 5-2) represents a lower energy configuration than a parallel spin alignment. This is consistent with the analogy of Fig. 5-23. In iron, the $3d$ band lies just inside the $4s$ valence band and contains six electrons. Since a full $3d$ band will have 10 electrons, the $3d$ band is only partially filled in iron, each level having one electron with spin up and one with spin down [Fig. 5-27(a)]. If this is visualized as two partial bands, one corresponding to all spin-up electrons and the other to all spin-

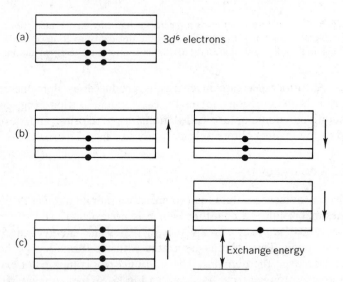

Fig. 5-27. Exchange energy in iron. (a) Band structure of incomplete subshell of iron with one electron of each spin per level; (b) decomposition of (a) into two subbands, one for each kind of spin; (c) rearrangement of (b) showing origin of exchange energy.

down electrons, it represents the unmagnetized state where the spin magnetic moments cancel in pairs [Fig. 5-27(b)]. A new state can be created in iron by having two spin-down electrons flip over and go into the spin-up band [Fig. 5-27(c)]. The energy corresponding to this new arrangement of electrons in the subbands is higher, since two electrons have moved from the middle of one partial band to the top of the other partial band. One band, however, has been reduced with respect to the other by an amount equal to the exchange energy. The exchange energy between adjacent domains, like magnetostatic energy, tends to minimize itself. The exchange energy, which is proportional to the number of domains, does exist.

5-9-8   ANISOTROPY ENERGY

In addition to the exchange energy, anisotropy energy (the other effect of wall motion) comes from the existence of easy, intermediate, and hard directions of magnetization (Fig. 5-24). The anisotropy energy will be a minimum when all magnetic moments are in the same direction. Thus rotation of entire domains increases the anisotropy energy above its minimum.

5-9-9   MAGNETOSTRICTION

Iron elongates in the direction of the induced magnetic polarization and contracts laterally. This phenomenon of magnetostriction has an associated mechanical energy which combines with exchange energy and anisotropy energy to oppose the magnetostatic energy.

5-9-10   FERRIMAGNETISM

Ferromagnetic materials are essentially metallic conductors and thus are subject to induced currents in variable magnetic fields, i.e., eddy currents. These eddy currents dissipate energy and dampen mechanical and electrical vibrations. Consequently, the usefulness of ferromagnetic materials in high frequency applications (TV tubes, high-speed switches, memory devices, etc.) is significantly reduced. Ferrimagnetic ceramics have a much lower electrical conductivity (order of 10^{-8}) and thus have much smaller energy losses from induced currents. This permits use in applications as indicated above but also in relatively novel circuit elements such as square hysteresis loop materials with rapid switching times. These are very useful for switching elements in digital computer magnetic memories. The ferrimagnetic materials are restricted, however, to those relatively few structures which contain ions with permanent magnetic dipoles, i.e., those ions with ferromagnetic properties.

The best known ferrimagnetic material is magnetite (Fe_3O_4). There are a number of other useful magnetic ceramics which are quite similar to Fe_3O_4

but with a higher resistivity. This class of materials has the general formula of AB_2O_4, of which the spinels are of prime interest. The spinels have the formula MFe_2O_4 in which M represents a metallic ion other than iron. In these structures, the O atoms are arranged in an (almost) FCC packing in a unit cell with 32 oxygen ions. This structure has 32 octahedral interstices (16 filled with B ions) and 64 tetrahedral interstices (8 filled with A ions). The cation arrangement is important in determining the ferrimagnetic behavior. In "normal" spinel, all 8 divalent ions are in tetrahedral (A) sites and the 16 trivalent ions are in octahedral (B) sites. In the "inverse" structure, the 8 tetrahedral sites are filled with trivalent ions and the 16 octahedral sites are equally divided between divalent and trivalent ions. Other ferrimagnetic materials include oxides such as $BaO \cdot 6 \, Fe_2O_3$, some magnetic perovskites, and rare earth garnets of the class $3R_2O_3 \cdot 5 \, Fe_2O_3$ where R is yttrium or some other rare earth.

In the case of magnetite, 8 Fe^{3+} ions are in tetrahedral sites while the other 8 Fe^{3+} and the 8 Fe^{2+} ions are in octahedral sites. The electron spin moments from the Fe^{3+} ions at the two different sites are opposite and cancel each other. Magnetic polarization is, therefore, determined solely from the spin moments of the Fe^{2+} ions. Similar reasoning applies to other divalent metals in the spinel structure. Combinations can be used to improve properties. For example, $ZnO \cdot Fe_2O_3$ mixed in 1:9 proportions with $MnO \cdot Fe_2O_3$ and fired into a single ceramic body gives a 10% increase in bulk magnetization. This occurs because some of the antiparallel Mn ions are replaced by Zn ions having no spin magnetic moment.

While magnetic properties are limited to materials that contain ions with ferromagnetic properties, this does not limit the physical form in which they may be used. Metallic magnets may be fabricated into a great variety of shapes and forms. This is also true for ceramic magnets. While we don't usually think of polymeric magnets, it is possible to make magnetic polymeric materials. This can be done by incorporating magnetic material as a filler during fabrication of the polymer. In fact, this has been done with some elastomers to produce a "rubber" magnet.

EXAMPLE 5-6

Calculate the intensity of magnetization in iron at room temperature when all the atomic dipoles are in perfect alignment. (Intensity of magnetization is defined as the magnetic moment per unit volume. Each atom in metallic iron has a magnetic moment of 2.2 Bohr magnetons.)

Solution:

$$M = \frac{2(2.2)(0.927 \times 10^{-20})}{(2.866 \times 10^{-8})^3} = 1730 \text{ gauss}$$

The measured value is 1714 gauss.

5-10 THERMAL PROPERTIES

In one sense, all properties of materials could be classified as thermal properties since they are affected, in varying degrees, by temperature. Among the properties generally classified as thermal properties are heat capacity, coefficients of expansion, thermal conductivity, etc.

Internal energy in solids is principally composed of vibrational energy of atoms about their mean lattice positions and the kinetic energy of free electrons. Thermal properties, therefore, depend upon energy changes of the atoms and free electrons. The thermal properties do not appear to be greatly affected by defects (Chapter 4) in contrast with the mechanical properties (Chapter 6). Thermal energy in lattice vibrations is visualized as a series of superimposed strain waves or sound waves with a frequency spectrum determined by the elastic properties. A quantum of elastic energy is called a *phonon*.

5-10-1 HEAT CAPACITY

Heat capacity of solids is zero at $0°K$ and rises rapidly with temperature to an approximately constant value at higher temperatures. Specific heat is the heat capacity per unit mass. The classical theory was proposed by Dulong and Petit. They postulated that specific heat is the same for all elementary solid materials, about 6 cal/mol-$°K$ or $3R$ per gram atom. R is the universal gas constant and is equal to $N_0 k$ where N_0 is Avogadro's number and k is the Boltzmann constant. This is based on $\frac{1}{2}k$ for each degree of freedom for kinetic energy and $\frac{1}{2}k$ for each degree of freedom for potential energy or $3k$ per atom.

Experimental results show that, although materials approach the value of $3R$, the law is not obeyed in all cases. The three curves, A, B, and C, in Fig. 5-28, are found to correlate with hardness, melting point, or some other parameter. Lead is like A, copper is like B, and diamond is like C. The Dulong-Petit law is generally correct at room temperature (and higher) for elements with atomic weight greater than 40. At low temperatures, the specific heats of all elements approach zero. A number of light, high-melting-point elements (B, Be, C, Si) have much lower room temperature values than predicted. For a number of highly electropositive elements (Na, Cs, Ca, Mg), the specific heat is considerably above $3R$.

The first two observations can be explained using quantum theory and by considering the solid to be a continuous vibrating medium, somewhat like a bowl of jelly. Thus the thermal vibration of atoms is regarded as a random mixture of phonons. The full explanation is much beyond our intended context. We might suspect that electrons would contribute $3k/2$

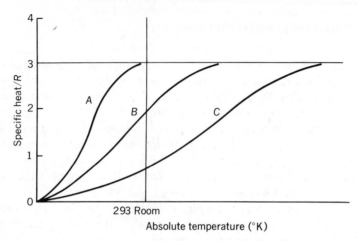

Fig. 5-28. Specific heat of solids.

(per free electron) to the specific heat. Actually, less than 1% of the valence electrons are responsible for the electronic specific heat in most circumstances. Electronic specific heat is important at very low temperatures as the lattice specific heat approaches zero and at high temperatures in electropositive metals.

5-10–2 THERMAL EXPANSION

Almost all materials expand on heating. This thermal expansion corresponds to an increase in the average interatomic distance. As temperature increases, the amplitude of lattice vibration increases up to about 12% of the interatomic spacing at the melting point. This is not sufficient, however, to account for thermal expansion. At any given temperature (with its associated vibration energy level), the atom can oscillate between the extremes indicated by the solid curve in Fig. 3-6. As the temperature increases, the energy level also increases, and the atom can oscillate between wider limits. Since the total interatomic potential is asymmetrical, it is obvious that the mean position shifts toward a position of greater interatomic spacing (for the case in Fig. 3-6).

Thermal expansion is "normal," i.e., volume increases as temperature increases, if the frequency of lattice vibration decreases as the volume increases in relatively isotropic structures. In anisotropic structures, this may not be true. Graphite (Fig. 4-17), for example, has a large thermal expansion along the c-axis but a small contraction along the a-axis. Such behavior is not limited to graphite but is found in Zn below 86°K and in a number of other materials at low temperature. Many of these have some directed covalent bonding. There are a number of ceramic materials which have essentially zero (or even negative) thermal expansion at room temperature.

EXAMPLE 5-7

When the length of a metal rod is measured in boiling water, it is found to be 0.4% longer than when measured in ice water. (a) What is the thermal coefficient of linear expansion in metric units? (b) In English units?

Solution:

(a)
$$\frac{\Delta L}{L} = 0.004 = \alpha_L(100 - 0)$$

$$\alpha_L = 40 \times 10^{-6}/°C = 40 \times 10^{-6} \text{ cm/cm} - °C$$

(b)
$$(40 \times 10^{-6})\left(\frac{1}{1.8}\right) = 22.2 \times 10^{-6}/°F$$

$$\alpha_L = 22.2 \times 10^{-6}/°F = 22.2 \times 10^{-6} \text{ in./in.} - °F$$

5-10-3   THERMAL CONDUCTIVITY

Thermal conductivity is the transfer of heat through a material by electronic conduction and by a traveling wave mechanism involving phonons. In pure metals, the electron contribution dominates; in disordered alloys and semiconductors, the electron and phonon contributions are about equal; and in normal insulators, the electron contribution is unimportant. In polymers, heat transfer occurs by molecular rotation, vibration, or translation. In metals, electrons can move readily and provide electrical conductivity. Since each charged particle can carry heat energy in addition to its electric charge, the thermal conductivity of a metal should be proportional to its electrical conductivity multiplied by the absolute temperature. This relationship was first discovered experimentally in 1853. At the same time, materials which have no free-moving charge carriers are excellent electrical insulators but are not necessarily good thermal insulators. BeO, for example, is an excellent electrical insulator and an excellent thermal conductor. In covalent or ionic materials, heat is transferred mainly by phonons. Although phonons travel with the speed of sound, they collide relatively often with each other and with lattice defects. As a result, the thermal conductivity is much lower than the phonon velocity would lead one to believe.

EXAMPLE 5-8

If one calculates the ratio of thermal conductivity to electrical conductivity, it is found to be in the range of 6 to 8 joule-ohm/sec-°C for many of the metals. Will one find the same ratio for ceramics and polymers?

Solution:

No. Most of the ceramics and polymers have no free electrons and thus have negligible electrical conductivity. Thermal energy can, however, be transferred by atomic vibrations although at a much lower rate than in metals. Thus one would not find the same ratio.

5-10–4   THERMOELECTRICITY

The fact that two pieces of wire, made from different materials, will show a voltage difference between two free ends when their common junction is raised to a higher temperature was discovered by Seebeck in 1822. This property, thermoelectricity, has been used extensively in thermocouples for the specific purpose of measuring temperatures. The fact that thermoelectricity is a property of semiconductors as well as metallic conductors makes it a potential and promising factor in direct conversion of heat into electricity.

If we visualize the upper bar in Fig. 5-29 as being made of an *n*-semiconductor and having a higher temperature at the right end than at the left,

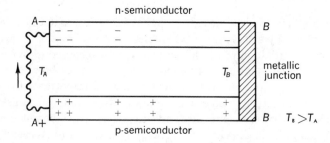

Fig. 5-29. Schematic diagram of a semiconducting thermoelectric generator.

then the average velocity of the excess electrons near the right end will be greater than the average velocity of the electrons near the left end. This would imply a net flow of excess electrons down the temperature gradient. Actually the first electrons arriving at the cold end have no outlet. As a consequence, a negative charge builds up and opposes the thermal emf. To produce a continuous flow of electricity, the circuit must be closed. Obviously, there will be no gain if the lower bar in Fig. 5-29 is identical to the upper one. If, however, the lower bar is a *p*-semiconductor, then a steady electric current can be generated. A thermoelectric generator operating as shown in Fig. 5-29 is capable of only about 3% efficiency with current semiconductors. Thermodynamic analysis shows that greater efficiencies can be achieved by increasing the thermal emf coefficient (thermoelectric power) at the expense of the ratio of thermal to electrical conductivity. To generate a continuous current with metals, two different metals must be used but the thermoelectric power is about one-tenth (at best) that of semiconductors.

Table 5-2   Comparison of thermal properties of some metals, ceramics, and polymers (at room temperatures)

Material	Specific heat, $cal/g/°C$	Coefficient of thermal expansion, $cm/cm/°C$ $(\times 10^{-5})$	Thermal conductivity, $cal/sec/cm^2/°C/cm$ $(\times 10^{-4})$
aluminum (1100)	0.21	2.36	5300
copper	0.092	1.65	9400
iron	0.11	1.17	1700
Type 304 stainless steel	0.12	1.66	390
alumina (Al_2O_3)	0.20	0.5–0.8	540
beryllia (BeO)	0.24	0.60	4200
fluorethylenes	0.22–0.25	4.5–10	1.4–6.0
polyamides (eg, Nylon)	0.4	10–15	5.2–5.8
polyethylenes	0.55	16–18	8
polytetrafluorethylene (eg, Teflon)	0.25	10	6
styrenes	0.32–0.35	3.4–21	1.0–3.3
vinyls (rigid)	0.3–0.5	7–25	3–4

QUESTIONS AND PROBLEMS

1. The text makes the tacit assumption that the conduction band in a covalent or ionic solid is empty. Why?

2. (a) The electron configuration for Na is $1s^2 2s^2 2p^6 3s$. Draw a schematic energy band diagram for one mole of solid sodium. Which levels do you expect to broaden? Indicate (perhaps by shading) the occupation of bands by electrons. Indicate the number of states in each band and the number of occupied states. Indicate conduction and valence bands.

   (b) Mg has one more electron than Na, i.e., the 3s band is full. Explain why Mg is also an electrical conductor.

   (c) Carbon has a $1s^2 2s^2 2p^2$ configuration. Bonding of carbon in diamond is covalent, converting the $2s^2 2p^2$ into sp^3 hybrid states with only four hybrid levels in the shell, per atom. Why then is diamond an insulator?

3. Some metals show a reduction in electrical resistivity with pressure and some show an increase (Li and Bi). How can this be explained in terms of band theory?

4. Why does the electrical conductivity decrease for a conductor with increasing impurity content and temperature? Why do insulators behave in the opposite fashion?

5. Pure Ge has an electrical conductivity of 0.025 ohm^{-1}-cm^{-1}. The number of negative carriers, n_n, equals the number of positive carriers, n_p. What fraction of the conductivity is due to electrons and what fraction is due to holes?

6. An n-type silicon semiconductor has a resistivity of 180 ohm-cm. Calculate the number of charge carriers.

7. Germanium is doped with boron to give a conductivity of 2.0 ohm^{-1}-cm^{-1}. Determine the Ge/B atomic ratio if there are 8 Ge atoms per unit cell and the unit cell volume is 5.66×10^{-8} cm^3.

8. What elements could be added to pure Si to make it a p-semiconductor? Draw a band diagram for Si, as follows: full valence band (four sp^3 hybrid states, filled) with the bottom of the empty conduction band 1.2 eV above the top of the valence band. Draw an acceptor state band with an 0.01 eV binding energy. If 1 ppm of Al is added to Si, how many states are there in the acceptor level per mole of Si?

9. What elements could be added to pure Si to make it an n-semiconductor? Rework the above problem but with a donor band of 0.01 eV. If 1 ppm of P is added to Si, how many states are there in the donor level per mole of Si?

10. Four semiconductors, their approximate energy gaps, and their appearance are listed:

Semiconductor	Energy Gap	Color
Diamond	5.6 eV	colorless
Silicon carbide	3.1 eV	blue-green
Sulfur	2.4 eV	yellow
Silicon	1.1 eV	opaque

Explain the observed colors.

11. Silicon has a density of 2.33 g/cm^3. (a) How many Si atoms per cm^3? (b) Enough phosphorus is added to Si to make an extrinsic semiconductor with a conductivity of 1 ohm^{-1}-cm^{-1}. How many P atoms are added per cm^3?

12. Aluminum is added to Si to give an extrinsic semiconductor with a conductivity of 2 ohm^{-1}-cm^{-1}. (a) How many Al atoms are added per cm^3? (b) How many Si atoms are there for each Al atom?

13. Ionic materials conduct electricity although not nearly as well as metals with conductivity of ionic materials increasing nearly exponentially with increasing temperature. Explain.

14. Compare electrical conductivities (or resistivities) of magnesium, Mg, and magnesia, MgO. Why the difference?

15. Compare the resistivities of conductors and semiconductors, especially in terms of similarities and differences.

16. Green light has a wave length of about 5250 A. What is the frequency? What is the energy of the photons?

17. Luminescence is caused by the 1.1 eV of energy released when an electron and hole recombine in silicon. Is this visible to the human eye?

18. The average energy of a C—Cl bond in polyvinyl chloride is 81,000 cal/mole. Does visible light (violet = 4000 A to red = 7000 A) have sufficient energy to break the average C—Cl bond?

19. Consider designing devices for detecting (a) visible radiation, (b) x radiation, (c) gamma radiation. What differences would you find in the devices?

20. Discuss the similarities and differences of light, x rays, and gamma rays.

21. What is meant by the term *metastable energy level*?

22. In order to obtain lasing, a population inversion is required in which the number of atoms in the excited state exceeds the number in the ground state. Explain why this must be so to obtain intense stimulated emission in a laser.

23. Define coherent electromagnetic radiation.
    (a) Distinguish between time and space coherency.
    (b) With what other devices than lasers can coherent radiation be obtained? How?

24. Diamagnetic specimens are repelled by the pole of a bar magnet, paramagnetic specimens are weakly attracted, ferromagnetic specimens are strongly attracted. Explain.

25. A "needle" of material is freely suspended at its center of mass and allowed to rotate in a uniform magnetic field. In what position will the needle come to rest, relative to the field, if the needle is (a) diamagnetic, (b) paramagnetic, (c) ferromagnetic?

26. A designer wants to produce a refrigerator door with a magnetic latch and a gasket for sealing against air flow. Can you suggest a way of achieving both objectives in the same material?

27. Susceptibility of a paramagnetic material is positive and decreases as temperature increases. Susceptibility of an antiferromagnetic material is also positive and increases as temperature increases. Why?

28. The work function, ϕ, for a cesium-coated oxide-silver photoemitter is 0.9 eV:
    (a) What is the longest wavelength which will eject photoelectrons?
    (b) What is the maximum energy of emitted electrons if incident light on the emitter is 3200 A?

29. Discuss the reasons that the Dulong-Petit "law" is incorrect at low temperatures. Why is the "law" valid, in most cases, at relatively high temperatures?

30. Show, to a first approximation, that the coefficient of volume expansion is three times the coefficient of linear expansion.

31. Explain why the coefficient of linear expansion approaches zero at low temperatures.

32. A mile-long bridge with a steel superstructure is subjected to seasonal temperature extremes of $-40°F$ to $100°F$. (a) How much total expansion joint space must be allowed to keep the superstructure free from thermal stresses? (b) If the superstructure were completely welded and stress free at $70°F$, what thermal stress is developed at $-40°F$? Is this stress a function of the length of the bridge structure?

33. A given metal rod is 0.17% longer in boiling water than in ice water. What is the coefficient of thermal expansion in (a) metric units, (b) English units? What metal might this be?

34. On the basis of volume at room temperature, lead has a volume of 1.02 (as a solid) at the melting point of $327°C$, while liquid lead has a volume of 1.06 at $327°C$. Assuming the radius of a hard-sphere model does not change with temperature, determine the atomic packing factors for solid and liquid lead at the melting point. Explain any difference.

35. Is the coefficient of thermal expansion isotropic in glass? Why? How does this compare with materials like iron, uranium, and graphite?

36. Why are the bottoms of 18:8 stainless steel cooking pans heavily plated with copper whereas those of aluminum pans are not?

37. Why do metals feel colder to the touch than ceramics or polymers although all are at the same temperature, e.g., room temperature?

38. Thermal conductivity is known to increase, remain essentially constant, or decrease with increasing temperature, depending on the particular material. (This is true even within one given class of materials, e.g., metals.) Explain.

39. Why does a porous material provide better thermal insulation than a solid material, even with the same material?

40. Assume you need a thermal insulator with a high electrical conductivity. Can you meet the requirements? What material would you use?

41. Assume you need an electrical insulator with a good thermal conductivity. Can you meet the requirements? What material would you use?

42. The shock parameter P is an empirical measure of the ability of a material to withstand a sudden temperature change, where $P = \sigma_T (TS)/E\alpha_L$ and σ_T is the thermal conductivity, TS is tensile strength, E is Young's modulus, and α_L is coefficient of expansion. What can you say about the thermal shock resistance of (1) metals, (2) polymers, (3) ceramics?

43. In a thermoelectric generator (such as Fig. 5-29), metal is used as an intermediary for both the hot and cold junctions. Why are direct p-n contacts

avoided? Why are p and n type versions of the same material often used in the same generator?

44. List possible applications for thermoelectricity in underdeveloped countries. Can you think of any possible extraterrestrial uses?

45. Discuss the similarities and differences between photons and phonons.

6

MECHANICAL PROPERTIES

6-1 INTRODUCTION

Mechanical properties are of major interest to the designer of operational components. Unfortunately, these properties cannot be deduced from structural or bonding considerations alone since most of them are "structure-sensitive." A piece of steel, for example, can have its tensile strength modified by a factor of three or four by differences in treatment. The properties developed also represent a compromise. Service conditions require structural materials which resist applied loading with limited deformation, yet many processes for fabricating metals require a capability of deforming in various ways with comparative ease. Determination of suitability of a material to meet both requirements cannot usually be made simultaneously by the same test. Thus two general types of testing, almost always destructive, due to the nature of the required measurements, are necessary. A third type of test is often used to determine the homogeneity or soundness of the component as a whole. These are usually nondestructive, e.g., radiographic and ultrasonic tests.

Tests may be divided into two classes: (1) those that attempt to simulate service conditions as nearly as possible, and (2) those that study the behavior of materials under specified conditions which are not necessarily related to a specific service condition. The first class is most satisfactory, but usually entails a slow and expensive procedure. The second class is based on the fact that, although the test conditions are not necessarily related to those encountered in service, the behavior in these tests depends on the same properties that determine usefulness in service. Service loading can then be related to test loading in various ways. Mechanical properties measured in standard laboratory tests are:

1. Properties of an "absolute" nature which can be used in design calculations.
2. Properties of a "relative" nature which cannot be used directly for design but which establish a material as being similar to one which has been found satisfactory in service.

Standard laboratory tests can be considered in three general categories:

1. Static tests in which the loads are applied slowly enough that quasi-static equilibrium of forces is maintained.
2. Cyclic tests in which loads are applied and (a) partly or wholly removed or (b) reversed a sufficient number of times to cause the material to behave differently than under static loading.
3. Impact tests in which loading is applied so rapidly that the material must absorb energy rather than resist a force.

Mechanical properties (such as tensile yield strength, tensile strength, percent elongation, compressive yield strength, shear yield strength, hardness, wear resistance, machinability, etc.), as measured and reported in the literature, are empirical in nature. Most of the tests are arbitrary, although they are standardized and performed under specified conditions. In many cases, reported values are nominal rather than "exact." For example, tensile strength is normally defined as the instantaneous load divided by the initial value of cross-sectional area, even though the area decreases as the test proceeds. The empirical nature of these mechanical properties does not destroy their usefulness to the designer. It does require exercise of his experience and judgment to translate these values into a good working design.

6-2   STRESS AND STRAIN

When external loading is applied, deformation occurs. Internal reaction forces are set up which transmit the applied loading through the component. The intensity of the internal force is called "stress" and has the dimensions of force per unit area. Stress is of two types: (1) a normal stress, σ, which acts perpendicular to a given area, and (2) shearing stress, τ, which acts in a plane, rather than normal to the plane.

As indicated, the loaded component is deformed. The amount of deformation in any direction depends on the magnitude and direction of the loading and on the condition of the material. The intensity of deformation, that is, the unit deformation, is called "strain." Strain is of two types: (1) linear strain, ϵ, which is the change of length per unit length, e.g., inches per inch, and (2) shearing strain, γ, which measures the change in angle from an originally right angle.

6-3   TENSION

A tension test, of which there are many varieties, is relatively simple to perform and is widely used to determine certain mechanical properties. In a
tensile test, a specimen (Fig. 6-1 shows one of several types) is subjected to a

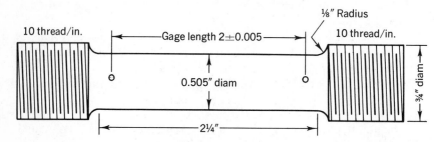

Fig. 6-1. One form of tensile test specimen, typically used in testing ductile
metals.

progressively increasing tensile force until it fractures. Up to fracture, the
material deforms as long as load increases. If the normal stress (tensile load
÷ original area),

$$\sigma = \frac{P}{A} \tag{6-1}$$

is plotted against strain (i.e., unit deformation, change of length ÷ gage
length),

$$\epsilon = \frac{\Delta l}{l} \tag{6-2}$$

a nominal or conventional "stress-strain" diagram is obtained as shown in
Fig. 6-2 for ductile metals.

In the early stages (Fig. 6-2), the stress developed is proportional to the
strain. This is an experimental confirmation of Hooke's law which applies
within the elastic limit. In this range, the elongation of the specimen is:

$$\delta = \frac{Pl}{AE} \tag{6-3}$$

where P/A is the stress (P is the applied load and A is the cross-sectional
area), l is the original gage length of the specimen, and E is the modulus of
elasticity or Young's modulus. In this linear range, the specimen returns to
its original dimensions when the load is released. Once the curve departs
from linearity, plastic deformation begins accompanied by a reduction in
cross-section known as "necking."

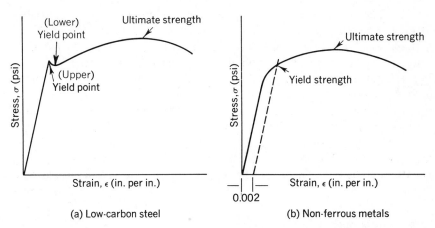

Fig. 6-2. Typical conventional stress-strain diagrams for ductile metals.

As deformation continues, the applied load increases until the *tensile strength or ultimate strength* (Fig. 6-2) is reached. This "strength" is the maximum point shown on the stress-strain curve. After fracture, the overall elongation and the reduction in area can be measured. Both elongation and reduction of area are used as criteria for determining relative plasticity, but neither measures any fundamental characteristic of the material.

In conventional stress-strain diagrams (Fig. 6-2), stress is calculated on the basis of original area and strain on the basis of original gage length. Obviously, stress and strain, as calculated, are not exact but approximate. Thus the conventional stress-strain diagram is somewhat fictitious, and data obtained therefrom are largely empirical. In addition, these values are a function of the specimen size and shape, temperature, and testing rate.

The data normally obtained in a tensile test are yield strength, ultimate strength (maximum load divided by original area), reduction of area (normally expressed in percent of original area), and elongation (usually expressed in percent of the original gage length, which, in a round specimen, is four times the original diameter). Although these data are empirical in nature, they are widely used for design purposes. The tensile test is also used extensively as a means for determining if a given lot of material meets specification values. Standard procedures for making tensile tests are given in American Society for Testing and Materials (ASTM) Standard E8 and other standards.

6-3–1 METALS

With a few exceptions, metals and alloys exhibit ductile behavior in standard tensile testing. Typical resulting stress-strain curves are shown in Fig. 6-2. In metals, work hardening (see Secs. 6-8-6 and 9-3) accompanies plastic deformation. The shape of the curve depends on the rate of work hardening. As loading proceeds beyond the elastic limit, we may find a continuing deformation, possibly with some decrease in loading, as shown in Fig. 6-2(a). This is a phenomenon often observed in plain carbon steel and in low-alloy steels. For most other metals, deformation increases with relatively small increases in load, as shown in Fig. 6-2(b). In the former case, the upper yield point can be used for a working determination of the elastic limit. In the latter case, it is necessary to resort to an arbitrary definition. *Yield strength* is defined as the stress required to produce an arbitrary permanent deformation. The deformation most often used is 0.2% although other values can be used. When yield strength is stated, however, the corresponding deformation must also be given.

6-3-2 CERAMICS

Most ceramics have very limited plastic deformation and fracture soon after exceeding the elastic range (Fig. 6-3). There are a few ceramics, however, such as viscous glasses, which do neck. Most ceramics have relatively low tensile strengths and are rarely used in applications in which they are subjected to tensile loading.

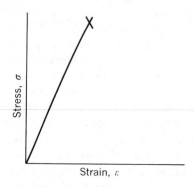

Fig. 6-3. Typical tensile stress-strain diagram for a brittle material.

6-3-3 POLYMERS

Polymers have a wide spectrum of mechanical properties, ranging from gel-like materials, through flexible elastomers and semirigid structures, through hard, tough materials, to hard, brittle materials. Because of this wide spectrum, it is not reasonable to either reject all polymers as structural materials or credit the excellent properties of certain types to all polymers.

One can generalize on the tensile properties of polymers as indicated in Fig. 6-4 and Table 6-1. At the same time, however, generalities about poly-

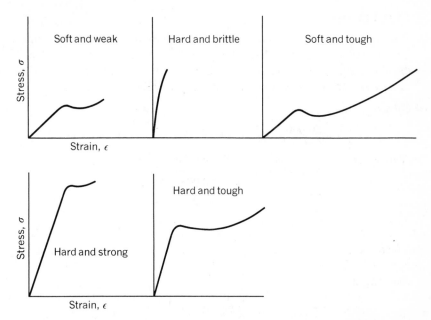

Fig. 6-4. Typical tensile stress-strain diagrams for polymeric materials.

Table 6-1 General tensile characteristics of polymers

Polymer type	Elastic modulus	Yield point	Elongation at fracture	Ultimate strength
Soft and weak	low	low	moderate	low
Hard and brittle	high	undefined	low	moderate to high
Soft and tough	low	low	high	moderate
Hard and strong	high	high	moderate	high
Hard and tough	high	high	high	high

mers can not safely predict performance. Therefore, polymers should be tested under conditions as close as possible to those of expected service if these are appreciably different from conditions used in standard tests.

Metals and polymers form a strength continuum ranging from about 2000 psi for low density polyethylene up to about 500,000 psi for cold drawn stainless steel. Polymers without reinforcement have only moderate tensile strengths and are thus not particularly attractive for structural use. Reinforced polymers have much greater strengths. Mat-reinforced polymers are about one-third to one-half as strong (on a volume basis) as aluminum alloys while cloth-reinforced polymers are about as strong as aluminum alloys.

EXAMPLE 6-1

Two rods, 1 inch in diameter, of 2024–T4 aluminum alloy must support a load of 50,000 lb. (a) What is the stress? (b) What is the strain? (c) If the aluminum rods are replaced by annealed magnesium and the strain cannot exceed that in the aluminum, what diameter is required?

Solution:

(a) $\sigma = \dfrac{P}{A} = \dfrac{50,000}{2[\pi(\frac{1}{2})^2]} = 31,800$ psi

(This is less than the yield strength given in Appendix B.)

(b) $\epsilon = \dfrac{\sigma}{E} = \dfrac{P}{AE} = \dfrac{31,800}{10.6 \times 10^6} = 0.0030$

(c) $A \geq \dfrac{31,800}{(6.5 \times 10^6)(0.0030)} = 2.56$

    $r^2 = \dfrac{2.56}{2\pi} \longrightarrow r = 0.64$ in

                     $d = 1.28$ in

checking the stress,

$$\sigma = \frac{50,000}{2.56} = 19,500 \text{ psi}$$

This is greater than the yield strength (18,000, Appendix B) and thus outside the elastic range. To stay within the elastic range and thus have the above equations apply,

$$\sigma \leq 18,000 \geq \frac{50,000}{A} \longrightarrow A \geq 2.78$$

$$r \geq 0.665 \text{ in} \quad \text{or} \quad d \geq 1.33 \text{ in}$$

6-4 "TRUE STRESS-TRUE STRAIN" TENSION TEST

The conventional tension test gives somewhat fictitious values, since they are based on original rather than instantaneous area. In some applications, we prefer to know the instantaneous value of load divided by area. This can be determined from a so-called "true stress-true strain" test. This type of test can be readily performed with specimens of circular cross-section. The diameter of the specimen is measured by a micrometer or dial gage for various increments of load up to rupture. Care is exercised to select the minimum diameter by moving the gage over the entire gage length of the specimen. The "true stress" and "true strain" can be determined from load and diameter measurements. Stress is obtained by dividing load by instantaneous area, whereas strain is determined from an assumption of constant volume. The equations are (where d is the instantaneous diameter at any load P, and d_0 is the initial diameter):

$$\sigma = \frac{P}{\pi d^2/4} \qquad (6\text{-}4)$$

$$\epsilon = \ln \frac{A_0}{A} = 2 \ln \frac{d_0}{d} \qquad (6\text{-}5)$$

The data, plotted on linear coordinates, give an essentially straight line in the plastic portion. Figure 6-5 shows a typical curve in comparison with a conventional stress-strain curve. Even this type of test gives somewhat empirical values because, once necking starts, the state of stress in the neck is triaxial (radial, tangential, and longitudinal) rather than uniaxial (longitudinal) tension.

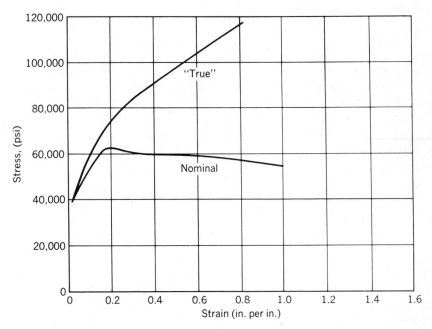

Fig. 6-5. "True" and nominal stress-strain diagrams for a semi-killed steel.

6-5 MISCELLANEOUS MECHANICAL PROPERTIES

6-5-1 MODULUS OF ELASTICITY

The modulus of elasticity, (E), or Young's modulus, is the proportionality constant between stress and strain for elastic materials. This modulus may be considered as the stress required to produce a strain of 1 (i.e., 100%) if the material were completely and indefinitely elastic. For steels, the value of the modulus is approximately 30×10^6 psi while for aluminum it is $10\text{--}11 \times 10^6$ psi. The modulus can be determined from a stress-strain curve obtained using a high-precision tensile test. The modulus of elasticity can also be determined using vibration techniques, although the results differ from those obtained from tensile testing.

Moduli of elasticity for polymers range from about 0.04×10^6 psi for nonrigid polyvinyl chloride to about 8×10^6 psi for filament-wound reinforced polymers. Since polymers are so much more flexible than metals, they are not suitable in applications where bending under load must be minimized. On the other hand, interference fits between polymers can be made easily and polymers can be molded in complex designs that can be snapped out of molds—an impossibility with metals.

6-5-2 COMPRESSION

A compression test is made in the same general fashion as a tensile test. A differently shaped specimen, one which shortens as the cross section increases, is used. In general, the only datum obtained is the compressive yield strength which is determined in the same manner as tensile yield strength (Fig. 6-6). Standard procedures for making a compression test are given in ASTM Standard E9.

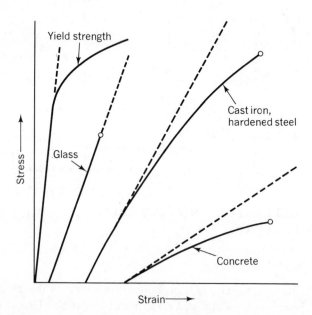

Fig. 6-6. Compressive stress-strain diagrams for various materials. (Schematic, not to scale.) Elastic lines are shown dashed.

6-5-3   DUCTILITY AND MALLEABILITY

Ductility is defined as the property that permits permanent deformation by stress in tension before fracture. The percent elongation and the reduction of area in tension are often used as empirical measures of ductility. Unfortunately, elongation is seldom uniform but varies along the specimen and is greatest at the neck. The reduction of area, however, is calculated from the minimum diameter of the neck and is probably a better indicator of ductility.

Malleability is defined as the property that permits deformation of a material subjected to rolling or hammering. The more malleable a material, the more easily it can be rolled. Ductility and malleability are frequently used interchangeably, but it is better practice to differentiate between them by remembering that ductility is considered a tensile property and malleability a compressive property.

6-5-4   BRITTLENESS

Brittleness is defined as a tendency to fracture without appreciable deformation and is therefore the opposite of ductility or malleability. If a material can be mechanically worked to a different size or shape without breaking or shattering, it is ductile or malleable; but if little or no change in the dimensions can be made before fracture occurs, it is brittle. The criterion of differentiation is rather arbitrary. An elongation of less than 5% in a two-inch gage length is often taken to indicate a brittle material. The appearance of the fracture is perhaps a better means of differentiation. Brittle fractures normally follow the grain boundaries (intergranular or intercrystalline), whereas ductile fractures normally occur through the grains (transgranular or transcrystalline).

6-5-5   TOUGHNESS

Toughness is a descriptive term often defined as the ability to absorb considerable energy before fracture, although it is nearly impossible to measure in absolute units. It is usually represented by the area under a stress-strain curve and thus involves both strength and ductility. We note that the total area under the stress-strain curve is the work expended in deforming one cubic inch of the material until it fractures. This work or energy is sometimes called the modulus of toughness. Since plastic deformation of a tensile specimen is not uniform, the modulus of toughness is highly empirical. This is true using either conventional or "true" stress-strain curves, although the latter may give a somewhat better measure. Comparative values of modulus of toughness apply only to tests involving the same type of specimen and method of loading. A change in either gives a different result. Toughness is related to impact strength, i.e., resistance to shock loading, but there is no definite relationship between the energy values obtained from static and impact testing.

6-5-6   RESILIENCE

Resilience is closely related to toughness and is defined as the capacity of a material to absorb energy in the elastic range. It is measured by the modulus of resilience, i.e., the strain energy developed per unit volume in straining material in tension from zero stress to yielding.

6-5-7   HARDNESS

No fully satisfactory definition for hardness exists. This is not surprising, since hardness tests, in commercial applications, do not measure an inherent mechanical property of the material tested, but are merely arbitrary tests performed under certain standard conditions. Each kind of hardness test measures a different combination of properties. A common definition is the one that states that hardness is a measure of the ability of a material to resist deformation, indentation, or abrasion.

Perhaps the following quotation from Dr. L. B. Tuckerman* best defines "hardness":

> " 'Hardness' in common parlance represents a hazily conceived conglomeration or aggregate of properties of a material, more or less related to each other. These properties include such varied things as resistance to abrasion, resistance to scratching, resistance to cutting, ability to cut other materials, resistance to plastic deformation, high modulus of elasticity, high yield point, high strength, absence of elastic damping, brittleness, lack of ductility and malleability, high melting temperature, magnetic retentivity, etc. This confusion under the one designation 'hardness' results from the fact that there is a rough parallelism in these properties in a large number of materials. The fact that 'hardness,' thus conceived, is a conglomeration of different, more or less unrelated properties makes it impossible to correlate any one definite measurable property with the current implications of hardness.
>
> This does not mean that under the hazy conglomeration of properties which are included in the common understanding of hardness, there are not included very important properties of the material."

The so-called hardness tests fulfill important functions in industry, even though these tests are empirical rather than fundamental. The various hardness tests are often used as inspection devices which are able to detect certain differences in materials, even though the inner qualities may be undefinable. Two pieces of material that have the same hardness may or may

---

*L. B. Tuckerman, Discussion of "Significance of the Hardness Test of Metals in Relation to Design," Proc. Am. Soc. for Testing Materials, Vol. 43, 1943, pp. 803–856.

not be alike, but if their hardnesses are different, the materials are certainly different.

There are some thirty hardness tests ranging from the scratch test made with a fingernail or pocket knife to more elaborate and reproducible laboratory methods, but only six methods are in general use. These are the Brinell, Rockwell, Vickers, Tukon, and, to a lesser degree, Scleroscope and file hardness tests. The Brinell, Rockwell, Vickers, and Tukon hardness tests are similar in that a standard load is applied to an indentor or penetrator resting on the surface of the material being tested. For instance, in the Brinell hardness test, a standard load (500 or 3000 kilograms) is applied through a hardened steel ball one centimeter in diameter. The diameter of the resulting impression is measured, and the surface area of the indentation is calculated. The Brinell hardness number is equal to the load divided by this area.

The accuracy of an indentation hardness test is influenced by (1) accuracy of the load, (2) rate of load application, (3) duration of load application, (4) condition of the indentor, (5) condition of the anvil on which the specimen rests, (6) method of measuring the impression, (7) thickness of the specimen, (8) surface condition of the specimen, (9) shape of the specimen, (10) location of the indentation on the specimen, and (11) uniformity of the material.

6-5-8   WEAR RESISTANCE

Wear resistance is defined as the ability to resist wear and abrasion. Measurement and evaluation are difficult since wear resistance depends on many factors such as the nature and hardness of the abrading material, corrosion, general service conditions, and the condition of the abraded material. Although no standards have been established for wear resistance, the scratch test can be used as a rather rough indication of wear resistance.

6-5-9   MACHINABILITY

Machinability is defined as the ability to be readily machined. No standard measurement has been devised since there is considerable difference of opinion as to what constitutes good machinability. Three criteria of machinability are often used:

1. Power required to remove a given quantity of material from a standard specimen. Under this criterion, the more machinable materials require less power.
2. Appearance of the machined surface. Under this criterion, the more machinable materials have a better appearance or a cleaner cut.
3. Length of tool life. Under this criterion, tools last longer between sharpenings with more machinable materials.

6-6 COMPARATIVE STRENGTHS OF MATERIALS

On the basis of bonding forces, we can postulate that materials with exclusively covalent bonding would be strongest and that those with exclusively ionic bonding would be nearly as strong. Metallic bonding would be clearly third and molecular bonding would be weakest. Measurements indicate this order is essentially correct, although there is some overlap between covalent and ionic bonding, depending on the materials. Since the strength of bonding determines the rigidity, symmetry, and strength of the crystal structure, we would expect plasticity to vary inversely with bonding strength, and this is generally true.

Tensile strengths determined by conventional tests are, at best, only an approximate indication of cohesive forces within a material. But covalent and ionic solids have essentially no measurable plasticity in room temperature tests. They rarely develop their true strength since they fail prematurely. At sufficiently elevated temperatures, however, they have enough plasticity for use in engineering structures and are superior to the metals from a strength viewpoint. Polymers can be deformed in varying amounts (depending on the specific polymer) as a function of stress and time.

Most metals are strong and also ductile and malleable; i.e., metals can be deformed under applied loading and can undergo considerable change of shape or dimension. In fact, extensive use is made of various mechanical working processes to change dimensions, properties, and/or surface conditions in metals. Most ceramics are strong but are not capable of undergoing extensive mechanical working.

Metals and most ceramics are solids, i.e., crystalline in nature, composed of a large number of individual grains. Each grain is made of atoms in an ordered geometrical arrangement, and cohesion in the solid is the resultant of attraction and repulsion between the atoms. The elastic and plastic properties of materials have their origin in these interatomic forces and the atomic arrangements.

6-7 ELASTIC DEFORMATION

If a material is loaded (within certain limits), a temporary deformation of the crystals takes place through displacement of the atoms. Upon removal of load, the atoms return to their "stable" positions, and the crystal recovers its original shape. This temporary deformation of the crystal is apparently permitted by an elastic displacement of the atoms in the structure; i.e., the applied loads are not great enough to cause permanent shifting between the atoms. This can occur to some degree in nearly all solids.

EXAMPLE 6-2

The tensile strength of a given material is 76,000 psi with a yield point of 42,000 psi. A bar 0.50×1.00 in. deforms to a cross section of 0.43×0.93 in. before breaking under a load of 34,000 lb. (a) What is the maximum load which can be carried without permament deformation? (b) What is the maximum load which can be carried? (c) What is the maximum stress before fracture? (d) What is the ductility?

Solution:

(a) $P = (42,000)(0.50)(1.00) = 21,000$ lb

(b) $P = (76,000)(0.50)(1.00) = 38,000$ lb

(c) $\sigma = \dfrac{34,000}{(0.43)(0.93)} = 85,000$ psi

(d) reduction in area $= \dfrac{(0.50)(1.00) - (0.43)(0.93)}{(0.50)(1.00)} = 20\%$

Note: Part (b) was calculated on the basis of the original area. This is conventional practice but the actual area is somewhat less since there will have been some yielding of the material at the ultimate strength. The assumption of uniform tensile stress is also not completely valid. Part (c) was calculated on the basis of the final area but assuming uniform tensile stress in the cross section. This is not quite the case.

6-8 PLASTIC DEFORMATION

If a material is stressed beyond the elastic range, it will not return to its original form. It will either fracture or it will undergo some change in shape due to "flow" of the material under the applied loading. For example, if a zinc wire, made from a single crystal, is stressed in tension, the crystal elongates slightly, a line appears around it showing displacement of one part of the wire relative to the other, and there is no further elongation. If the load is increased, another line appears, parallel to the first, indicating another displacement and again no further elongation. Each successive extension or elongation requires greater applied load and is accompanied by the appearance of another line.

In Chapter 4 we found that atoms in the various crystal structures lie in a series of families of parallel planes. The interplanar spacing varies with the particular sets of planes, and there are one or more sets of planes which have the greatest interplanar spacing. Interplanar spacing, arrangement of atoms in the planes, and type of bonding between atoms are most important in determining whether or not plastic deformation can occur. Deformation does not occur unless there are dislocations (see Sec. 6-8-7).

6-8-1 DEFORMATION BY SLIP

Plastic deformation by slip occurs as a result of relative movement of lamellae in the crystal, where movement is concentrated in a succession of planes or in very thin blocks of planes as in Fig. 6-7. Intervening blocks are left undeformed, much like movement of cards in a pack when the pile is distorted. Displacement takes place on specific crystallographic planes (slip or glide planes) and in specific crystallographic directions (slip directions). In general, slip planes are atomic planes of greatest interplanar spacing, since the crystal yields by shearing of atomic bonds between crystallographic planes. We expect the planes with greatest interplanar spacing to have the weakest bonds between them. We also find that these same planes have the greatest planar density of atomic packing. The direction of movement on these slip planes is not random but is usually restricted to a crystallographic direction of greatest linear density of atomic packing within the plane; i.e., the slip direction coincides with one of the most closely packed rows of atoms.

In FCC metals such as Al, Cu, Ag, Au, Ni, etc., the slip plane is usually a {111} plane. The perpendicular distance between these planes is greater than for any other set of planes in the crystal, and the atomic population in these planes is denser than in any other family of planes. Thus planes of easiest slip are also planes of densest packing. The slip directions, i.e., the lines of greatest linear packing, are the ⟨110⟩ family (face diagonals).

The {111} planes in an FCC crystal are the octahedral planes (implying eight). There are, however, effectively only four sets of parallel planes so there are only four different planar orientations. Each plane, however, has three possible slip directions. Four possible planes with three possible directions in each plane gives twelve directions of potentially equally easy slip in an FCC crystal. These combinations of planes and directions are called slip systems. In an FCC structure, therefore, there are twelve slip systems.

We recall that the (111) plane of the FCC structure and the (0001) or basal plane of the HCP structure are identically arranged. In HCP structure, however, there is only one set of basal planes [rather than the four (111) planes in FCC] so there are only three systems of easiest slip in HCP structure. This difference in the number of slip systems has a pronounced effect on differences in mechanical properties and behavior of FCC and HCP structures.

In BCC structures, several families of planes have almost the same density of packing. In α-iron (ferrite), the {110} or dodecahedral planes have the greatest density of packing, but the {112} and {123} planes are almost as densely packed, and all three sets of planes can operate during slip. In all these planes, the $\langle 111 \rangle$ directions are the most densely packed. Thus BCC α-iron has 48 possible slip systems. In other BCC metals such as Mo, W, and Na, slip normally occurs on $\{112\}$ planes in $\langle 111 \rangle$ directions. In K, however, slip usually occurs on $\{123\}$ planes, again in $\langle 111 \rangle$ directions.

Slip systems indicate the modes of slip which are potentially possible. It has been found that, of all possible modes, slip occurs on a plane in the direction which lies closest to the direction of the maximum resolved shear stress acting within the crystal. In other words, whether the crystal is compressed or stretched, deformation actually takes place by shear.

Consider a single-crystal wire or rod supporting an axial tensile force P, applied as in Fig. 6-8. If the area normal to the axis is A, then the area of the slip plane, whose normal is at an angle ϕ to the axis, is $A/\cos \phi$. If the slip direction makes an angle λ with the axis, then the component of force in the plane in the slip direction is $P \cos \lambda$. This shearing force must be divided by the area to obtain the resolved shearing stress, τ, i.e.,

$$\tau = \frac{P}{A} \cos \phi \cos \lambda. \tag{6-6}$$

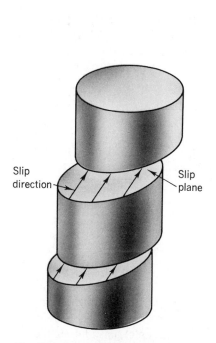

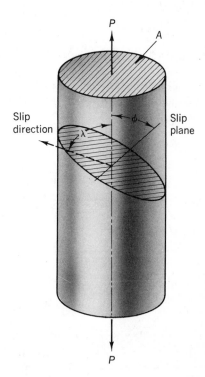

Fig. 6-7. Schematic representation of plastic deformation in a single crystal.

Fig. 6-8. Coordinates for resolving stresses.

The component of force normal to the slip plane is $P \cos \phi$. Normal stress is determined by dividing this force by area, i.e.,

$$\sigma = \frac{P}{A} \cos^2 \phi \qquad (6\text{-}7)$$

A crystal does not slip or flow at an appreciable rate until the resolved shear stress has reached a clearly defined value. In other words, slip starts in the most favorably oriented slip system only when the shear stress equals a threshold value, i.e., the critical resolved shear stress (CRSS). At lower values of resolved shear stress, the rate of strain is very slow. In this range of loading, the flow is known as creep, which will be discussed in Chapter 13. At, or above, the critical shear stress, plastic flow is readily measured.

CRSS depends on composition, purity, extent of prior deformation, and temperature. The value of CRSS usually increases with increasing alloy content, with increasing impurity content (soluble impurities being more effective than insoluble ones), and with increasing amounts of prior deformation. It generally decreases with increasing temperature and drops abruptly to zero at the melting point (or the solidus temperature in alloys). The CRSS is not influenced by the stress normal to the slip plane.

EXAMPLE 6-3

The critical shear stress in the [1$\bar{1}$0] direction on the (111) plane of a single crystal of an FCC metal is 0.15 kg/mm² (213 psi). What magnitude of stress along the cube edge [100] is required to produce slip on the (111) plane? What is the normal stress on the (111) plane?

Solution:

Referring to Fig. 6-8,

$$\cos \phi = \frac{\text{edge of unit cell}}{\text{long diagonal of unit cell}} = \frac{a}{a\sqrt{3}} = 0.577$$

$$\cos \lambda = \frac{\text{edge of unit cell}}{\text{short diagonal of unit cell}} = \frac{a}{a\sqrt{2}} = 0.707$$

from Eq. 6-6

$$0.15 = \frac{P}{A}(0.577)(0.707) \longrightarrow \frac{P}{A} = 0.37 \text{ kg/mm}^2$$

from Eq. 6-7

$$\sigma = (0.37)(0.577)^2 = 0.123 \text{ kg/mm}^2$$

Slip effectively consists of a relative displacement of parallel slip planes through the shearing action of the resolved shear stress, and a simultaneous rotation of the active slip system toward the direction of tension (i.e., the slip plane becomes more nearly parallel to the direction of tension). Although slip resembles simple relative gliding of two parallel planes, it differs in that:

(1) slip is restricted to specific planes and directions within the planes which are determined by internal directional forces, (2) the extent of slip is limited to an interatomic distance or an integral multiple of that distance, (3) resistance to continued slip increases (but is essentially constant in gliding), and (4) part of the energy of slip is stored in the crystal as potential (strain) energy in contrast to complete dissipation (as heat) of the energy of gliding.

Deformation, on a microscopic scale, is not uniform but is localized on slip planes and within slip bands (Fig. 6-9). The spacing between slip planes in a slip band may be of the order of 200 A, whereas the separation of adjacent bands may be 100 times as great. The slip on each plane may be of the order of several microns. The minimum force necessary to produce slip is several orders of magnitude less than we expect from our knowledge of crystal structures and interatomic forces. In a typical metal, the theoretical CRSS is about 10^{11} dynes/cm² while the experimental measurement is about 10^7 dynes/cm². Moderate deformations, even at low temperatures, do not drastically alter properties, and there is no indication of a breaking-up of the lattice. As the magnitude of applied stress increases, the number of active slip planes and the distance of slip along these planes increase.

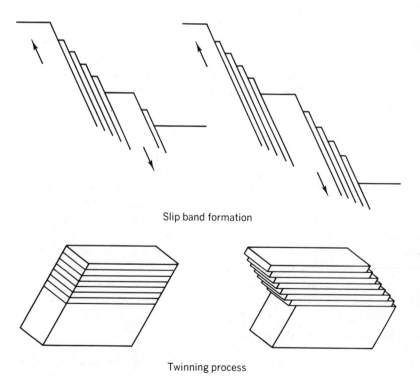

Slip band formation

Twinning process

Fig. 6-9. Schematic representation of slip and twinning.

Duplex slip sometimes occurs, particularly in FCC metals. Slip is accompanied by rotation of the slip plane toward the direction of applied tensile stress, and a continued increase in the stress is required for continued deformation. This rotation of the active slip plane toward the axis of tension causes rotation of other slip systems. If one of these systems rotates into a position in which the resolved shear stress in the system is greater than the CRSS but less than that required to continue slip on the first system, slip will then take place in the new system. As load increases, slip may occur in both systems by alternating between the two.

6-8-2 DEFORMATION BY TWINNING

Some crystals deform by twinning rather than by slip. Twinning occurs when the atoms in a layer within the crystal are rearranged during deformation in such a manner that the layer develops a mirror image relationship to an adjacent layer of the parent crystal. The twinning plane is a plane of symmetry relating the twinned layer to the parent crystal. The difference between slip and twinning is shown schematically in Fig. 6-9.

Although twinning is not actual rotation of planes, it is convenient to visualize the twinned portion of a crystal as if it were rotated about the plane of symmetry. Movement during twinning is really shear of adjacent planes. All planes involved shift in the same direction, and each plane shifts an amount proportional to its distance from the twinning plane. Thus it is possible to regard twinning as deformation by slip in which *every* plane of atoms in the twinned layer participates, in contrast to normal slip in which several inactive planes may be found between active planes. In twinning, as in slip, there is a threshold or critical value of resolved stress which must be exceeded for twinning to take place.

In twinning, each plane of atoms shifts a small but definite fraction of the interplanar spacing with respect to the adjacent planes. This shift results in an effective reorientation in the twin. This is in contrast to no change of orientation in slip. The twinned section of the crystal may participate in deformation in two ways: (1) twinning itself may accomplish an extensive change in shape; or (2) it may facilitate further slip by placing planes of potential slip in more favorable positions with respect to the applied stress. The latter is generally more effective in producing deformation.

The critical resolved shear stress for twinning, as is the CRSS for slip, is a function of composition, purity, prior deformation, and temperature.

6-8-3 POLYCRYSTALLINE MATERIAL

Deformation in polycrystalline material is far more complicated than in a single crystal although the process is basically the same. Complicating factors come from grain boundaries and constraints imposed on flow of a given grain by its neighbors and by the flow of the aggregate. In other words, stresses from external loading are transmitted to individual grains by surrounding grains, and the change in shape of an individual grain must conform to the change in shape of its neighbors. One obvious effect of polycrystallinity is to increase strength. We can also infer that decreasing grain size in a given volume increases strength and hardness.

Loading in any given grain can be resolved into shear components for each possible slip system and normal components. Plastic deformation in any slip system will occur when the shear component in that system is greater than the critical resolved shear stress. During deformation, each grain rotates so that the active slip directions become more nearly parallel to the axis of tension with the result that the grains become elongated in the direction of flow. The deformation of polycrystalline metal is the integrated result of combined deformations of individual grains.

Deformation of polycrystalline metal involves appreciable movement along grain boundaries as well as along slip planes. This is not a simple condition, since slip normally changes direction at a grain boundary.

6-8-4 PREFERRED ORIENTATION

Each grain in polycrystalline material generally has a different crystallographic orientation than its neighbors. Considering the entire piece of material, orientations of individual grains can be randomly distributed with respect to a frame of reference. During slip, the active slip planes tend to rotate and make the slip direction more nearly parallel to the axis of tension. This means a departure from a random orientation to one in which orientations of individual grains tend to cluster, to some degree, about some particular orientation or orientations. This condition of nonrandom distribution of crystal orientations is known as *preferred orientation* or *texture*.

Preferred orientation developed by slip and twinning is called deformation texture to differentiate it from preferred orientations developed by various heat treatments and those which are found in nonmetallic aggregates such as rocks, fibers, and various organic and inorganic substances. A common example of deformation texture is found in cold-drawn wires in which the individual grains are so oriented that a given crystallographic direction $[uvw]$ in most of the grains is parallel (or nearly parallel) to the axis of the wire. Another prominent example is cold-rolled sheet in which most of the grains are oriented with a certain plane (hkl) approximately parallel to the surface of the sheet and with a certain direction $[uvw]$ in that plane approximately parallel to the direction of rolling.

Texture is normally stated in terms of Miller indices. For instance, we find cold-rolled FCC metals usually have $\{110\}$ planes parallel to the rolling plane with $\langle 112 \rangle$ directions parallel to the rolling direction. This texture is written as $(110) \parallel R.P. [\bar{1}12] \parallel R.D.$ or more simply, $(110)[\bar{1}12]$. The principal texture in cold-rolled BCC metals is $(001) [110]$. The wire texture of FCC metals is usually a double texture with $[111]$ and $[100]$ parallel to the axis; i.e., the grains have either $[111]$ or $[100]$ directions parallel to the wire axis and have random planar orientation around the axis. The relative number of grains having these two orientations depends on the metal. Aluminum has only $[111]$ texture, but copper has about 40% $[100]$ and about 60% $[111]$. BCC metals have a simple $[110]$ wire texture.

Texture ranges from none in a completely random specimen to nearly complete and uniform orientation for a very heavily cold-worked metal. Preferred orientation is a crystallographic condition and has nothing to do with grain shape as seen under a microscope. It is true that grain shape is affected by the same forces which produce texture; i.e., grains become flattened and elongated, but a flattened grain is not direct evidence of preferred orientation.

Preferred orientation is important because of the effect, often very marked, which it has on the overall macroscopic properties of materials. For example, the dimensional stability of uranium and its alloys is highly sensitive to texture. Individual crystals are anisotropic. In a specimen having random crystal orientation, there is an effective cancellation or neutralization with a net result of macroscopic isotropy. As preferred orientation develops, the effectiveness of the cancellation diminishes, and the specimen shows directionality in its macroscopic properties. Directional properties are often objectionable, although they are preferred in some applications; for example, in steel sheet for transformer cores, a high magnetic permeability is desired in the direction of the applied field.

A common example of undesirable directionality is "earing," which sometimes develops in deep drawing of sheet metal. If cups or similar shapes are drawn from cold-rolled sheets of FCC metal, four symmetrically arranged "ears" may form on the upper rim. Each ear represents a direction in which the resistance to deformation was relatively low. The height of the ears varies directly with the degree of preferred orientation that existed in the blank. These greater than average elongations in the region of the ears are accompanied by greater than average reductions in thickness so that there is a variation in wall thickness of the drawn shape. The directionality may be so pronounced that splitting will occur during subsequent forming.

6-8-5   EXTENT OF PLASTIC DEFORMATION

While the foregoing discussion in Sec. 6-8 has been in general terms, it applies primarily to metals in which plastic deformation is usually extensive.

Nonmetals, on the other hand, are severely limited in deforming plastically. This difference depends primarily on differences in crystal structure and interatomic forces in materials.

Binding forces are strongly directional in ionic and covalent bonding. The covalent bond comes from the mutual attraction of pairs of atoms for specific pairs of shared electrons, such that each pair is confined to a limited region between two atoms. A slight angular displacement between the two atoms results in a weakening of this atom-electron attraction which becomes relatively small after even a limited amount of displacing motion. Therefore, in shear, the covalent bond is destroyed before a given atom reaches a position where it can form a similar bond with a new neighbor. Upon loading such materials, we find some elastic deformation followed by fracture and little or no intermediate plastic deformation. Most ionic materials are as brittle as covalent materials, but there are a few, e.g., MgO and NaCl, in which slip can be made to occur in specific slip systems.

In metals, however, cohesion between atoms is obtained by the attraction between an electron "gas" and a relatively uniform electrostatic force field surrounding the atoms. Consequently, the bond strength is not particularly affected by small relative angular displacements of atoms. Upon application of shear, a given atom will be attracted to a new equilibrium position before there is a major decrease in the forces holding it to its original position. In other words, during slip, each atom is continuously bound to its immediate neighbors.

An excellent example of the effect of type of bonding on plastic deformation is given by tin. The normal structure of this element is body-centered tetragonal. In this form it is known as white or β-tin, is definitely metallic, and is capable of extensive plastic deformation. When it is cooled below 13°C, a polymorphic transformation occurs from body-centered tetragonal to a tetrahedral cubic (diamond structure). This change in structure also implies a change to covalent bonding. This new form is known as gray or α-tin and is brittle.

All polymers are viscoelastic to some degree, i.e., they have mechanical properties which are characteristic of both elastic solids and viscous fluids. Consequently, analysis of mechanical response is very complicated since time and rate of loading are important variables. Plastic deformation can occur in polymers. For example, most glassy polymers can undergo appreciable plastic deformation at room temperature under moderate strain rates. This can occur by shear yielding or by normal stress yielding, depending on the stress conditions and ambient temperature.

Deformation of amorphous polymers is analogous to viscous flow in a fluid. Upon application of a shearing load, uniform flow begins throughout the material until complex entanglements and molecular interactions resist further deformation. With crystalline polymers, the situation is much more complicated since some regions in the structure are much more resistant to flow than others.

Polymers are isotropic as long as the chains are randomly arranged even when there is a relatively high degree of crystallization. If the material is stretched, the chains tend to align and develop along a common fiber axis which leads to anisotropic characteristics.

6-8-6 STRAIN HARDENING

Many properties of metals change appreciably while undergoing plastic deformation. The most noticeable are increased tensile strength, yield strength, and indentation hardness (but not scratch hardness), and decreased plasticity and formability (ductility). If we bend a piece of wire between our fingers, it becomes harder and stronger at the bend. If we continue to bend the wire back and forth, the hardness increases, it becomes more difficult to make the bend, and, after a few reversals, the wire breaks. This increase of strength due to mechanical working is known as *strain hardening* or *work hardening*.

The ability of metals to strain harden is commercially important. Pitch, for example, pressed into a die, yields at one point, becoming thin and tearing. In metals, however, the thinner part becomes stronger and shifts the deformation to thicker parts, thereby causing a reasonably uniform deformation instead of a localized one. Without this property of strain hardening, extensive mechanical working would be impossible. At the same time, not all changes produced by cold plastic deformation are necessarily improvements. For example, if a severely deformed metal is placed in service, it may be unable to sustain additional plastic deformation due to an overload. Thus it will be relatively brittle and likely to fail. In addition, preferred orientation with its accompanying directionality is developed. The metal may also be more susceptible to corrosion.

Structure-insensitive properties are little affected by cold forming. For instance, density increases very slightly during deformation. On the other hand, changes in structure-sensitive properties can be very important. For example, specific electrical resistance sometimes increases by several percent. Practically all the mechanical properties, with the exception of modulus of elasticity, are greatly affected by deformation. Yield strength, for example, can often be doubled.

Experimental evidence has established the following: (1) shortly after deformation starts, the lattice "shatters" into "fragments" with dimensions of the order of 10,000 times the interatomic distance. As deformation continues, however, there is no evidence of additional "break-up"; (2) as slip proceeds, preferred orientation develops; (3) as the slip plane becomes more nearly parallel to the line of applied force, the shear component diminishes, and a larger applied force is required to continue slip; and (4) crystallographic dimensions are essentially unchanged, regardless of the amount of deformation. Distortion of structure cells is elastic.

Strain hardening is not common to all polymers although it does occur to a marked degree when the material can be oriented into a fiberlike structure.

6-8-7 DISLOCATIONS

Slip depends on dislocations which propagate through the crystal upon load application. The edge dislocation (Fig. 4-39) is the simplest type. It can be visualized as a case in which $n + 1$ atoms on the upper side of the slip plane are opposite n atoms on the lower side. This is considered a positive dislocation. If the relative positions are reversed, it is a negative dislocation. In either case, application of shear along the slip plane will shift one portion a distance of one atomic spacing with respect to the other. When the dislocation reaches the edge of the crystal, it will have produced unit slip. The unit slip or the displacement of the line defect can be represented by a vector which is one interatomic distance in length and points in the direction of motion. This "slip vector" or "Burgers vector" is normal to the edge dislocation.

A dislocation cannot end inside a crystal but ends only at a grain boundary or a surface of a crystal. It is possible for a dislocation to change its character within the crystal; i.e., an edge dislocation can be converted to a screw dislocation. The screw dislocation (Fig. 4-40) is essentially a spiral surface formed by the atomic planes around the dislocation line. In this case the Burgers vector is parallel to the dislocation line. It is also possible for components of edge and screw dislocations to join together to form a closed dislocation loop within the crystal. The material within the loop is considered to have slipped on the specified slip plane relative to the material around it.

The total length of dislocation line per unit volume is called the dislocation density. In "soft" crystals this is about 10^8 cm/cm^3, which indicates an average distance between dislocations of a few thousand atoms. If each dislocation produced only unit slip, this small number of dislocations could not produce macroscopic slip of the magnitude observed. Thus the dislocations must multiply and thereby increase their effectiveness by orders of magnitude.

The Frank-Read mechanism for effective multiplication is shown schematically in Fig. 6-10 and by the photograph in Fig. 6-11 In Fig. 6-10(a), an edge dislocation extends only part way through the crystal. The dislocation lines, Bz and $B'z'$ are unable to slip, either because their slip vectors are not in the slip plane or because the lines are pinned by the presence of foreign

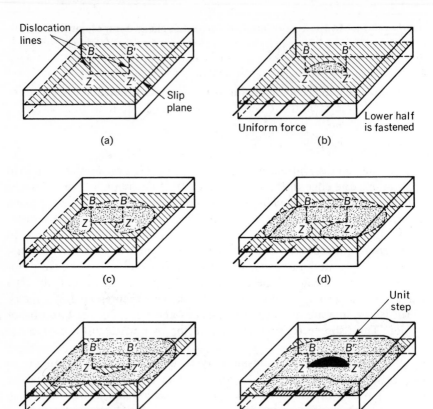

Fig. 6-10. Frank-Read mechanism for extensive slip from a single source.

atoms. When a force is applied to the upper half of the crystal [Fig. 6-10(b)], the dislocation $zz'\ BB'$ bulges and begins to move forward, gradually traversing the entire slip plane by a sweeping motion [Fig. 6-10(c)–(f)] around the anchors Bz and $B'z'$. This results in the material above the slip line moving forward one atomic spacing, thus forming a protrusion on the front and leaving a step in the wake. This is shown partly formed in Fig. 6-10(f), but it eventually extends across the entire width.

The net effect is to move the upper half of the crystal, together with the anchor lines and the dislocation (which is still present), forward by one atomic spacing relative to the lower half. If application of the force is continued, the process is repeated [indicated by the black area in Fig. 6-10(f)] until the Bz and $B'z'$ anchors reach the back edge of the lower half of the crystal. At that time, the dislocation disappears, and slip ceases.

The portions of the sweeping dislocation closest and farthest from the reader are of the edge type, whereas the portions on the sides are of the screw

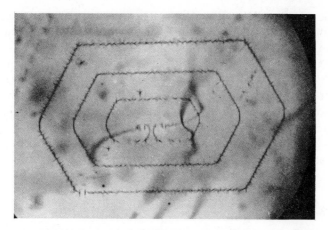

Fig. 6-11. Dislocation loops spreading out from a Frank-Read source in a silicon crystal.

type. Both types move in the direction of the slip vector; the edge type moves perpendicularly to the dislocation line, and the screw type moves parallel to it.

Under these conditions, a limited number of anchored dislocations can produce extensive slip. In fact, one anchored dislocation is sufficient to produce considerable slip since the free end of the dislocation can sweep the slip surface like the hand of a clock.

When a generator operates as shown in Fig. 6-10 and the area to be swept is large compared with the length of *BB'*, the "tail ends" of the expanding bulge come so close to each other that the two screw elements (of opposite sense) are mutually attracted and annihilate one another, leaving a completely closed dislocation ring [Fig. 6-10(d) and (e)], which also results in lower total strain energy. The closed ring continues to expand until it disappears on leaving the crystal. Several such rings can exist simultaneously as shown in Fig. 6-11.

Progressive strengthening with increasing deformation (strain hardening) comes from impedance of dislocation movement. As a dislocation moves through a crystal, it may encounter an imperfection such as an impurity atom, a precipitated particle, a vacancy, an interstitial atom, or another dislocation. The portion of the dislocation which is at the local barrier is held back and exerts an effective backward pull on the remainder of the dislocation which tends to continue its forward motion under the applied load. An additional load is required to overcome the "back stress" thus developed before additional motion is possible.

Strain hardening is also possible from interaction between dislocations on intersecting slip planes. Depending on whether the intersecting dislocations are of the edge or screw type, a line of vacancies or interstitialcies may be created, or part of the dislocation may be displaced into a slip plane parallel to the initial slip plane. This discontinuity is called a "jog" and is produced only by expending energy. These jogs can move easily only in the direction of the slip vector. If forced to move otherwise, a jog will usually leave a trail of vacant lattice sites at further expenditure of energy. Since the number of points of interaction presumably increases with increasing strain, this can be a source of strain hardening.

While it is known that dislocations and other defects exist in polymeric crystals, study of them has been very limited.

6-9   FRACTURE

Plastic deformation begins imperceptibly but, if continued long enough, can end with relative suddenness and finality in fracture. Fractures are usually considered to be ductile or brittle, based either on gross energy absorption or shape change accompanying fracture. In single crystals, fracture may be by cleavage or by shear. In polycrystalline material, fracture may be intercrystalline or transcrystalline.

The theoretical strength of an ideal crystal has been calculated on the basis of the energy required to form two new surfaces at the break and is 100 to 1000 times the strengths normally observed. This discrepancy is customarily attributed to imperfections, principally cracks. These cracks can be microcracks which exist as a result of the prior history of solidification or mechanical processing. An initially sound material, however, can develop cracks on an atomic scale. If a large enough stress is applied, Frank-Read sources can generate a succession of dislocations on favorably located slip planes. A large barrier, such as a grain boundary, can stop the first dislocation and cause succeeding ones to pile up against the first one. Such a pile-up creates a stress concentration which is conducive to formation of the nucleus of a crack. From the dislocation viewpoint, one of two phenomena, the antitheses of each other, can occur: (1) extremely limited dislocation motion which permits separation of material across a "single" plane under action of a normal stress on that plane; or (2) extremely extensive dislocation motion on a "single" plane which results in separation of the material by a slipping apart of the two halves across the plane. The end result is a brittle fracture since the two halves are essentially undeformed, and energy absorption is relatively small.

Ductile fracture is less well understood although shear *strain* appears to have a major role in this type of fracture. In fact, it appears that excessive shear strain alone is sufficient to cause fracture in compression and torsion. In a tensile test, it is impossible to tell whether shear alone is sufficient or whether there is a simultaneous effect of tensile and shear stresses.

Fracture in metals does not occur instantaneously but is progressive, and usually occurs through formation and propagation of a crack that grows across the section during some finite and measurable time period. This period may be extremely short; i.e., the propagation rate may be rapid. Velocities of crack advance as high as 6600 feet per second have been measured in brittle impact failures in steel. In ductile impact fractures, however, the crack may advance at rates on the order of 20 fps. In any case, fracture begins with formation of a small crack which extends "spontaneously" until complete separation occurs.

Most ceramics fracture in a brittle manner, i.e., with little or no plastic deformation. Noncrystalline ceramics, such as glass, are always completely brittle below the softening temperature. In crystalline ceramics, brittle fracture usually takes place by cleavage on specific crystallographic planes. At high temperatures, crystalline ceramics can fail along grain boundaries when shearing takes place with cracks opening up between the grains, thereby causing a local stress concentration and ultimate separation.

A process similar to that of crystalline ceramics at high temperatures occurs in polymers with high molecular weights. In these cases, sliding of structural elements leads to formation of pores with fracture occuring some time after load application. It has also been shown that shear bands in glassy polymers are rather similar to slip bands in metals in that both have very high stress concentrations at intersections of bands. These concentrations act as nuclei for crack formation leading to fracture.

6-9-1 CLEAVAGE STRESS

For those cases in which a critical normal stress must be exceeded to initiate or propagate cracking, it appears that the critical stress is independent of shear stress and is little affected by prior plastic deformation. It is influenced by temperature, composition, crystal structure, and the crystallographic plane under consideration. Planes of easy cleavage are generally planes of low indices, i.e., planes of easy slip, since these are the planes of greatest atomic density within the plane (greatest *intra*planar bonding) and the planes with the largest spacing between them (least *inter*planar bonding).

6-10 CRITICAL STRESSES FOR SLIP, TWINNING, AND FRACTURE

Under applied load, a piece of material may slip or twin, it may fracture, or it may deform somewhat and then fracture. A critical value of shear stress must be exceeded before slip begins, and a similar situation exists for twinning. In like manner, a critical value of tensile stress must be exceeded for cleavage to take place. When load is applied, a crystal will slip, twin, or fracture depending on whether the resolved shear stress in a slip system, or the resolved shear stress in a twinning system, or the resolved tensile stress normal to a cleavage plane is the first to exceed its respective critical value.

If a single-load application is carried to the point of fracture, three strength phenomena are exhibited: (1) a shear yield strength (shear stress above which plastic flow occurs); (2) a shear ultimate strength (shear stress at which ductile fracture occurs); and (3) a cohesive strength (normal stress at which brittle fracture occurs). Tensile yield strength, often measured and reported for design criteria, is actually the magnitude of the normal stress at the shear yield strength of the material. It is not, therefore, a fundamental characteristic. Likewise, the (ultimate) tensile strength (maximum load divided by the original cross-sectional area) is often measured and reported. Although this is often used as a control check for consistency in various lots of material, it is not a fundamental property. Upon load application, there is an elastic reaction proportional to the load, however large. If a sufficiently large load is applied to ductile materials, flow will occur by slip or by twinning whereas fracture will occur by shear or by cleavage. In truly brittle materials, no flow develops, and fracture is by cleavage.

In a tensile test of many ductile metals, fracture is often initiated at the center of the specimen (due to development of a triaxial stress condition), followed by a shear failure of an annular region along a surface at 45°. This results in a typical "cup and cone" fracture as shown in Fig. 6-12.

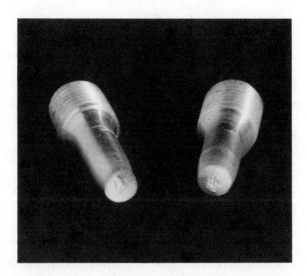

Fig. 6-12. Typical cup-cone fracture of a ductile material tested in tension. Note the reduction of the cross-section in the fracture region (known as the neck).

6-11   EFFECT OF STRAIN RATE

An increase in the stress at which a material deforms is produced by an increase in the rate of straining. At moderate temperatures, the effect of strain rate changes is small in metals. For example, an increase in strain rate by a factor of 10^4 or more may be required to double the flow stress. The resistance increases as the logarithm of strain rate, characterized by a greater increase in soft metals than in hard ones.

At higher temperatures, the effect is more pronounced; the strength increases markedly with the strain rate. The effects of changing strain rate and temperature generally oppose. For example, a *decrease* of 100°F would give a metal about the same increase in strength as about a 10^5 *increase* in strain rate.

The effect of strain rate in ceramics is not clear although Al_2O_3 is an interesting case since alumina is important in high-strength and high-temperature applications. At temperatures above about 1000°C, plastic deformation in Al_2O_3 shows (1) a strong temperature dependence, (2) a large strain rate dependence, and (3) a definite yield point in constant strain rate tests or an equivalent induction period in long-time low-stress tests.

Since most polymers are viscoelastic, there is a pronounced strain rate dependence. For example, a true elastomer appears rubbery under an alternating stress with a frequency as low as 100 cycles/sec. At 10^6 cycles/sec, however, the same elastomer behaves as if in a glassy state.

6-12   FATIGUE

The term fatigue, applied to materials, refers to failure under action of repeated stresses. Materials can fail under the action of cyclic stress even when the maximum cyclic stress is far below the static breaking strength. As a result, many service failures can be traced to fatigue action. We tend to think only of rotating components, e.g., axles and crankshafts, in connection with fatigue, but other structures, e.g., ship hulls and floor beams of bridges, are also subject to fatigue. Piping and other equipment, often considered statically loaded, can also be subject to fatigue through temperature variations and consequent cycling of thermal stress.

Fatigue failures begin at points of high stress or stress concentration such as abrupt changes of cross-section, small cracks, pieces of dross or slag, or small nicks or dents (caused by careless machining or subsequent handling). These can be on the surface or internal. A crack spreads fanwise from this starting point by slip along certain crystallographic directions. Under sufficiently high repeated stresses, this slip apparently causes local crystal fragmentation. This rupture of atomic bonds forms submicroscopic cracks which grow by intermittent stages into macroscopic cracks. Formation of successive zones of fracture continues until the remaining section is unable to sustain the applied loading, at which point it fractures.

Fracture surfaces produced by fatigue failure have a characteristic appearance. As the crack spreads, the opposing surfaces rub each other giving a smooth, burnished appearance with "clamshell" markings indicating the successive boundaries of the crack. The final brittle failure of the reduced section produces a granular appearance in the remaining portion of the fractured surface. Typical fatigue failures are shown in Fig. 6-13.

Unfortunately, trying to assess fatigue data, we find much uncertainty both as to the actual mechanism and as to the specific effect of each of the large number of factors which affect fatigue strength. These factors include: (1) types of stress action (bending, direct, shear, or combined stresses); (2) mean stress of the cycle; (3) frequency of stress application; (4) testing temperature; (5) specimen size; (6) form of the specimen; (7) specimen surface condition (e.g., ground or polished); (8) environment; (9) whether the specimen is smooth or notched; (10) chemical composition; and (11) method of fabrication. The order of listing does not indicate the relative order of magnitude of effect.

Cyclic stresses occur in a wide variety. Two typical examples are shown in Fig. 6-14. Figure 6-14(a) represents completely reversed stress, whereas Fig. 6-14(b) represents a fluctuating stress. In either case, the range of stress (σ_a) is one-half the algebraic difference between the maximum stress (σ_{max}) and the minimum stress (σ_{min}). The mean stress (σ_m) is zero for case (a) and is one-half the algebraic sum of the maximum and minimum stresses in case (b). Many variations of the stress range may occur, and mean stress may be either positive or negative. We often consider stress variation to be sinusoidal. In practice, however, many other forms of cyclic variation are encountered.

The variation in stress can come from many different sources in practical applications. In laboratory testing, the most usual procedure is to use simply-supported, round, rotating beam specimens. This gives a sinusoidal stress variation [Fig. 6-14(a)]. Less common are rotating cantilever beams (round specimen) and oscillating cantilever beams in which the "free" end of a sheet-type cantilever specimen is displaced through a given deflection about the static equilibrium position. Still less common are tensile types of tests in which direct axial loading is applied to a specimen. These last tests have the advantage that either tension or compression or a combination of both can be applied to obtain a curve of the type shown in Fig. 6-14(b).

For ferrous metals and alloys, the strength of the metal under repeated stresses is usually given by the *endurance* or *fatigue limit*, which is defined as the maximum stress that can be applied repeatedly an infinite number of times without fracture. If we plot the maximum stress in a rotating beam test versus

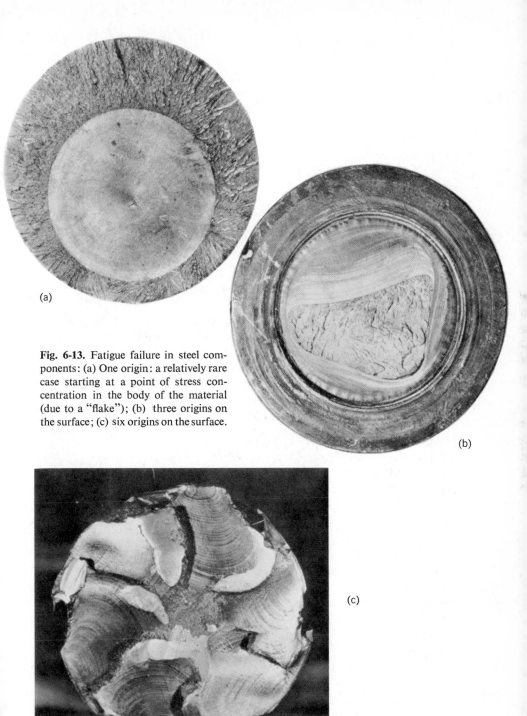

(a)

(b)

(c)

Fig. 6-13. Fatigue failure in steel components: (a) One origin: a relatively rare case starting at a point of stress concentration in the body of the material (due to a "flake"); (b) three origins on the surface; (c) six origins on the surface.

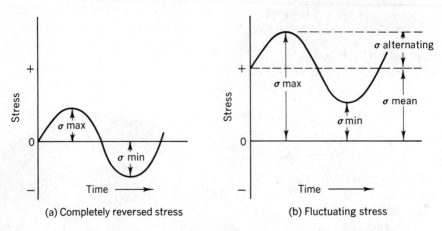

Fig. 6-14. Typical examples of stress reversal.

the logarithm of the number of cycles of repeated stress (Fig. 6-15), we find a rather sharp "knee" for many ferrous metals and a constant stress line with increasing number of cycles. The stress indicated by this horizontal line is the *endurance limit*. Most nonferrous metals do not exhibit such behavior; i.e., they show no such constant stress line. For these metals, an *endurance* or *fatigue strength* is defined as the maximum stress that can be applied repeatedly for a specified number of stress cycles without producing fracture. The number of stress cycles quite commonly selected is 500,000,000.

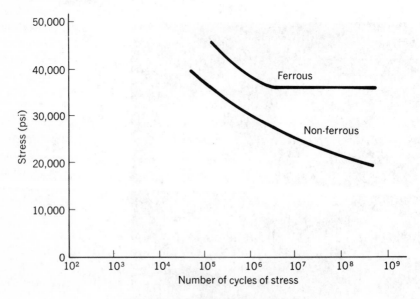

Fig. 6-15. Schematic fatigue curves.

EXAMPLE 6-4

It is sometimes assumed that if failure does not occur in a fatigue test in 10^8 cycles, then the stress is less than the fatigue limit. If a rotating beam machine runs at 10,000 rpm, how long will it take to get to that number of cycles?

(a) Extrapolate to the time required to develop a fatigue curve, e.g., Fig. 6-15 when ten or more data points are required.

(b) If the material is hot rolled AISI 1040 and the stress is 34,000 psi, is the assumption reasonable?

(c) If the material is aluminum alloy 2024-T4 and the stress is 20,000 psi, is the assumption reasonable?

Solution:

$$\text{Time} = \frac{10^8 \text{ cycles}}{(10^4 \text{ cyc/min})(1440 \text{ min/day})} = 6.95 \text{ days}$$

(a) If any tests are run to 500×10^6 cycles, as is commonly done with many alloys, especially aluminum, one test would thus require about 35 days. Establishment of the complete curve, using only one machine thus might require several months.

(b) The assumption is fairly reasonable for AISI 1040 (see Fig. 6-15 and Appendix B).

(c) The assumption is very questionable for 2024-T4 (see Fig. 6-15 and Appendix B).

Note: Values for fatigue limit or fatigue strength, such as in Appendix B, are average (or nominal) values. Thus there may be significant deviation in stress from such values (for a given number of cycles) or in the number of cycles to failure (for a given stress). In other words, data on fatigue failures show much scatter. Therefore, average values should be used with much caution.

Fatigue curves obtained using notched specimens lie well below those obtained using smooth specimens. These differences in fatigue strengths emphasize the marked effect of geometry on fatigue strength. Fatigue strengths usually given are obtained from smooth specimens. Data from notched specimen tests are relatively scarce despite the obvious greater applicability to design. Unfortunately, there is no standard notch fatigue specimen, and data obtained from notched specimens depend on the notch geometry. In addition, there is no good correlation between smooth specimen fatigue strengths and notched specimen fatigue strengths. In fact, two metals or alloys which differ markedly in smooth specimen fatigue strengths may have notched specimen fatigue strengths (using a sharp notch) which are nearly the same.

There is no method known by which we can determine the remaining fraction of service life for a specimen in use. If a crack has started, obviously a replacement part is necessary. Unfortunately, detection of the very tiny initiating crack, normally located at some stress raiser, is often difficult if not impossible. Prior to formation of a crack, however, no method exists which will determine the damage, for there is no perceptible effect upon the microstructure. Good design, minimizing notches and stress raisers, is most essential for long fatigue life.

Fatigue failure in metals occurs under repeated cyclic stresses by growth of a crack nucleated in a surface region which has been extensively cold worked. Such failure is rare in ceramics. Cyclic stresses in polymers limit allowable loading to a small fraction (perhaps 20–30 %) of that permitted under static loading. Since polymers are viscoelastic, most of them have a rather large hysteresis loop under cyclic loading. This gives an energy loss, which is a measure of the internal heat generated under cyclic stressing, and is a measure of damping capacity. High damping capacity permits dissipation of vibrations which produce noise and stresses in other components, but it may also produce high temperatures leading to rather rapid degradation in the polymer.

6-12-1 STATIC FATIGUE

Static fatigue, also known as delayed fracture, sometimes occurs in ceramics and polymers. This appears to be failure under relatively low static stresses applied over an extended period. This phenomenon is not creep which is discussed in Chapter 13. Delayed fracture is highly sensitive to environmental conditions. In general, preferential stress corrosion occurs at the root of a crack, and under static stress, fracture develops some time after load application. In some silicate glasses, for example, moisture in the air may be sufficiently "corrosive" to cause static fatigue. In polymers static fatigue is sometimes known as "solvent cracking." In general, however, the liquids or vapors involved are not solvents for the polymer or even strong swelling agents. For example, liquids such as vegetable oils and butterfat can lead to delayed fracture at low stress levels in polymers which are not detectably affected by these liquids in the absence of stress.

6-13 IMPACT PROPERTIES

Impact or shock loading differs from static and cyclic loads in two respects: (1) load is applied rapidly, that is, with appreciable speed, and (2) loading is seldom repeated, since failure often occurs on the first application, if it occurs at all. An impact test normally determines the energy absorbed in fracturing a test piece under high-speed loading.

In the tensile impact test, a variable-speed flywheel delivers a tensile impact loading to a smooth specimen similar to a tension specimen, or to a somewhat similar specimen with a circumferential notch. Such a test permits study of the impact strength of a material under a unidirectional stress, and at various speeds. Energy of rupture is a measure of the impact strength.

Charpy and Izod tests are considered bending impact tests, but this name is misleading since the rate of load application is relatively low, and the specimens have a stress raising notch. These tests, therefore, do not indicate impact properties as much as the effect of the notch. In both Charpy and Izod machines a weighted pendulum is swung through an arc to fracture a carefully prepared specimen. This type of machine is shown schematically in Fig. 6-16 as well as the types of specimen. The number of foot-pounds of energy required for fracture is considered a measure of the impact strength. Obviously, the two tests use different types of support and thus give different results, although a keyhole or V-notch may be used with either type of test. Standardized procedures for making impact tests are given in ASTM Standard E23.

One of the most significant uses of the Charpy and Izod impact tests is in determining transition temperature, i.e., the change from a ductile failure to a brittle one as a function of decreasing temperature. Figure 6-17 shows a typical energy absorption curve. Low energy absorption (low temperature) corresponds to brittle fracture whereas high energy absorption corresponds to ductile fracture. Since evidence of limited ductility has been observed in otherwise completely brittle fractures in some specimens tested at low temperatures, it is difficult to clearly determine a specific transition temperature. Therefore, an arbitrary value (15 ft-lb is very common) is used to determine transition temperature. Various materials can be compared for transition temperature only by using identically dimensioned specimens since the data cannot be reduced to any common units. Measured transition temperature depends, to some extent, on the shape of the notch. Higher transition temperatures have been observed with sharper notches.

The ductile-brittle transition is found in a large number of ferrous alloys and some refractory metals, such as molybdenum, which are BCC. It is not usually observed in FCC metals, such as aluminum, copper, and nickel. If the transition temperature is close to or above the anticipated operating temperature, there is a strong potential for catastrophic failure in service. Molybdenum, for example, has a transition temperature above room temperature. Certain steels, exposed to neutron radiation, show a shift of transition temperature from below to above room temperature.

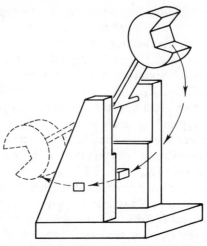

(a) Typical notched bar impact machine

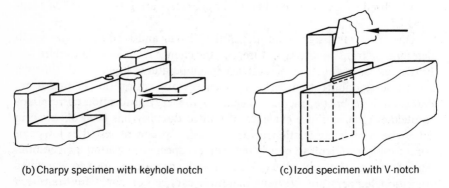

(b) Charpy specimen with keyhole notch   (c) Izod specimen with V-notch

Fig. 6-16. Notched bar impact tests.

Impact "strength," as measured by Izod or Charpy tests, is used as a measure of toughness. While data from these tests have questionable quantitative significance, they do provide a comparative basis for estimating relative shock resistance. Metals, in general, are relatively resistant, although the ductile-brittle transition must be kept in mind. Essentially all ceramics are susceptible to failure under impact loading. In polymers, two different and opposite characteristics impart good impact resistance. Polymers with high tensile strength and moderate elongation usually have good impact resistance. Flexible polymers, while relatively weak, also have good impact resistance because they absorb the energy and distribute it through a relatively large volume. Many polymers do not fracture under the first impact loading but

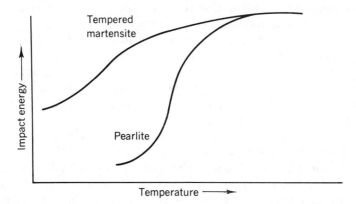

Fig. 6-17. Comparative impact properties of pearlitic and martensitic steels.

the ability to absorb energy is greatly reduced by repeated impact. Selection of polymers becomes especially critical if they are to be used at low temperatures which make many of them relatively brittle.

6-14 NOTCH SENSITIVITY

Materials differ greatly in the degree of damage introduced by stress raisers. Different lots of the same material, made by the same process, and having the same tensile properties, frequently differ when notched specimens are used in testing. Some materials are severely damaged in that the notch develops into a crack with very little straining. Stress is so localized and intensified that only a small volume of material is plastically deformed, and the energy absorption is low. Such materials are notch sensitive or notch brittle. Most nonmetals, especially ceramics, are markedly notch sensitive. Other materials, primarily metals, may be deformed to such a considerable degree in the region of the notch that the radius of curvature of the notch increases. Thus, a larger volume of material deforms plastically before development of a crack. Since a greater amount of energy can be absorbed, such materials are said to have a high notch toughness. In other words, they are less sensitive to stress raisers which may be introduced during design, manufacture, or service.

Notch sensitivity is defined as the reduction of nominal strength, static or impact, by the presence of a stress concentration and is usually expressed as the ratio of the notched to the unnotched strength. It is often determined by conducting tensile tests using both smooth and notched specimens of the same material. Unfortunately, no one notch contour, size, or surface finish has been accepted as standard. Consequently, data in the literature are apt to be misleading or of little value unless all test conditions are clearly stated.

Notched bar impact tests are often used to evaluate notch sensitivity although they do not give a ratio of notched to unnotched strength. They are also subject to criticism because the energy measured by the test applies only to specific size and shape of specimen, contour and finish of the notch, temperature, and method of loading. It is also difficult to obtain reasonably identical specimens.

Notch sensitivity cannot be predicted, even approximately, unless we have specific knowledge of the behavior, obtained experimentally, of different classes of materials. For example, good ductility in tension does not necessarily mean low notch sensitivity. There is no adequate correlation between yield or tensile strengths and fatigue notch sensitivity. Likewise, there is no good correlation between notched-specimen fatigue strength and the energy absorbed in the single-blow impact tests.

6-15 SUMMARY COMMENT

None of the mechanical properties (in contrast to mechanisms, Sec. 6-8) described in this chapter can be considered fundamental properties. Essentially all of them are structure-sensitive. They are properties derived from experimental data which are taken in a standardized manner and analyzed by standard procedures. This gives data whose major virtue is reproducibility. In essence, there is nothing wrong with such empiricism, and much useful information comes from the standardization. At the same time, no data are obtained which are quantitatively applicable to the underlying basic mechanisms.

No attempt is made here to give representative values of mechanical properties of specific materials although some selected data are given in appendices. A great quantity of published mechanical property data exists and is readily available.

QUESTIONS AND PROBLEMS

1. What two aspects of mechanical properties must be kept in mind from the practical viewpoint? What kinds of tests may be used to determine mechanical properties? What are the relative merits and drawbacks of each? What kinds of results can be expected from these tests?

2. Define stress and strain. What kinds of stress and strain are there? What is the difference between them? What is the relationship between stress and strain? In mathematical terms, which of these is the independent variable?

3. A steel bar and an aluminum bar are each under a 1000 lb load. If the cross-section of the steel bar is one square inch, what area must the aluminum bar have for the strain to be the same in both bars? (E for steel is 30×10^6 psi; E for aluminum is 10×10^6 psi.)

4. What does 0.2% yield strength mean? What does 0.002 offset in a tensile test mean?

5. A structural tension member 10 ft long is subjected to a load of 120,000 lb. There is a choice between an aluminum alloy at 45 cents per lb. and structural steel at 21 cents per lb. Select the more economical material based on

   (a) ultimate strengths with a factor of safety of 5 if

   $$\sigma_{\text{ult-Al}} = 45{,}000 \text{ psi} \quad \text{and} \quad \sigma_{\text{ult-St}} = 90{,}000 \text{ psi}$$

   (b) a limiting elongation of 0.20. Values of moduli are

   $$E_{\text{Al}} = 10.5 \times 10^6 \text{ psi} \qquad E_{\text{St}} = 30 \times 10^6 \text{ psi}$$

6. The following data were obtained from a tension test for an aluminum base alloy. The specimen was of circular cross-section with 0.500 in. diameter, and it had a 2 in. gage length.

Load (pounds)	Elongation (inches)	Load (pounds)	Elongation (inches)
0	0	7200	0.0070
1000	0.0010	8000	0.0080
2100	0.0020	8300	0.0090
3200	0.0030	8600	0.0100
4200	0.0040	8700	0.0120
5200	0.0050	8750	0.0140
6200	0.0060	8800	0.0180
		(fracture load)	

   (a) Plot the stress-strain diagram.
   (b) Determine the yield strength for 0.2% offset.
   (c) Determine the ultimate strength.
   (d) Determine the modulus of elasticity.

7. The following data were obtained from a tension test of an AISI 1020 steel. The initial diameter is 0.505 in., the final diameter is 0.391 in., and the gage length is 2 in.

Load (pounds)	Elongation (inches)	Load (pounds)	Elongation (inches)
600	0	8,500	0.0058
1200	0.0002	8,500	.0068
1800	.0004	8,400	.0080
2400	.0006	8,200	.0098
3000	.0008	8,200	.020
3600	.0010	8,400	.030
4200	.0012	9,400	.050
4800	.0014	10,600	.075
5400	.0016	11,400	.100
6000	.0018	12,300	.150
6500	.0020	12,800	.200
7000	.0023	13,500	.300
7400	.0026	13,700	.380
8000	.0032	13,100	.450
8300	.0038	11,500	.500
			(fracture)

(a) Calculate the stresses and strains from the given data and plot two stress-strain diagrams, one using all the data and one for elongations up to 0.0100 in.

Determine:

(b) upper yield point.

(c) lower yield point.

(d) yield strength at 0.2% offset.

(e) modulus of elasticity.

(f) tangent modulus of elasticity at 65,000 psi.

(g) ultimate strength.

(h) nominal fracture strength.

(i) "true" fracture strength.

(j) percentage elongation.

(k) percentage reduction in area.

(l) Sketch your estimate of the "true stress-true strain" curve.

8. A "true" stress-strain diagram for a 1020 semi-killed steel was obtained in tension, using a specimen of circular cross-section. For the plastic range the readings of load and diameter were:

Load (pounds)	Diameter (inches)	Load (pounds)	Diameter (inches)
3500	0.352	5800	0.316
4300	0.350	5700	0.302
4800	0.347	5600	0.290
5175	0.344	5300	0.270
5400	0.340	5050	0.260
5650	0.334	4950	0.255
5775	0.324	4650	0.235
		(rupture)	

The original diameter of the specimen was 0.364 in. Determine the "true" stress and "true" strain values and plot the "true" stress-strain diagram.

9. Four tension members have 2 sq in. cross-sectional area. These members are a steel alloy, an aluminum alloy, a magnesium alloy, and a laminated glass plastic. The ultimate strengths are 70,000, 68,000, 44,000 and 40,000 psi, respectively. The moduli of elasticity are 30×10^6, 10.6×10^6, 6.5×10^6, and 1.5×10^6 psi, respectively. The densities are 0.284, 0.100, 0.065, and 0.060 lb/in^3, respectively. Determine the strength-weight ratio based on:
   A. (a) ultimate strength.
      (b) allowable strain of 0.005
   B. Discuss the applicability of these results.

10. Assume that a body can be elastically deformed without any change in volume. What must Poisson's ratio be for this to be possible?

11. A tensile stress of 1500 psi is applied to a cubic crystal, e.g., aluminum, along the [100] direction. What shear stress is developed in the [111] direction on the (110) plane?

12. A load of 6000 lb is to be supported by a cable with a minimum weight. No plastic deformation is permitted. If steel (YS = 140,000 psi), copper (YS = 60,000 psi) and aluminum (YS = 50,000 psi) are available, which one will you use?

13. Solve problem 9 as a compression problem based on yield strength if the yield strengths are 35,000, 46,000, 18,000, and 35,000 psi, respectively.

14. An 1100-H18 aluminum bar (3 in.^{2}) is surrounded by a hot-rolled AISI 1040 steel sleeve (1 in.^{2}) of the same length. What is the maximum compressive load which can be supported without yielding on the part of either metal?

15. Assume twisting of a circular shaft (no bending) to failure. Sketch the appearance of the break if
   (a) the material is ductile, e.g., copper, steel, etc.
   (b) the material is brittle, e.g., cast iron, chalk, etc.

16. Discuss the similiarities and differences of ductility, malleability and brittleness.

17. Define hardness and discuss its practical significance.

18. Discuss the similarities and differences of the various hardness tests.

19. Thermosetting "plastics" do not soften appreciably at elevated temperatures; thermoplastics soften and flow easily. Describe differences in structure which might account for this difference in behavior. What relative advantages (and disadvantages) can you see between the two types?

20. What is slip? Of what significance is it? How does it operate?

21. On what crystallographic planes and in what directions is slip most likely to occur in BCC, FCC and HCP metals?

22. Most metals and solid solution crystals have a simple (relatively so) structure cell. Intermetallic compounds have complex structure cells. How and why does this suggest that intermetallic compounds are relatively brittle?

23. What is a slip system? How many slip systems are there in FCC and HCP metals?

24. Demonstrate the largest perpendicular interplanar distance in FCC crystal structures is between (111) planes. Demonstrate that the {111} planes are the most densely populated.

25. Repeat problem 24 for the (0001) planes in HCP crystal structures.

26. What is the meaning of "critical resolved shear stress"? On what factors does it depend and how? What is the distinction between "resolved shear stress" and "critical resolved shear stress"?

27. Discuss the differences and similarities between slip and twinning.

28. What is duplex slip? What effect does it have on mechanical working?

29. Show how two edge dislocations of opposite sign on the same slip plane can annihilate each other. Is it possible for two screw dislocations of opposite sign to do likewise?

30. Titanium is an HCP metal with c/a of less than the ideal ratio of 1.633. Slip occurs in the $\langle 11\bar{2}0 \rangle$ direction on either the (0001) plane or $\{10\bar{1}0\}$ planes. Sketch a hexagonal prism and show both planes and the common slip direction. For a single crystal of Ti, the stress-strain curve looks more like that of FCC-Cu than that of HCP-Mg. Why?

31. Why can a jog in an edge dislocation glide through a lattice along with the line, while a jog in a screw dislocation cannot glide?

32. What is the difference between slip in single crystals and slip in polycrystalline metals? Of what practical significance is this difference?

33. What is strain hardening? Is it of commercial significance? Why?

34. What is the mechanism of strain hardening?

35. What is preferred orientation? What is texture? How is either obtained? What are the principal effects?

36. Iron, brass or copper sheets which have been annealed sometimes show localized thinning or "earing" when cupped. Why? How may this be avoided?

37. Why are there four rather than some other number of ears in cupping FCC metal?

38. What are the basic different types of "dislocations"? In what respects are they similar and in what respects do they differ? What effect do they have in strain hardening?

39. What is fracture? In what respects is it similar to slip and in what respects is it different? What are the effects of cracks on fracture strength?

40. Why do cubic ionic crystals such as MgO, NaCl, and LiF cleave on cube faces? On what planes do BCC transition metals cleave? On what planes do FCC metals cleave?

41. A single crystal of a certain substance having an FCC crystal structure is machined in the form of a right circular cylinder with a cross-sectional area of 0.200 sq in. An axial tensile force of P lb is applied. Slip can occur in $\langle 110 \rangle$ directions in $\{111\}$ planes. Twinning can occur in $\langle 11\bar{2} \rangle$ directions in $\{111\}$ planes. Cleavage frature can occur between $\{111\}$ planes. The relationship between slip and twinning directions in the (111) plane is shown in the sketch.

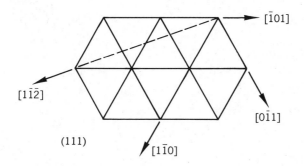

Cleavage fracture, critical normal stress,   450 psi
Twinning, critical resolved shear stress,   360 psi
Slip, critical resolved shear stress,   300 psi

(a) Consider the case (refer to Fig. 6-5) when $\phi = 45°$ and $\lambda_{slip}[1\bar{1}0] = 59°40'$ and $\lambda_{twin}[1\bar{1}\bar{2}] = 50°$.
    i) If $P = 150$ lb, what happens to the crystal?
    ii) If P is increased to 170 lb, what happens to the crystal?
(b) Consider the case when $\phi = 0$.
    i) If the force $P = 80$ lb, what happens to the crystal?
    ii) If P is increased to 100 lb, what happens to the crystal?

42. If a certain steel fails in a brittle manner in an impact test at 0°C, does this mean that any structure made of this steel will also fail in a brittle manner? Why?

43. Define fatigue. Discuss the practical significance of fatigue, the measures of fatigue life, and its influence on mechanical design.

44. Why is a shiny, smooth area surrounded by a relatively rough, dull area characteristic of a fatigue failure?

45. Endurance or fatigue strength is often taken as the maximum stress to produce no failure in 5×10^8 cycles. (a) If R. R. Moore machines run at 10,000 rpm, how long will it take to test just one specimen to that number of cycles? (b) If a test machine is directly coupled to an electric motor that runs at 1800 rpm, how long will it take? (c) Assume you have four Moore machines available, estimate how long it might take to establish a reasonable fatigue curve for a given material.

46. A steel ball weighing 1.5 lb must be dropped from a height of 3 ft onto a sheet of plate glass to break it. What height is required for a 0.25 kg ball to also break the glass? How much energy is involved?

7

MATERIALS IN EQUILIBRIUM

7-1 INTRODUCTION

A common statement of the concept of equilibrium in a system is: "Equilibrium is a state in which properties do not change with time." This concept is insufficient and is unacceptable as a definition. The words "change with time" are not precise, and the statement can apply to metastability as well as to equilibrium. Before defining equilibrium, however, we should define other pertinent terms.

A *system* is any physical body (or group of bodies) isolated for consideration. A more precise definition is: A *system* is a substance (or group of substances) so isolated from its surroundings that it is unaffected by these and is subjected to changes in overall composition, temperature, pressure, or total volume only to the extent allowed by the investigator. A system may be composed of gases, liquids, solids, or any combination of them and may involve metals and nonmetals, either separately or in any combination.

The *state* (of a system) is the physical condition as defined by any combination of quantities which uniquely fixes the condition. State is analogous to geometric form. For example, a given triangle has a specific unique form defined by the angles and sides; i.e., the "state" of the triangle is defined by the "quantities" angles and sides. In a system, the quantities fixing the condition are either *intensive* (nonadditive, e.g., pressure, temperature, density, etc.) or *extensive* (additive, e.g., mass, volume, etc.).

Equilibrium in a system is the state of minimum free energy under any specified combination of overall composition, temperature, pressure, and total volume. Free energy can be defined in two ways:

$$1. \quad A = E - TS \tag{7-1}$$

or

$$2. \quad F = E + PV - TS \qquad\qquad \text{(7-2a)}$$
$$= H - TS \qquad\qquad\qquad \text{(7-2b)}$$

The Helmholtz free energy (A) is given by Eq. 7-1, and the Gibbs free energy (F) is given by Eq. 7-2, where E is the internal energy (the sum of all potential and kinetic energies), P is the pressure, V is the volume, H is the enthalpy, T is the absolute temperature, and S is the entropy. For a system to be in equilibrium, the pressure and temperature must be uniform throughout and the chemical potential or vapor pressure of each component must be the same in every phase.

Once equilibrium is attained, any change, however slight, in amount, composition, volume, or state of any substance within the system means an increase in free energy. Thus it is impossible to have spontaneous changes in a system at equilibrium. This thermodynamic concept of equilibrium is concerned only with the macroscopic aggregate and not with individual atoms.

Attainment of equilibrium is not instantaneous and can require extremely long periods of time, in which case the system is in a metastable state. Metastability is often of more practical interest than equilibrium. Hardened steels and other heat-treated metals, for instance, are in metastable conditions. Nevertheless, we must know the equilibrium state in order to prevent its occurrence in some cases or to speed its appearance in others. Some knowledge of kinetics or rate processes (Chapter 8) is necessary for studying the approach of a system to equilibrium.

The number of *components* in a system is the smallest number of individual substances that must be listed to completely describe the chemical composition of the system (or any part of it) at any temperature and/or pressure. The components may be chemical elements, compounds, or polymers. In metallic systems, the components are normally the chemical elements present.

Components in a system combine to form various *phases* in the system. A *phase* is a substance, or portion of matter, which is homogeneous in the sense that its smallest mechanically separable parts are indistinguishable from one another. A more elaborate definition is: "The number of phases present in a system is the number of different substances that exist in it, each of which (at equilibrium under a given temperature and pressure) is chemically and structurally homogeneous within itself, is physically separated by definite boundary surfaces from all dissimilar substances also present, and is distinguishable from those other substances by some detectable difference in composition, physical state, crystal structure, or properties." A given phase may be continuous throughout the system but this is not necessary. For example, the ice cubes in a drink are one phase and the drink is a second phase.

Since all gases mix with one another in all proportions, there can be only one gas phase in a system. Two liquids may dissolve in each other to form one phase which may show wide variations in composition depending on other conditions of the system. Two liquids which are essentially insoluble in each other (e.g., water and oil), or which have limited solubility, will separate into two distinct liquid phases. In solids, each different type of crystal present is a separate phase.

One additional concept is that of *degrees of freedom*, which are defined as the number of independent variables whose values must be specified in order to exactly define the state of a system. It is obvious that the variables for any given system must be separated into independent and dependent variables so that we can determine the number of degrees of freedom for the system.

The number of components and the overall composition are dependent variables, since these are fixed when the system is specified. Temperature and pressure are usually considered independent variables. The number of phases present is a dependent variable, since this is uniquely determined by the other conditions of the system. The amount of each phase is not a degree of freedom since equilibrium relationships are concerned with the number of phases and their individual compositions, not the amount. Concentration, i.e., composition of individual phases, not the overall composition, is generally considered independent. In an n-component system, there are $n - 1$ possible degrees of freedom due to concentrations. Specific volume may (or may not) be considered an independent variable. If temperature, pressure, and concentrations are fixed, then the specific volume is also fixed. If specific volume is taken as an independent variable, then either temperature or pressure becomes a dependent variable.

Obviously, we have some choice as to which variables are taken to be independent. The usual practice is to consider temperature, pressure, and concentrations as the independent variables, i.e., the degrees of freedom.

7-1-1 GIBBS PHASE RULE

All equilibrium relationships in metals, ceramics, and polymers conform to a general law called *Gibbs phase rule*. This rule prescribes the number of phases present in a given system of specific composition in equilibrium at specified temperature and pressure. The phase rule does *not* indicate the time required for a system to reach equilibrium. The usual statement of the phase rule is:

$$F + P = C + 2 \tag{7-3}$$

where F is the number of degrees of freedom,* C is the number of components, and P is the number of phases in equilibrium.

For systems under constant pressure, one degree of freedom is eliminated, and Gibbs phase rule then becomes

$$F + P = C + 1 \tag{7-3a}$$

The major limitation on the phase rule is that it applies only to equilibrium states which require homogeneous equilibrium within each phase and heterogeneous equilibrium between phases. A system in equilibrium always obeys the phase rule. The reverse is not always true, i.e., conformation with the phase rule is not conclusive proof of equilibrium. Nevertheless, one principal use of the phase rule is to assess whether or not equilibrium has been reached. It is not possible to have less than zero degrees of freedom (except in very unusual circumstances†). Therefore, a binary system in equilibrium at *constant pressure* will not have more than three phases present. That is, when $C = 2$, $F = 3 - P$ (from Eq. 7-3a), or $P = 3$ when $F = 0$. Although the three phases may be any combination of gas, liquids, and solids, the principal practical cases are one liquid and two solids, two liquids and one solid, or three solids (in a binary system). *The presence of four phases in a binary system under constant pressure is positive indication of incomplete equilibrium.*

EXAMPLE 7-1

Consider an alloy of iron, chromium, nickel, manganese, and carbon.
(a) Determine, under equilibrium conditions:
1. the minimum number of phases
2. the minimum number of degrees of freedom
3. the maximum number of degrees of freedom
4. the maximum number of phases
5. the number of degrees of freedom with three phases
(b) Determine the same items if the system operates under constant pressure and in equilibrium.
(c) If the system is not under equilibrium conditions, is it possible to have more phases present than indicated by Gibbs rule? Is it possible to have fewer phases present than indicated by Gibbs rule?

Solution:

The number of components is 5. Rewriting Eq. (7-3), $F + P = 7$.
(a) 1. There must be at least one phase present in any real system.

---

*The symbol F here should not be confused with the same symbol that is used for Gibbs free energy in equation (7-2a).

†H. F. Halliwell & S. C. Nyburg, "The Phase Rule: The Significance of Negative Degrees of Freedom," *J. Phys. Chem.* 64, 855 (1960).

2. The least number of degrees of freedom is zero.
3. 6
4. 7
5. 4

(b) Eq. (7-3a) applies, i.e., $F + P = 6$.
1. 1
2. 0
3. 5
4. 6
5. 3

(c) The answer to both questions is *yes*. Whether there are more or less phases present than indicated by Gibbs phase rule for equilibrium depends on the specific system and what has been done to it in the sense of heating rate, holding time, and/or cooling rate. For example, a system may be cooling but have more than the equilibrium number of phases (see end of Sec. 7-1-1). If a system is single phase at some elevated temperature and two phase at a lower temperature, under equilibrium, then very rapid cooling from the higher to the lower temperature may retain the single phase as a supersaturated phase (see Sec. 9-5-1).

7-1-2   UNARY (ONE COMPONENT) SYSTEMS

In a one-component system, the phase rule becomes:

$$F + P = 3 \tag{7-3b}$$

This implies that the maximum number of phases in equilibrium is three (when $F = 0$). These phases can be vapor, liquids, or solids including polymorphs of liquids and solids. This is illustrated in Fig. 7-1 which has four regions, i.e., vapor, liquid, and two solids. In any one of these regions, temperature or pressure may be varied with no change in phase; i.e., there are two degrees of freedom. Two phases coexist along any of the three lines separating one phase from another; i.e., there is only one degree of freedom. In other words, for two phases to continue to coexist, *only* temperature *or* pressure can be varied independently. Three lines of phase separation intersect in a common point, known as the "triple point." At this point (and only at this point), with specific temperature and pressure ($+0.0098°C$, 4.58 mm Hg for H_2O), all three phases, i.e., solid, liquid, and vapor, coexist in equilibrium. If temperature or pressure is changed, at least one phase must disappear if equilibrium is to prevail.

     Good examples of one-component systems are the development of high-temperature, high-pressure methods for making synthetic diamonds and

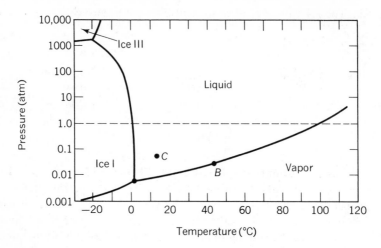

Fig. 7-1. Temperature-pressure diagram for H_2O.

hard boron nitride. Silica (SiO_2) is also a one-component system of wide interest. There are five phases which occur under equilibrium. The temperatures
for solid-solid transitions under one atmosphere pressure are shown in Fig.
7-2.

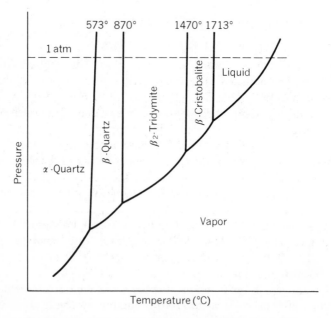

Fig. 7-2. Schematic temperature-pressure diagram for SiO_2.

7-2 PHASES IN MULTICOMPONENT SYSTEMS

It is obvious that one, two, or three phases can coexist in equilibrium in a one-component (unary) system. If we have two or more components in a system, more phases can be present, and their compositions can change with the state of the system. As the number of components increases from one (unary) to two (binary) to three (ternary) to four (quaternary) and so on, the number of possible combinations increases tremendously. Although pure materials are of commercial interest (germanium for semiconductors is produced with an impurity content of 1 part in 10^{10}), systems with two or more components are of much greater interest.

To keep the discussion relatively simple, we shall use binary metallic alloys for illustration but the principles apply equally to other types of systems and to systems with more than two components.

7-2–1 ALLOYS

A substance which is composed of two or more chemical elements such that metallic atoms predominate in composition and the metallic bond predominates is called an alloy. The element which is present in the largest proportion is called the base metal, and all other elements present are called alloying elements. Physical, chemical, and mechanical properties of the base metal can be changed (often very drastically) by the presence of alloying elements. The type and extent of change of properties depends on whether the alloying elements are insoluble in, dissolve in, or form a new phase with the base metal.

7-2–2 INSOLUBILITY

In principle, complete insolubility is thermodynamically impossible. In practice, however, solubility can be extremely limited. Lead, for example, is essentially insoluble in iron. From the practical viewpoint, the alloy of Fe and Pb is an intimate mechanical mixture of the components where each component retains its own identity, properties, and crystal structure. Leaded bronze bearings, for example, have small globules of lead trapped in a bronze matrix which supports the load while the lead acts as a lubricant.

7-2–3 SOLID SOLUTIONS

Components in an alloy may be soluble in each other in amounts ranging anywhere from rather limited solubility to complete intersolubility. Two kinds of solid solutions are formed: (1) *substitutional* solutions in which some of the atoms of base metal are replaced in their normal lattice sites by solute atoms as indicated in Fig. 7-3(a), and (2) *interstitial* solutions in which solute atoms are found in the "holes" or interstices between solvent atoms as indicated in Fig. 7-3(b). A further distinction is often made between

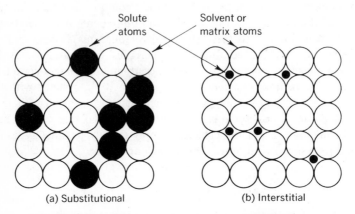

Fig. 7-3. Schematic representation of solid solutions.

solutions formed upon adding the first solute atoms (those which retain the crystal structure of the solvent) and solutions that develop after extensive alloying (those which normally have a different crystal structure than either component). The former are known as *primary* or *terminal solutions* and the latter as *secondary* or *intermediate solutions*.

7-2-4 SUBSTITUTIONAL SOLUTIONS

It seems obvious that extensive solubility in a substitutional solution occurs only if solute and solvent atoms are nearly the same size. This has been confirmed by evidence that solid solubility of one element in another becomes very restricted if the atomic diameters of the two elements concerned differ by more than about 15%. Within this 15% range the two components have a favorable size factor. We should remember, however, that an atom in solution seldom has the same radius that it has as an elemental metal. Three other factors, chemical affinity, relative valence, and crystal structure, also influence the extent of solubility.

Two metals with great chemical affinity for each other will not form an extensive solid solution. In other words, strongly electronegative and strongly electropositive elements will tend to form a compound rather than a solution.

Thus metals which have similar electronic structures can form extensive solutions in each other. The relative valence effect is similar to the chemical affinity effect in that solubility of one metal in the other is considered a function of electron solubility. We find that a metal of high valence can dissolve only a small amount of a lower valence metal, although the lower valence metal may have reasonable solubility for the higher valence metal. Crystal structure is determined mainly by bonding forces, and we would expect metals of similar crystal structure to have greater potential solubility. We find, in fact, that a continuous series of solid solutions is formed only between elements of the same crystal structure and then only when the sizes of the atoms differ by less than about 8%. Cu-Ni and Ag-Au-Ni are examples of binary and ternary systems with complete solid solubility.

A qualitative estimate of solid solubility of one component in another can be made by considering the above four factors. An unfavorable size factor alone indicates a low solubility. If the size factor is favorable, the other three factors must be considered. For example, in the Cu-Zn system, Cu has a wide range of solubility for Zn, but there is very limited solubility of Cu in Zn.

Ideally, solute and solvent atoms are randomly distributed on lattice sites, with a statistically uniform distribution. In certain cases, however, substitution is regular instead of random. This can occur in binary systems with $1:1$ or $1:3$ atomic ratios. The $1:1$ system often has a BCC structure with atoms of one type on the corners and atoms of the other type at the center. These regular arrangements are known as superlattices. These superlattices are formed upon slow cooling of random substitutional solutions. At elevated temperatures, there is a random distribution of solute and solvent atoms (disordered solution) which can be preserved at room temperature by rapid cooling. During slow cooling, however, the atoms will begin to take up preferred positions when a critical temperature is reached, and the solution becomes "ordered." A classic example of this is the copper-gold alloy, $AuCu_3$, which, at high temperatures, is disordered FCC. On slow cooling (or holding) below about 750°F (390°C) the FCC structure remains, but we find the gold atoms at cube corners and the copper atoms at face centers. If we consider the AuCu alloy in the same system, it is disordered FCC at high temperatures. On cooling below 410°C, the structure becomes an ordered tetragon with a c/a ratio of 0.93. In this arrangement alternate (001) planes have only Au or Cu atoms.

Substitutional solid solutions occur in ceramics as well as in metals. For example, a complete series of solid solutions exists in the Al_2O_3—Cr_2O_3 system (rubies are 0.5 to 2% Cr_2O_3 in Al_2O_3). Magnesia (MgO) dissolves significant amounts of NiO or FeO with Ni^{2+} or Fe^{2+} ions substituting for Mg^{2+} ions in the crystal structure.

EXAMPLE 7-2

Small quantities of an element in an alloy are often considered to be impurities with negligible effect on the system. While this is true for many engineering applications, it is not always true. Compute how far apart copper (Cu) atoms will be in an alloy of 99.99 a/o nickel (Ni) and 0.01 a/o Cu. Assume a homogeneous distribution of Cu atoms in Ni with a lattice parameter of 3.52 A.

Solution:

Both Ni and Cu are FCC and dissolve in each other in substitutional solid solutions.

For every 10,000 atoms, 9999 will be Ni and 1 will be Cu. These 10,000 atoms are equivalent to 2500 unit cells. Assuming three-dimensional homogeneity, then Cu atoms are separated from each other by a distance of $(2500)^{1/3} = 13.6$ cells. The linear distance is $13.6 \times 3.52 = 47.9$ A. Thus even though the Cu atoms are only 1 in 10,000, they are close enough together to have non-negligible effects on electrical properties, optical properties, etc. (see Sec. 7-2-11).

7-2-5 INTERSTITIAL SOLUTIONS

Since solute atoms in an interstitial solid solution occupy the spaces between solvent atoms, these solute atoms must be small relative to the solvent atoms. The elements H, C, N, B, and O all have radii less than 1.0 A and form interstitial solutions. Interstitial solid solutions normally have very limited solubility and generally have limited importance. Carbon in iron is a major exception and forms the basis for hardening steel which will be discussed in Chapter 10.

Iron atoms in γ-iron (austenite) are on FCC sites and the carbon atoms occupy interstitial positions. The largest interstices or "holes" in this lattice are at the midpoints of the structure cell edges and at the cube center. These positions are crystallographically equivalent. Smaller interstices exist which are surrounded by a tetrahedral arrangement of 4 solvent atoms. The largest interstices in iron have a radius of 0.52 A whereas the next smaller interstices have a radius of only 0.28 A.

The largest of these interstices can accommodate a carbon atom (0.8 A radius) or a nitrogen atom (0.7 A radius) only by expansion of the lattice, while the next smaller interstices are essentially inaccessible to solute atoms. In α-iron (BCC) the largest interstices have a radius of 0.36 A. The next smaller "hole" has a radius of only 0.19 A. Therefore, interstitial solution of carbon in α-iron should be more limited than in γ-iron. This is the case as we will see in Chapter 10.

Interstitial solid solutions depend on the size, valency, and chemical affinity factors—as do substitutional solid solutions—but not on the type of crystal structure. With ceramics, addition of ions in interstititial sites requires some adjustment to maintain electrical neutrality. This can be done by forming vacancies or substitutional solid solutions or by changing the electronic structure. If YF_3 or ThF_4 is added to CaF_2, Y^{3+} or Th^{4+} substitute for Ca^{2+} while F^- ions go into interstitial positions to maintain electrical neutrality.

EXAMPLE 7-3

The largest interstitial hole in FCC has $r/R = 0.414$ while the largest interstitial hole in BCC has $r/R = 0.291$.

(a) If the atomic radii of some elements are as given, rank them in the order of possible solubility in both BCC and FCC iron.

Atom	Radius, A
H	0.46
O	0.60
B	0.97
C	0.77
N	0.71

(b) What can you say about the relative solubility in BCC and FCC iron?

(c) Is it possible to have interstitial and substitutional solubility simultaneously?

Solution:

(a) BCC iron at 20°C has an atomic radius of 1.24 A. FCC iron at 950°C has an atomic radius of 1.29 A. Reading from top to bottom of the table, the potential solubility decreases. With the exception of H in FCC Fe, all of the atoms are too big to fit into the interstices without any distortion of the surrounding crystal structure.

	r/R	
Atom	BCC	FCC
H	0.371	0.356
O	0.485	0.465
N	0.572	0.550
C	0.620	0.596
B	0.781	0.751

(b) Since the radius ratio of each atom relative to FCC Fe is closer to the radius ratio of the largest hole than for BCC Fe, the solubility in FCC Fe is greater than in BCC Fe. Solubility is limited in both FCC and BCC Fe.

(c) Both interstitial and substitutional solubility can exist simultaneously. Iron, for example, can simultaneously dissolve manganese substitutionally and carbon interstitially.

7-2-6 PROPERTY CHANGES IN SOLID SOLUTIONS

Addition of solute in an interstitial solution increases the lattice parameter. In a substitutional solution, the lattice of the base metal is usually expanded by solute atoms which are larger than the solvent atoms and contracted by solute atoms which are smaller.

Addition of solute also causes changes in properties of the solvent. Some properties change according to the law of mixtures; i.e., the properties of the alloy are directly proportional to the properties and the relative amounts of each component. Other properties may change much more drastically. Tensile, compressive and shear moduli of elasticity, coefficient of thermal expansion, specific heat, density, and lattice parameter normally obey the law of mixtures. The strength, hardness, formability, and electrical and thermal resistance usually vary much more widely with alloy content than predicted by the law of mixtures.

For two elements which are completely intersoluble in all proportions, e.g., Cu and Ni, we find strength and hardness are a maximum at about 50 a/o as indicated in Fig. 7-4. This strength increase is accompanied, however, by a decrease in formability. Electrical and thermal conductivities in such an alloy system decrease rapidly with alloy addition as indicated in Fig. 7-4.

Introduction of solute atoms automatically produces distortion of the crystal structure of the base metal. The first small additions of solute only slightly alter the average lattice parameter of the solvent, but each solute atom usually introduces a rather severe local crystallographic disturbance. This noticeably interferes with slip in the crystal (Chapter 6), thus causing a marked increase in strength and hardness accompanied by a marked decrease in formability. As more solute is added, the number of distorted regions increases without increasing the intensity of effect of each region. Each successive increment of solute is thus somewhat less effective in altering properties than the preceding increment.

In alloys with limited solid solubility, the general effect of alloy additions on properties is similar to that shown in Fig. 7-4, for relatively small amounts of solute. Should the alloy consist of two phases, strength, hardness, and electrical and thermal resistance can generally be determined by using the law of mixtures. Ordering of a solid solution has the effect of adding peak cusps to the electrical conductivity, thermal conductivity, and magnetic properties with corresponding minima in strength and hardness.

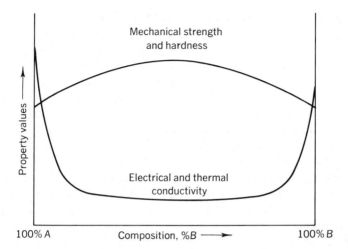

Fig. 7-4. General effect of composition of solid solution alloys on some properties.

7-2-7 INTERMEDIATE PHASES

If the alloying element is added in excess of the limit of solid solubility, a second phase appears along with the primary solid solution. This second phase may be a primary solid solution of base metal in the alloying element, or an "intermediate" phase which differs in both crystal structure and properties from either primary solid solution. This intermediate phase may be a secondary solid solution, or it may be one of a variety of phases loosely classified as intermetallic compounds. These intermediate phases may have either narrow or wide ranges of homogeneity, and may or may not include a composition having a simple chemical formula. For instance, the phase, "CuAl," exists in a homogeneity range that does not include the actual composition CuAl. Intermediate phases may range between the ideal solid solution on one hand and the ideal chemical compound on the other. An alloy may be a true solid solution at high temperature and yet may resemble a compound at low temperatures if it becomes an ordered superlattice. It is also possible that the tendency to form an ordered superlattice may be so weak that ordering cannot be detected or so strong that the alloy melts before thermal agitation is able to produce detectable disorder. A variety of intermediate phases are formed in ceramic systems in much the same manner as in metallic systems.

7-2-8 INTERMETALLIC COMPOUNDS

Intermediate phases are true intermetallic compounds only if they have: (1) a narrow range of homogeneity, (2) simple stoichiometric proportions, and (3) atoms of identical kinds occupying identical sites in the lattice. The intermetallic compound thus has a characteristic lattice which is generally more complicated than the lattices for pure metals or solid solutions. The intermetallic compounds have characteristic properties which often differ drastically from those of any constituent element. Some intermetallic compounds can form rather extensive solid solutions with their component elements. Many properties obtainable in industrial alloys depend on these intermetallic compounds and solid solutions. The intermetallic compound $CuAl_2$ and the related Al-rich primary solid solution provide an important precipitation hardening process (Chapter 9).

Interstitial compounds can be considered a special case of intermetallic compounds since they form between the transition elements and C, N, B, and H, upon exceeding the solubility limit of the interstitial element. These compounds fulfill the three requirements listed above. The compounds are metallic in nature, yet have high hardnesses and high melting points, which indicate very strong bonding. [Iron carbide (Fe_3C), which has an extremely important role in steels (Chapter 10), is a good example of an interstitial compound.] The metal atoms are normally located at lattice points of BCC, FCC, or HCP structures with the C, N, B, or H atoms contained interstitially. We should note that interstitial compounds are very different from interstitial solutions. In interstitial solutions, the solute atoms (or ions) are not in regular patterns but are randomly distributed throughout the solvent. In interstitial compounds, there is a regular pattern in the form of a crystal structure which is characteristic of the specific compound. These compounds, particularly the carbides, are often complex, since there may be substitution of other metal atoms in the lattice and also substitution for the interstitial element. Iron atoms in Fe_3C, for example, may be partly replaced by Cr, V, Mo, or W atoms and some of the C may be replaced by N.

We customarily write intermetallic compounds in the same manner as true chemical compounds. This can be misleading. Structurally speaking, TiC, for example, is not one atom of carbon bonded to one of titanium but rather six atoms of carbon bonded to six of titanium. Although Ti_6C_6 might be more correct it is still incomplete, for it is the scheme of repetition, rather than the number of atoms, that is significant.

7-2-9 VALENCY COMPOUNDS

It is rare to find alloy systems in which intermediate phases obey normal valences. At the same time, two chemically dissimilar metals tend to form phases which show ordinary chemical valence. Since bonding tends to be ionic or covalent, the properties are substantially nonmetallic, and the "compounds" are relatively brittle. The intermediate phases which tend to obey normal valence laws are found in a variety of crystal structures. These phases are most often oxides, fluorides, hydrides, carbides, and combinations of metals and metalloids such as S, Se, Te, As, Sb, and Bi.

7-2-10 ELECTRON COMPOUNDS

Some intermediate phases which occur in various binary systems show remarkable similarities. These are called electron compounds, because they depend on the ratio of valence electrons to atoms regardless of the specific alloy system. One set of these phases (β) exists at or near compositions (in each alloy system) which have a ratio of three valence electrons to two atoms. Such a structure is found in AgCd, for example, where one electron is contributed by each atom of silver and two electrons by each atom of cadmium, giving three electrons for two atoms. The same structure occurs in Cu_3Al, where three valence electrons from aluminum and one from each copper atom give six electrons for four atoms. This β phase occurs in Cu_5Sn, where we find a ratio of nine electrons to six atoms (four from tin, one from each copper). Another set of phases (γ) has an electron-to-atom ratio of twenty-one to thirteen, and a third set of phases (ϵ) has an electron-to-atom ratio of seven to four. The crystal structure of these compounds can be rather complicated. The β phases, with an electron-to-atom ratio of three to two, have a crystal structure which is either BCC or complex cubic with twenty atoms per unit cell. The γ phases have a complex cubic structure, while ϵ phases are HCP. (The β, γ, and ϵ phases shown in Fig. 7-24 are electron compounds.)

7-2-11 IMPURITIES

Whether an alloy is composed of only two (or more) metals or of one metal and one (or more) nonmetal, the components may be combined in an infinite variety of proportions. If the secondary element is present in a relatively small fraction, it is normally considered an impurity, although the real distinction between an impurity and an alloying element is that an alloying element is intentionally added, whereas the impurity is usually a residue from refining or fabrication.

Whether or not the presence of the foreign atoms is intentional, their presence still has an effect on the properties. Usually, the magnitude of the effect is small when the proportion of foreign element is small. At the same time, in many cases, as little as 0.01 % of the second element may profoundly change some or all of the properties of the base metal. For example, oxygen-free copper has remarkable toughness in a twisting test, yet this copper differs from ordinary copper only by the removal of 0.03–0.08 % of oxygen. A few hundredths of 1 % of calcium in lead readily increases that metal's strength, hardness, and endurance limit. Gold which contains as little as 0.005% of lead is unworkable because of a brittle eutectic which forms at the grain boundaries. The magnetic properties of iron are profoundly altered by the presence of 0.01 % of carbon. Platinum sponge becomes unusable as a catalyst if it contains 1.3×10^{-7} % of HCN.

Nuclear usage, for example, has established new concepts in purity requirements for large-scale applications. In conventional usage, impurities of several hundred ppm (parts per million) may have no practical effect on the desired properties. In nuclear reactors, however, a few hundred ppm of certain elements may be enough to inhibit successful operation. For instance, in reactor grade aluminum, boron cannot exceed 10 ppm without seriously affecting operation.

7-3   PHASE DIAGRAMS

In the above discussion, we considered the principal phases which can be developed between two or more components in an alloy. If we were required to describe each alloy that can be formed by varying the proportions of the alloying elements, giving the number of phases, composition and structural arrangement of each phase at room temperature, we would need a lengthy description or tabulation. If we were also required to describe changes due to temperature. such a description or tabulation would become practically impossible. Fortunately, the number of phases and their compositions can be incorporated in a simple diagram after proper exploration of an alloy system. This diagram is called an *equilibrium, phase* or *constitution diagram.* It is essentially a graphical application of Gibbs phase rule.

The phase diagram is the most successful method yet devised to accomplish the task of systematically recording, condensing, and presenting the vast body of information which has been accumulated concerning phase relationships and phase changes in materials in equilibrium.

We stated the phase rule, Eq. 7-3, and applied it to a one-component system in Fig. 7-1. For alloys, the components, C, are the base metal and the alloying additions, and the degrees of freedom, F, are the independent changes which the alloy can undergo, namely, changes of temperature, pressure, and concentration. Pressure changes and the vapor phase are usually neglected in applying the phase rule to liquids and solids under normal conditions. *Neglecting pressure effects*, the phase rule can be reduced to:

$$F + P = C + 1 \tag{7-3a}$$

The degrees of freedom are simultaneously reduced to temperature and concentrations. Since any transformation, such as a polymorphic change, which is accomplished by pressure alone, requires several thousand atmospheres, no significant error is made in neglecting pressure as a degree of freedom. This is fortunate, since neglect of pressure allows construction of a complete binary phase diagram in two dimensions, i.e., temperature and concentration of components.

7-3-1 CONSTRUCTION OF A SIMPLE EQUILIBRIUM DIAGRAM

To construct an equilibrium diagram, it is a necessary and sufficient condition that the boundaries of the one-phase regions be known. In other words, the equilibrium diagram is a plot of solubility relations between components of the system. It shows the number and composition of phases present in any system under equilibrium conditions at any given temperature. Construction of the diagram is often based on solubility limits determined by thermal analysis, i.e., using cooling curves. Changes in volume, electrical conductivity, crystal structure, and dimensions can also be used in constructing phase diagrams.

For the sake of simplicity, our discussion will be based on a binary (two-component) system under constant pressure. The principles, however, are general.

Solubility can range from essential insolubility to complete solubility in both liquid and solid states. Water and oil, for example, are substantially insoluble in each other while water and alcohol are completely intersoluble. Let us visualize an experiment on the water-ether system in which a series of mixtures of water and ether in various proportions is placed in test tubes. After shaking the test tubes vigorously and allowing the mixtures to settle, we find only one phase present in the mixtures of a few percent of ether in water or water in ether, whereas for fairly large percentages of either one in the other there are two phases. These two phases separate into layers, the upper layer being ether saturated with water and the lower layer being water saturated with ether. After sufficiently increasing the temperature, we find, regardless of the proportions of ether and water, that the two phases become one. If we plot solubility limit with temperature as ordinate and composition as abscissa, we have an isobaric [constant pressure (atmospheric in this case)] phase diagram as shown in Fig. 7-5. This system exhibits a "solubility gap."

If pure metal A in the molten state is cooled by removing heat at a constant rate, the resulting cooling curve will be similar to that in Fig. 4-33. The horizontal line represents the transition from the liquid to the solid state or the only temperature at which both solid and liquid can exist simultaneously for an indefinite period. At higher temperature, the solid disappears, and at lower temperature, the liquid disappears. This is one application of the phase rule to a unary system.

In a binary system, however, the cooling curve may appear somewhat different. Consider a binary system in which the two components are completely intersoluble in both the liquid and solid states (i.e., an isomorphous

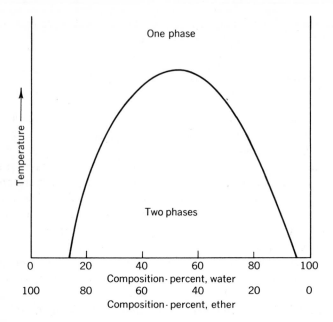

Fig. 7-5. Schematic representation of the solubilities of ether and water for each other.

system). Cooling curves for such a system are shown schematically in Fig. 7-6. As composition changes, the temperature at which solidification starts is different from that of either pure element, and the alloy solidifies over a temperature range rather than at one temperature.

Disregarding the time required for solidification, it is evident that the temperature changes shown on the cooling curves may be indicated in a two-dimensional plot in which ordinates represent temperature and abscissae represent the composition of various alloys. This is shown in Fig. 7-7, which is a phase diagram for this *A-B* system.

In this case, we find that addition of *B* to *A* decreases the melting point of *A*. This is a commonly observed phenomenon; for example, we can depress the freezing point of water by adding common salt. At the same time, however, we find addition of *A* to *B* increases the melting point of *B*. This seems ususual, and an explanation is in order.

We must first recognize that vapor pressures of a solid and a coexisting liquid are equal at the melting point. The explanation of change of melting point lies in the relative effect of addition of solute atoms on the vapor pressures of the liquid and solid phases. If solute atoms are soluble in the liquid but insoluble in the solid, it is clear that in the liquid, the attraction of solvent for solute atoms must be greater than that of solvent atoms for themselves. This greater attraction means that fewer solvent atoms escape

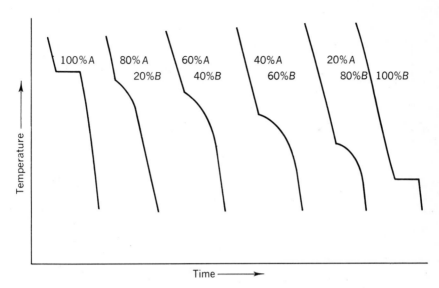

Fig. 7-6. Idealized schematic cooling curves for a series of alloys in the *A-B* system.

per second at the surface of the liquid, and thus, the vapor pressure of the liquid phase is lowered, but the vapor pressure of the solid phase is unchanged. Thus the melting point is lowered. When the solute atoms dissolve in the solid, as well as in the liquid, then the vapor pressure of the solid is also lowered. The melting point may then either rise or fall, depending on the relative degree of lowering in each phase. It should be emphasized that the above reasoning and results hold equally well for other phase changes, including polymorphic changes, as for melting and freezing.

Returning to Fig. 7-7, we find that the complete phase diagram makes it possible to predict the state of any alloy in the system at any temperature included in the diagram, if sufficient time is allowed for the system to reach equilibrium. In normal operation, however, it is fairly rare that a system reaches (or sometimes even approaches) true equilibrium, especially at lower temperatures. The equilibrium diagram is still, even then, of considerable importance, since it can serve as a guide in predicting the behavior of the system under nonequilibrium conditions.

In Fig. 7-7, we see two lines connecting the melting points of the pure components, *A* and *B*. The upper of these lines is a locus of points at which solidification starts; i.e., the area above this line represents a temperature range in which the entire system is molten. This line is called the *"liquidus."* The lower of these lines, called the *"solidus,"* is a locus of points at which solidification is complete; i.e., the area below this line represents a temperature range in which the entire system is solid. The region between the two

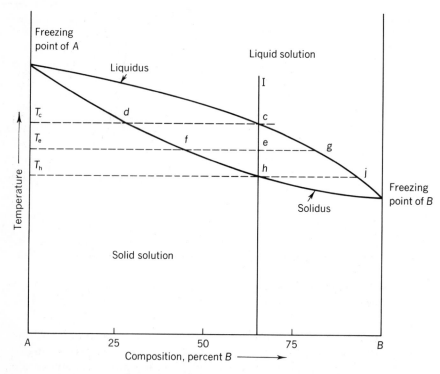

Fig. 7-7. Schematic equilibrium diagram of a binary system showing complete intersolubility in the liquid and solid states, i.e., an isomorphous system.

lines is a two-phase region in which liquid and solid solutions exist simultaneously.

Consider the behavior of Alloy I in Fig. 7-7, initially completely liquid, which is cooled to room temperature. As temperature decreases, there is no discernible change until the liquidus is reached, i.e., T_c. At any lower temperature, solid must appear. The composition of the solid is given by the solidus line (point d) at that temperature. Thus the first solid is much richer in A than the liquid, which is still essentially the composition of Alloy I, although it is slightly richer in B than before. The liquid now has a slightly lower melting point. As more heat is removed, additional solid is formed with a composition following the solidus, whereas the liquid becomes increasingly richer in B with composition following the liquidus. A decrease in temperature to T_e causes the melt to precipitate solid of f composition which both encases the existing crystals and forms new, separate crystals. For equilibrium at this temperature, diffusion (see Chapter 8) must take place between the d

cores and the *f* encasements. In addition, for the entire solid to be of *f* composition, some *B* atoms from the liquid must diffuse into the *A*-rich center of the crystals, since diffusion between *d* and *f* compositions can only result in a composition intermediate to *d* and *f*, not in composition *f*. A schematic representation of these solid and liquid combinations is given in Fig. 7-8.

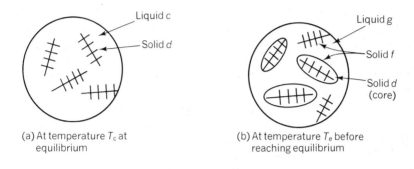

(a) At temperature T_c at
equilibrium

(b) At temperature T_e before
reaching equilibrium

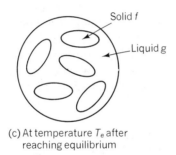

(c) At temperature T_e after
reaching equilibrium

Fig. 7-8. Schematic representation of the physical condition of alloy in Fig. 7-7.

If diffusion keeps pace with crystal growth (i.e., if true equilibrium is constantly maintained), melt composition moves downward along the liquidus, and solid composition moves downward along the solidus. Under the microscope, the separating solid appears the same as a pure metal. This continues until temperature T_h is reached. At that temperature, under equilibrium conditions, solidification is complete, and, on further cooling, there are no further discernible changes in the solid. If the temperature of the alloy is raised from room temperature, constantly maintaining equilibrium, to the point where the alloy is completely molten, the behavior is exactly the reverse of the cooling behavior.

7-3-2 INTERPRETATION OF PHASE DIAGRAMS

From the above discussion, we can draw two useful conclusions which are the only rules necessary for interpreting equilibrium diagrams of binary systems:

Rule 1. Phase Composition: To determine the composition of phases which are stable at a given temperature, we draw a horizontal line at the given temperature. The projections (upon the abscissa) of the intersections of the isothermal line with the liquidus and the solidus give the compositions of the liquid and solid, respectively, which coexist in equilibrium at that temperature. For example, draw a horizontal temperature line through temperature T_e in Fig. 7-7. The T_e line intersects the solidus at f and the liquidus at g, indicating solid composition of $f\%$ of B and $(100 - f)\%$ of A. The liquid composition at this temperature is $g\%$ of B and $(100 - g)\%$ of A. This line in a two-phase region is known as a *"tie" line*, because it connects or "ties" together lines of one-fold saturation; i.e., the solid is saturated with respect to B, and the liquid is saturated with respect to A.

Rule 2. The Lever Rule: To determine the relative amounts of the two phases, erect an ordinate at a point on the composition scale which gives the total or overall composition of the alloy. The intersection of this composition vertical and a given isothermal line is the fulcrum of a simple lever system. The relative lengths of the lever arms multiplied by the amounts of the phase present must balance. As an illustration, consider Alloy I in Fig. 7-7. The composition vertical is erected at Alloy I with a composition of $e\%$ of B and $(100 - e)\%$ of A.

This composition vertical intersects the temperature horizontal (T_e) at point e. The length of the line feg indicates the total amount of the two phases present. The length of line fe indicates the amount of liquid while the length of line eg indicates the amount of solid. In other words:

$$\left.\begin{array}{l} \dfrac{eg}{fg} \times 100 = \% \text{ of solid present} \\[3mm] \dfrac{fe}{fg} \times 100 = \% \text{ of liquid present} \end{array}\right\} \tag{7-4}$$

These two rules give both the *composition* and the *relative quantity* of each phase present in a two-phase region in any system in equilibrium regardless of physical form of the two phases. *The two rules apply only to two-phase regions independently of the number of components in the system.*

EXAMPLE 7-4

Consider the portion of a binary equilibrium phase diagram as shown. Further consider a material having a composition of 80% A and 20% B.

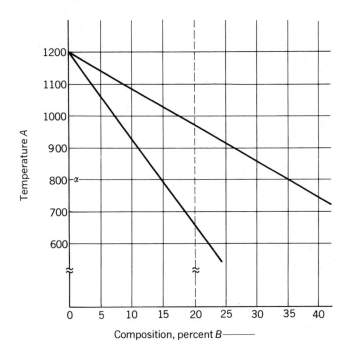

Composition, percent B———

(a) What are the phases present at 1000°?
(b) What are the phases present at 800°?
(c) What are the phases present at 600°?
(d) Can one obtain a material containing 92.5% A and 7.5% B from the material as given? How?

Solution:

(a) At 1000°, there is a single liquid phase containing 80% A and 20% B.
(b) At 800°, there is solid containing 85% A and 15% B coexisting with liquid containing 65% A and 35% B (Rule *1*, Sec. 7-3-2). From the lever rule (Rule *2*, Sec. 7-3-2)

$$\text{solid} = \frac{35 - 20}{35 - 15} = 75\%$$

$$\text{liquid} = \frac{20 - 15}{35 - 15} = 25\%$$

(c) At 600°, there is a single solid phase containing 80% A and 20% B. This is a primary solid solution but it is not a saturated solution, i.e., there could be more B in solution than there actually is.
(d) This can be done. Filter the solid out of the two-phase material at 800°. Heat this solid to 1000° and filter out a solid containing 92.5%

A and 7.5% *B*. In many situations, several steps like this may be necessary. This process of *liquation* is one method (among many) for obtaining materials of relatively high purity.

7-3-3   ISOMORPHOUS SYSTEMS

An isomorphous system is one in which there is complete intersolubility between the two components in the vapor, liquid, and solid phases as shown in Fig. 7-9. This system has been discussed and illustrated in Figs. 7-7 and 7-8 and no further discussion is required. The Cu-Ni system is both a classical and a practical example since the monels, which enjoy extensive commercial

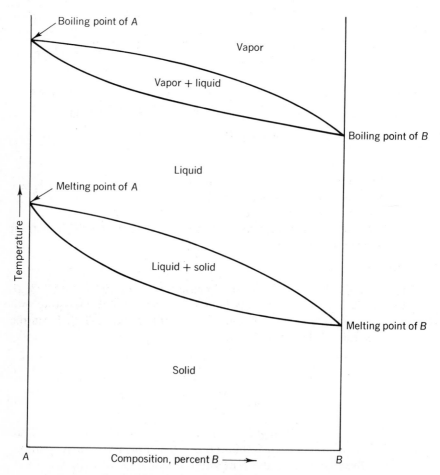

Fig. 7-9. Schematic phase diagram for a binary system, *A-B*, showing complete intersolubility (isomorphism) in all phases.

use, are Cu-Ni alloys. Many practical materials systems, both metallic and ceramic, are isomorphous.

7-3-4   AZEOTROPIC SYSTEMS

In some binary isomorphous systems the region between the liquidus and the solidus is not shaped as in Fig. 7-7 or Fig. 7-9, but the liquidus and solidus meet tangentially at a minimum temperature which lies below the melting points of both components. This is illustrated in the upper portion of Fig. 7-10. Examples of this are found in the Au-Cu, Au-Ni, and Cr-Fe systems as well as in many others. The alloy with the composition corresponding to this minimum (or maximum, e.g., Pb-Tl system) is known as an *azeotrope*. This implies a specific composition and a specific temperature at which the transformation takes place. The azeotropic alloy will behave much like a pure metal for it melts and freezes isothermally with the reaction $L \rightleftharpoons \alpha$.

7-3-5   INCOMPLETE SOLUBILITY

In many systems there is a tendency toward incomplete intersolubility below the transformation temperature, and we find a solubility gap (e.g., Cu-Pd system) similar to that in Fig. 7-11. If there is a greater tendency toward incomplete intersolubility, we may find a larger gap as in Fig. 7-10. Diagrams similar to this are found in the Au-Cu and Au-Ni systems and others. Devel-

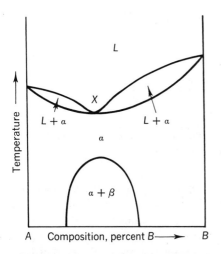 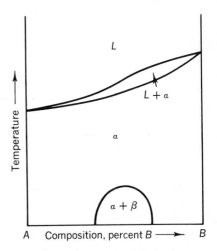

Fig. 7-10. Phase diagram of *A-B* system showing a larger solubility gap than Fig. 7-11 and a minimum liquidus tempera-ture, i.e., an azeotrope at point *X*.

Fig. 7-11. Phase diagram of *A-B* system showing a small solubility gap.

opment of a solubility gap, however, does not necessarily imply an azeotrope. The Au-Pt and Cd-Mg systems, for example, have a "cigar-shaped" region between the liquidus and solidus and a solubility gap at lower temperature.

If the tendency for formation of a solubility gap increases beyond that in Fig. 7-10 we might find the singular, but possible, case in which the maximum of the solubility gap and the minima of the liquidus and solidus have a common horizontal tangent. At the point of tangency, three phases (α, β, and liquid) would be in equilibrium without violating the phase rule.

7-3-6   EUTECTIC

If intersolubility decreases further, implying an increased solubility gap, then the upper transformation curves effectively overlap the gap, and we obtain a diagram of the kind shown in Fig. 7-12. This is a schematic representation of the eutectic type of diagram. All lines in Fig. 7-12, except the

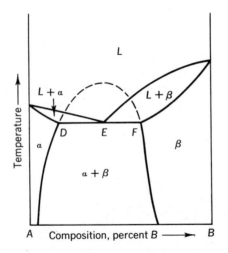

Fig. 7-12. Phase diagram of *A-B* system showing a large solubility gap. This is the origin of the eutectic type phase diagram.

horizontal line *DEF*, are one-phase region boundaries (i.e., transformation curves or solubility limit curves). The line *DEF* is part of the solidus.

The word *eutectic* (coming from the Greek, meaning "most fusible") has at least four different, but related, meanings, a fact which sometimes causes considerable confusion: (1) eutectic means the reversible, isothermal reaction of a liquid which forms two different solid phases (in a binary system) upon cooling, i.e., $L \rightleftharpoons \alpha + \beta$; (2) eutectic means the composition of the alloy which undergoes this reaction (Alloy III, Fig. 7-13); (3) eutectic

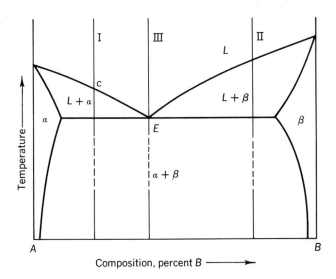

Fig. 7-13. Phase diagram of the *A-B* system showing a typical simple eutectic diagram.

means the temperature at which the eutectic reaction takes place, i.e., the lowest liquid-solid transformation temperature in the system; and (4) eutectic also refers to one of the specific microstructures formed by the eutectic reaction.

Figure 7-13 shows a typical eutectic diagram. If Alloy I (a *hypoeutectic* alloy, i.e., one having less *B* than the eutectic composition) is cooled under equilibrium conditions, solid starts to precipitate (or separate) from the liquid when the temperature is that of the intersection of the composition vertical with the liquidus line (point *c*). The solid phase is not pure *A*, but is a primary solid solution of *B* in *A* exactly as discussed for Alloy I in Fig. 7-7. The behavior of the two systems is identical until Alloy I (Fig. 7-13) is cooled to T_E. In other words, the composition of the solid moves downward along the solidus and the composition of the liquid moves downward along the liquidus until the liquid composition is that indicated at point *E*. Alloy II (a *hyper*eutectic alloy, i.e., one richer in *B* than the eutectic composition) will follow an analogous procedure until the liquid composition is that of the eutectic (point *E*).

In either case, or if we start with the eutectic composition (Alloy III), we arrive at point *E* on the diagram. At this temperature, the liquid is saturated with respect to both *A* and *B* and must, therefore, precipitate α and β phases. This occurs by alternate precipitation of crystals of alpha and of beta at various points in the remaining liquid. This takes place as rapidly as the heat of fusion released by solidification can be removed. The process continues until all the liquid has solidified. The temperature and composition (but not the relative quantities) of all three phases (liquid, α, β) remain constant during solidification.

Representative microscopic structures for these three alloys are shown in Fig. 7-14. Alloy I has primary crystals of α surrounded by eutectic. Alloy II has primary crystals of β surrounded by eutectic. Very crudely speaking, these two structures resemble pebbles in concrete. *The eutectic structure,* itself, whether in Alloy III, or in either of the above alloys, *is an intimate mechanical mixture of α and β. It is not a single phase structure.* This structure is shown in Fig. 7-14(c) and (g) as alternate platelets of α and β.

The lamellar, or alternate platelet, type of eutectic structure is most commonly found. One of three additional structures is often found when the eutectic composition is close to the composition of one of the solid solutions composing the eutectic. These structures are the "Chinese script," acicular (needlelike), and nodular (spheroidal) structures. A *divorced eutectic* is sometimes also found. In this case, there is no distinct boundary between the primary solid solution and the portion of the same solid solution in the eutectic. In other words, there is a tendency for the two phases to separate into large crystals so that there is little or no resemblance to one of the normal types of eutectic structure. Eutectics are found in many metallic and ceramic systems.

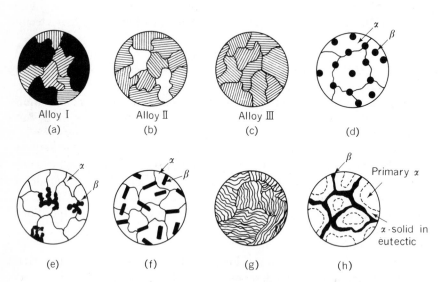

Fig. 7-14. Schematic representation possible in eutectic structures. (a), (b), and (c) are alloys shown in Fig. 7-13 (d) nodular (e) Chinese script (f) acicular (g) lamellar (h) divorced (a) is hypoeutectic, (b) is hypereutectic, (c) and (g) are somewhat different representations of the same structure; (c) through (h) are all eutectic compositions.

7-3-7 PERITECTIC

In Fig. 7-13, we see that two two-phase equilibria exist above the three-phase equilibrium, and one two-phase equilibrium exists below it. In Fig. 7-15, the arrangement is reversed. This often occurs when the melting point of one component is much lower than that of the other component. In the eutectic (Fig. 7-13), the liquid transforms into two solid phases. In the peritectic (Fig. 7-15), α and liquid phase combine to form β phase. We can thus consider the peritectic as an "upside-down" eutectic, and the peritectic reaction is written as $L + \alpha \rightleftharpoons \beta$. The Sb-Sn system is one example of a simple peritectic system.

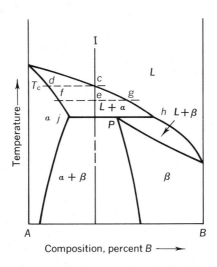

Fig. 7-15. Phase diagram of *A-B* system showing a typical peritectic.

Assume Alloy I in Fig. 7-15 is molten. As temperature is reduced no change occurs until the temperature is that of the intersection of the composition vertical with the liquidus, T_c, at which point the liquid is saturated with *A*. Upon further reduction of temperature, the alpha phase is precipitated. This enriches the liquid in *B* and, in turn, causes a continuous reduction of the transformation temperature in exactly the same fashion we saw earlier. Enrichment of the liquid in *B* continues until the temperature is that of *P*, at which time the liquid is saturated with *B* as well as with *A*. At a temperature just below T_P, the liquid phase disappears, and β phase appears upon completion of the transformation. In this transformation, the liquid must "react" with the α in the melt. This "reaction" takes place all around the surface of each crystal of primary α where the liquid touches it. Shortly thereafter, a layer of β is formed around every α crystal, and this layer obstructs further access of liquid to the material inside the layer or wall. Further reaction is slow since it must wait for diffusion through the peritectic wall. Because of this wall, an excessive time is required for complete diffusion, so that in practice, peritectic reactions are often incomplete in structures in a nonequilibrium condition. A schematic representation of these structures is given in Fig. 7-16.

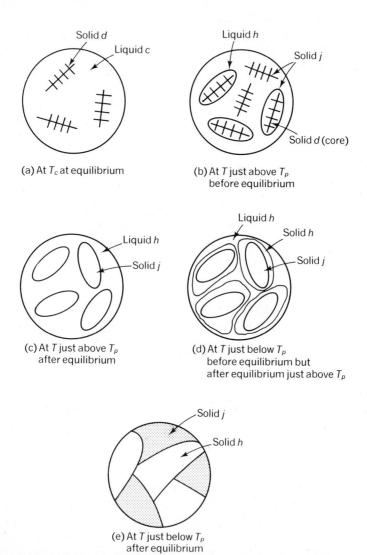

(a) At T_c at equilibrium

(b) At T just above T_p
before equilibrium

(c) At T just above T_p
after equilibrium

(d) At T just below T_p
before equilibrium but
after equilibrium just above T_p

(e) At T just below T_p
after equilibrium

Fig. 7-16. Schematic representation of physical condition of hypoeutectic alloy I in Fig. 7.15.

7-3-8 EUTECTOID

So far, we have considered only simple systems in which a liquid, whose two components are completely intersoluble, transforms into solids in which there is either complete or limited intersolubility. The principles hold equally as well for solid-solid transformations as for liquid-solid transformations. For instance, we merely substitute the words "solid solution" for the words "liquid solution" or "melt" in the diagram for the eutectic (Fig. 7-13), and we have, instead of a eutectic reaction, a transformation of a solid solution into two other solid solutions. This solid-solid transformation is called the *eutectoid* to distinguish it from the liquid-solid eutectic. A eutectoid is shown schematically in Fig. 7-17.

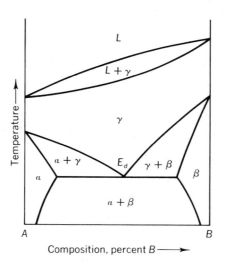

Fig. 7-17. Phase diagram of *A-B* system showing complete solubility at high temperatures and partial solubility at low temperatures; a eutectoid-diagram.

The eutectoid reaction is written as $\gamma \rightleftharpoons \alpha + \beta$. Figure 7-17 can be interpreted as a combination in which a diagram such as Fig. 7-7 is drawn above a diagram such as Fig. 7-13. In other words, Fig. 7-17 shows a system which is isomorphous at high temperatures. Both components undergo polymorphic or allotropic transformations at lower temperatures and have only partial intersolubility in the solid state at low temperatures. A practical example of the eutectoid reaction occurs in the iron-carbon system. In this case, austenite (a solid solution of carbon in γ-iron) decomposes to form pearlite (the eutectoid structure, an intimate lamellar mixture of α-iron and iron carbide). This reaction is the basis for much of the heat treating of steel and its variety of applications (see Chapter 10). The eutectoid reaction is found in many systems such as Cu-Al, Cu-Zn, Cu-Sn, Cu-Be, Al-Mn, and so on.

7-3-9   PERITECTOID

The peritectoid reaction is the transformation of two solids into a third solid in the same fashion as the peritectic reaction is the transformation of a liquid and a solid into a second solid. A representative diagram is shown in Fig. 7-18. The reaction equation is written $\alpha + \gamma \rightleftharpoons \beta$. The diagram in Fig. 7-18 can be regarded as a combination in which a diagram such as Fig. 7-7 is drawn above a diagram such as Fig. 7-15. Since the peritectoid is a solid-state reaction, it will proceed still more slowly than the peritectic so that there is a great possibility of finding the reaction incomplete. We called the peritectic an upside-down eutectic, and it follows that the peritectoid is an upside-down eutectic. Peritectoid reactions are found in Ni-Zn, Fe-Nb, Cu-Sn, Ni-Mo, and many other systems.

7-3-10   MONOTECTIC

The eutectic, peritectic, eutectoid, and peritectoid all depend on a gap in solid solubility. Solubility gaps are also possible in liquids. This can produce

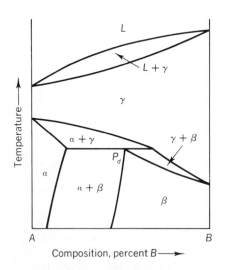

Fig. 7-18. Phase diagram of *A-B* system showing isomorphism at high temperature and partial solubility at low temperature; a peritectoid diagram.

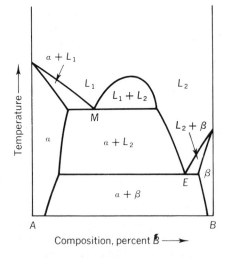

Fig. 7-19. Phase diagram of *A-B* system showing partial solubility in liquid and solid phases; a monotectic diagram with a eutectic.

a diagram similar to Fig. 7-19, i.e., a *monotectic*. This is similar to a eutectic except that one liquid, on cooling, transforms into a solid and a second liquid. The reaction is written $L_1 \rightleftharpoons \alpha + L_2$. This reaction is found in Zn-Pb, Cu-Pb, Cu-Cr, CaO-SiO_2, FeO-SiO_2, MgO-SiO_2, and other systems. The monotectic occurs only when there is a solubility gap of the type shown in Fig. 7-19. The two liquids are similar but different in composition and properties and thus are two distinct phases.

7-3–11   MONOTECTOID

Many alloy systems show solubility gaps in the solid state (Fig. 7-20) which have an appearance similar to the monotectic. This is called a *monotectoid* for essentially the same reasons that the terms eutectoid and peritectoid are used. The use of the term is not widespread, however, and we could argue that it is a form of eutectoid since the reaction equation is of the same form, i.e., $\gamma_1 \rightleftharpoons \alpha + \gamma_2$. Reactions of this type are found in Al-Zn and other systems.

7-3–12   SYNTECTIC

The monotectic and monotectoid reactions are of the eutectic type, but there is one more reaction of the peritectic type. This is the *syntectic*, shown in Fig. 7-21, which reacts according to the equation $L_1 + L_2 \rightleftharpoons \beta$. This reaction

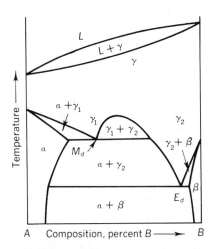

Fig. 7-20. Phase diagram of *A-B* system showing isomorphism at high temperature and partial solubility at low temperature; a monotectoid type diagram with a eutectoid.

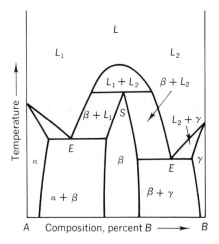

Fig. 7-21. Phase diagram of *A-B* system showing partial solubility in liquid and solid phases; a syntectic type diagram with two eutectics.

is relatively rare, and no useful alloys occur within the range of any known syntectic reaction. The best known example of this reaction occurs in the Na-Zn system.

7-3-13 INTERMEDIATE PHASE SYSTEMS

We know some systems form intermediate phases between the two components. An example is given in Fig. 7-22, in which the components, A and B, combine to form phase A_mB_n. As shown, this intermediate phase is stable and forms eutectics with A and B. This system can also be treated as two separate systems, i.e., A-A_mB_n and A_mB_n-B.

Phase diagrams showing intermediate phases are by no means limited to eutectic reactions. For example, Fig. 7-23 illustrates the case of formation

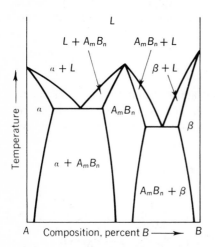

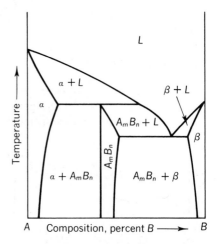

Fig. 7-22. Phase diagram of A-B system showing the existence of an intermediate phase A_mB_n which has limited intersolubility with the α and β phases.

Fig. 7-23. Phase diagram of A-B system showing the existence of an intermediate phase A_mB_n which is precipitated by a peritectic transformation.

of an intermediate phase by a peritectic type of reaction between a solid solution and a liquid.

We know several possibilities exist for intermediate phases, including electron compounds. An excellent example of a system forming electron compounds is the commercially important Cu-Zn system, Fig. 7-24, with all three electron compounds. This system has an order-disorder reaction in the β phase, five peritectic reactions, and one eutectoid reaction.

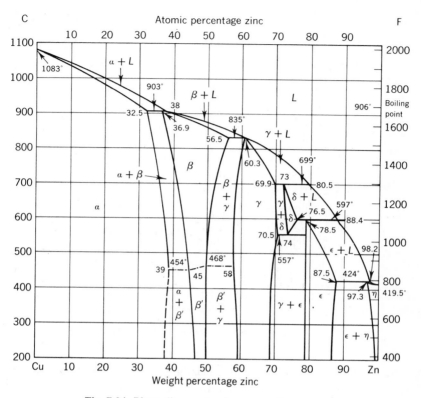

Fig. 7-24. Phase diagram for the copper-zinc system.

7-3–14 SUMMARY OF BINARY SYSTEMS

We have approached equilibrium reactions and corresponding phase diagrams in sequence, dealing only with simple diagrams, yet the phase diagrams
for some systems appear complex. In fact, simple diagrams are found much
less frequently than complex ones which, however, are combinations of the
simple invariant reactions discussed above. For example, we find a eutectic,
a eutectoid, a peritectic, and a peritectoid in the Ni-Zn system. The Cu-Zn
system, Fig. 7-24, is an example of a system with apparent complexities.
The Cu-Al system (of commercial importance), Fig. 7-25, has a monotectic,
eutectic, peritectic, eutectoid, and peritectoids.

Regardless of the apparent complexity of a phase diagram, its interpretation is relatively simple. In all cases, we can use the two rules for interpreting
phase diagrams, keeping in mind that the lever rule applies *only* to a two-
phase region. In addition, most diagrams can be considered to be composed
of a series of simple diagrams such as Figs. 7-7, 7-13, and 7-15.

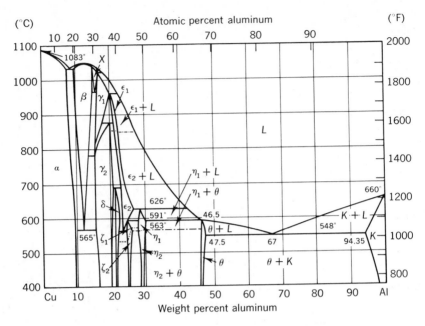

Fig. 7-25. Phase diagram for the aluminum-copper system.

By way of emphasis, we should remember that each simple reaction, i.e., azeotrope, eutectic, peritectic, eutectoid, peritectoid, monotectic, monotectoid, and syntectic takes place at constant temperature in an alloy of specific composition. The cooling curve for an alloy of the composition of each of these reactions will appear similar to that of a pure material (Fig.

Table 7-1. Invariant isothermal fixed-composition equilibrium reactions in binary systems

Reaction	Reaction equation	Diagram	Actual system (simple)
Pure metal	$L_1 \rightleftharpoons S_1$		
Azeotrope	$L_1 \rightleftharpoons S_1$	Fig. 7-10	Zr-Ti
Eutectic	$L_1 \rightleftharpoons S_1 + S_2$	Fig. 7-13	Al-Si
Eutectoid	$S_1 \rightleftharpoons S_2 + S_3$	Fig. 7-17	U-Al
Peritectic	$L_1 + S_1 \rightleftharpoons S_2$	Fig. 7-15	U-Mo
Peritectoid	$S_1 + S_2 \rightleftharpoons S_3$	Fig. 7-18	U-Zr
Monotectic	$L_1 \rightleftharpoons L_2 + S_1$	Fig. 7-19	U-Th
Monotectoid	$S_1 \rightleftharpoons S_2 + S_3$	Fig. 7-20	Th-Zr
Syntectic	$L_1 + L_2 \rightleftharpoons S_1$	Fig. 7-21	Na-Zn

4-15). The microstructure, however, will be different for each reaction. Each one of these terms, applied to a single reaction, implies a specific reversible isothermal reaction occurring in a material of a specific composition at a specific reaction temperature with a resulting characteristic microstructure formed as a result of the reaction. These reactions have been summarized in Table 7-1. Figure 7-26 shows the phase diagram of a hypothetical binary system having all these reactions.

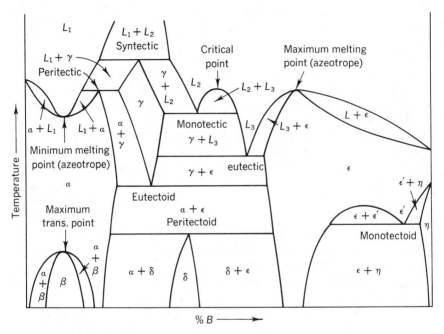

Fig. 7-26. Equilibrium phase diagram showing all invariant reaction points in a hypothetical binary system at constant pressure.

The discussion of binary systems has been in terms of metallic systems but it should be emphasized that the principles apply equally well to any system with two components whether metals, a metal and a gas (e.g., a metal and oxygen), liquids, a metal and a liquid, ceramics, or any components which form a binary.

7-4 TERNARY ALLOYS

An improvement in qualities or properties of a binary alloy is frequently gained by adding a third element. The addition of nickel to steel (Fe-C) to improve toughness and lead to brass (Cu-Zn) to improve machinability are only two of a host of commercial examples. Whether the presence of the third element is deliberate (alloying) or accidental (an impurity), its presence will affect the alloy and the phase diagram, possibly to a marked degree.

In principle, use and interpretation of ternary phase diagrams is no different than use and interpretation of binary diagrams. The phase rule, the composition vertical rule, and the lever rule are equally applicable, although we must remember that *the lever rule applies only to isothermal lines in two phase regions.* In practice, however, ternary systems are more complex than binaries and require more skill (plus the capacity to visualize in three dimensions) to use and interpret.

Because four variables (pressure, temperature, and two concentrations) are involved, we have no simple means of representing a ternary system. Even a constant-pressure section, analogous to the binary diagrams previously discussed, is three-dimensional for a ternary system. Since a three-dimensional diagram is not convenient to handle and is sometimes hard to visualize, we much prefer two-dimensional diagrams printed on paper. We customarily obtain two-dimensional diagrams by sectioning the constant pressure ternary diagram. The sections are usually: (1) isothermals (constant temperature) or (2) isopleths (constant concentration of one component, or constant ratio of two components). The isothermal sections show the true compositions existing in equilibrium at the temperature under consideration. The vertical sections do not give the true composition but are useful for predicting changes expected on heating or cooling.

The usual procedure for representing isobaric ternary systems is to place the three components at the corners of an equilateral triangle as shown in Fig. 7-27. Any given point within the triangle represents a definite proportion of the three components, since each corner is equal to 100% of each component as indicated. The amount of each component in a particular ternary alloy, X (Fig. 7-27), is determined by drawing a line through X parallel to the side of the triangle opposite the component. The point of intersection of that line and either side of the triangle indicating the constituent is its percentage composition. Thus, the percent of A in the ternary alloy X is determined by the intersection of the line a-a' with AB or AC, where the line a-a' is drawn parallel to CB.

The sides of this triangle represent the bases of the three binary systems (Fig. 7-28). If these three binary diagrams are placed upright so that temperature becomes an ordinate, perpendicular to the triangle, we have a three-dimensional solid with a triangular base. A vertical line arising from any point X in the base triangle represents the phase behavior of a given alloy of specific composition at a series of temperatures.

In general, if we can visualize the single-phase regions, then the rest of the three-dimensional phase diagram follows "automatically." In constructing ternary diagrams, it is fairly standard practice to use a basal plane pro-

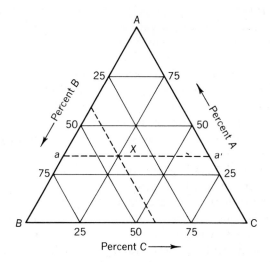

Fig. 7-27. Basal isothermal plane of the ternary equilibrium diagram.

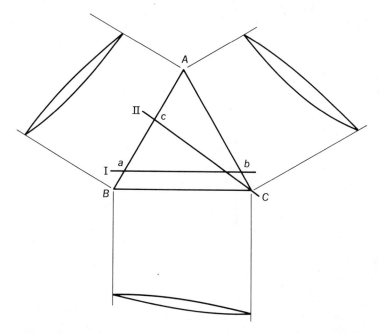

Fig. 7-28. Base plane and the three binary phase diagrams of a ternary system *A-B-C* which is completely intersoluble in both the liquid and solid states.

jection. This projection, in effect, collapses the entire three-dimensional isobaric ternary diagram onto the base plane, which is normally the room temperature isothermal. Among the items of information which may be displayed on a basal plane projection are: the base plane, the three binary diagrams, intersections of liquidus surface (with arrowheads to indicate direction of temperature decrease), lines of two-fold saturation, ternary eutectic triangles, ternary eutectic points, and threefold saturation points. Although any (or all) of this information may be found on a given basal plane projection, it is not customary to show all of it on one diagram because of possible confusion in reading the diagram.

To gain some understanding of ternary diagrams, consider the simple isomorphous system in which all three components are completely inter-soluble at all temperatures. The base plane and the three binary systems are shown in Fig. 7-28, and a sketch of the three-dimensional diagram is found in Fig. 7-29. For this system only three phases exist: one vapor, one liquid, and one solid; and we will neglect the vapor phase. The liquidus surface and the solidus surface, similar to liquidus and solidus lines in a binary diagram, enclose a region in which liquid and solid coexist.

Isothermal sections of this ternary system at temperatures above and below those of transformation will contain a one-phase region only. A typical isothermal section which intersects the one-phase region boundaries is given in Fig. 7-30. The curved lines in this figure are the intersections of the iso-

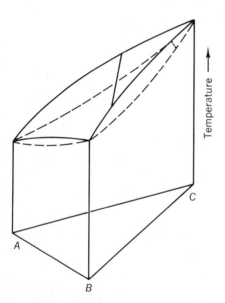

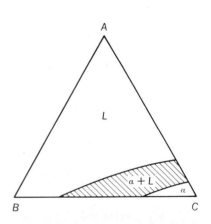

Fig. 7-29. Phase diagram of ternary system *A-B-C* shown in Fig. 7-27.

Fig. 7-30. Isothermal section of the *A-B-C* system shown in Figs. 7-27 and 7-28.

thermal plane with the boundary surfaces. Other isothermal sections at temperatures within the transformation range differ in appearance only in the position of the $\alpha + L$ region. Isopleth I (vertical section) in Fig. 7-31 is for the case of a constant composition of A; the location of this section is indicated in Fig. 7-28. Figure 7-32 shows an isopleth (vertical section II in Fig. 7-28) with a constant ratio of A and B. We should reemphasize that the *lever rule cannot be applied in vertical sections* but only in isothermal sections. *Vertical sections are for qualitative use* only.

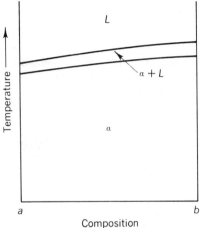

Fig. 7-31. Vertical section I of the *A-B-C* system as shown in Figs. 7-28 and 7-29.

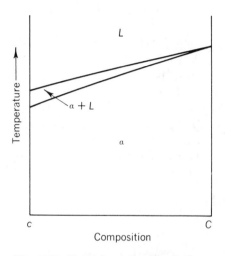

Fig. 7-32. Vertical section II of the *A-B-C* system as shown in Figs. 7-28 and 7-29.

This isomorphous ternary system illustrates all the basic ideas involved in ternary diagrams. At the same time, it is obvious that many more combinations can be found in ternary systems than in binary systems. The difference is, however, basically in degree rather than in kind, and only practice will give facility in use and understanding. Some examples of the types of systems which can be found are:

1. A completely isomorphous system (Figs. 7-28 through 7-32).
2. Two isomorphous binaries and a eutectic binary.
3. Three eutectic binaries which form a ternary eutectic.
4. Two peritectic binaries and an isomorphous binary.
5. Two eutectic binaries and an isomorphous binary.
6. Two eutectic binaries and a peritectic binary.

In short, relatively simple ternary systems can be formed from any combination of simple isomorphous, eutectic, and peritectic binary systems. Additional complications are involved if there is an allotropic transformation, eutectoid, peritectoid, monotectic, monotectoid, or syntectic reaction in one or more of the binary systems.

In binary systems, a system with an intermediate phase can usually be treated as two separate systems. In a ternary system, this is sometimes, but not always, possible. Obviously, however, if an intermediate phase is formed in one or more of the binary systems, the complexity of the ternary diagram will increase. As an example, consider the Al-Cu-Zn system. The binary diagrams for Al-Cu and Cu-Zn are given in Figs. 7-25 and 7-24, respectively. The Al-Zn system has a eutectic and a monotectoid. While the ternary system will obviously be quite complex, the use and interpretation of the diagram will differ only in degree, not in kind, from the simpler ternaries and binaries, since the basic rules of interpretation apply to all phase diagrams.

In addition to the complications deriving from the various possible combinations of binary systems in a ternary system, there is a possibility of formation of a ternary "intermetallic" compound. A true ternary compound is, however, rarely encountered in metallic systems with the possible exception of double carbides. The existence of a ternary compound or ternary intermediate phase introduces nothing new, in principle, in use and interpretation of the phase diagram.

7-5 SYSTEMS WITH MORE THAN
THREE COMPONENTS

In principle, systems with four or more components are no different than unary, binary, or ternary systems. A quaternary system (or one with any greater number of components) is as amenable to thermodynamic treatment as a unary system. From the practical viewpoint, however, an isobaric ternary system exhausts the available dimensions of space. Representation of a quaternary system may be made by constructing isothermal-isobaric tetrahedra whose faces are equilateral triangles. This of course requires considerable understanding of, and proficiency with, binary and ternary systems. One alternative method which may be helpful is to use projections or sections of sections.

For quinary, sexinary, septenary, etc., systems, graphical representation is impractical and analytical methods are used. An alternative practical approach, in cases where some components are present in relatively small amounts, is to construct binary or ternary diagrams as modified by constant small fractions of those components. This is sufficiently accurate to permit prediction of heat-treating temperature, precipitation hardenability, and other processes of practical interest.

QUESTIONS AND PROBLEMS

1. Discuss the differences and similarities of equilibrium and metastability.

2. Define the following terms: system, state, phase, component, constituent, degrees of freedom, equilibrium.

3. What is an alloy? How many components are found in an alloy? What is meant by base metal? What are alloying elements? How do you distinguish between alloying elements and impurities?

4. Consider a system composed of Al_2O_3 and UO_2. Is this a binary or a ternary system? Which way would you treat it? Is this system an alloy in a strict sense? Is this system an alloy in a broad sense? Does the phase rule apply?

5. If a beaker contains NaCl and H_2O at room temperature, how many phases are present?

6. What is the major distinction between a binary metallic system and a binary ceramic system?

7. Is an alloy possible between two metals which are essentially insoluble in each other? How?

8. Discuss the similarities and differences between substitutional and interstitial solid solutions.

9. Consider the following metals (see Table 4-4). Co is HCP up to 390°C and FCC above that temperature. Ti is HCP up to about 900°C and BCC above that temperature. In the absence of further information (except perhaps the periodic table), what would you predict about the possibilities of having isomorphous systems in the following pairs: W-Ta, Pt-Pb, Co-Ni, Zn-Co, Ti-Ta?

10. Why is complete intersolubility possible in a substitutional solid solution but not in an interstitial solid solution?

11. What are secondary solid solutions? Intermetallic compounds? Electron compounds? Discuss the similarities and differences among them. How do secondary solutions differ from primary solutions?

12. Calculate the radius of the largest atom which can enter interstitially into a face-centered cubic crystal without causing distortion. State the answer as a fraction of the radius of the larger atoms.

13. Repeat problem 12 for a body-centered cubic metal.

14. Repeat problem 12 for a hexagonal close-packed metal.

15. When does a second element constitute an alloying element, and when does it constitute an impurity? Of what significance is the difference? What is the difference in the effect on the various properties?

16. What is meant by solid solution hardening? What general effects does it have on the various mechanical and physical properties? How does it operate to cause these effects?

17. What is meant by the "order-disorder reaction"? What are the general effects on the various properties? How is the reaction made to proceed?

18. What is the lattice parameter of 75 a/o nickel 25 a/o copper if the lattice parameters of nickel and copper are 3.517 A and 3.608 A, respectively?

19. Define "phase." What different kinds of phase are possible?

20. Assume you are told that four distinct phases are observed in a laboratory specimen of a binary alloy. Would you believe this statement? Explain.

21. "Derive" or justify the lever rule.

23. Can one apply the lever rule at the temperature of an invariant reaction? Why?

23. When substitutional atoms do *not* occupy random sites in a crystal structure, an *ordered structure* results. For example, although Cu and Au show complete isomorphism at elevated temperatures, ordering takes place at lower temperatures at about 25 a/o Au and at about 50 a/o Au. Sketch an FCC unit cell and designate the most likely sites for the Au atoms in the two situations.

24. Five w/o LiBr is dissolved in NaCl. What percent of the ions are (a) Br? (b) Li?

25. Discuss the differences and similarities of:
    (a) a minimum point and an azeotrope.
    (b) an azeotrope and a pure metal.
    (c) an azeotrope and a eutectic.
    (d) a eutectic and a peritectic.
    (e) a eutectic and a eutectoid.
    (f) a peritectic and a peritectoid.
    (g) a eutectoid and a peritectoid.
    (h) a monotectic and a eutectic.
    (i) a monotectic and a syntectic.
    (j) a monotectic and a monotectoid.
    (k) a monotectoid and a eutectoid.
    (l) a *hypo*eutectic and a *hyper*eutectic.

26. "Eutectic" implies four different but related meanings. Do these same meanings apply to monotectic, peritectic, eutectoid, etc.? If so, how? If not, why not?

27. What is the normal microstructure for a eutectic? A peritectic? A eutectoid? A peritectoid? Sketch these typical microstructures.

28. What is meant by a divorced eutectic? What sort of structure does it have? What is the significance?

29. Discuss the differences and similarities of homogeneous and heterogeneous equilibrium.

30. It has been noted that in any constitution diagram a two-phase region is bounded by a single-phase region or a three-phase line (or three-phase region). Is this a requirement of the phase rule? Why?

31. Silver melts at 961°C and copper at 1083°C. A eutectic forms at 780°C with a composition of 72% Ag. At the eutectic temperature each metal dissolves 8% of the other; at room temperature, only 1%. Draw the phase diagram assuming straight lines. For the alloys containing 5%, 20%, 28%, and 80% Cu,
   (a) Draw typical time vs. temperature cooling curves, showing the temperatures of initial crystallization and final solidification, and label all changes.
   (b) Sketch the final microstructure of each; label all structures.
   (c) What proportion of eutectic does each contain at 400°C and at room temperature?

32. Why must the liquidus and solidus of Fig. 7-11 be tangent to each other at point X rather than meet to form a single line?

33. If air contains nitrogen, oxygen, argon, carbon dioxide, and water vapor, how many components are present? How many phases?

34. A simplified phase diagram for H_2O-NaCl is shown. A system with 5 w/o NaCl is at −15°C. What phases are present? In what relative amounts? What is the composition of each phase present? Can you suggest a method of obtaining freshwater from seawater or brackish water?

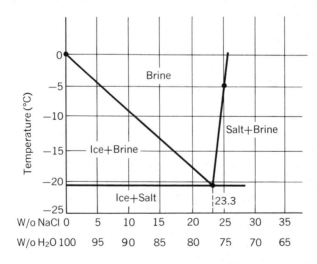

35. Sidewalks can be cleared of ice by spreading NaCl on the ice. (a) How effective will this be at −15°C? (b) at −25°C?

36. Consider the peritectic occurring in the iron–iron carbide diagram in Fig. 10-2. For equilibrium cooling of an alloy containing 0.25 w/o carbon fill in the following table.

Temperature	Phases present	Compositions of phases	% of Total weight for each phase	Sketch of microstructure
2720°F				
2715°F				
2700°F				
2600°F				

37. A plumber prefers a 65 Pb-35 Sn solder whereas a radio repairman prefers a 60 Sn-40 Pb solder. Why?

38. Is insolubility ever exhibited among gases? Why? What physical phenomenon is necessary to demonstrate insolubility?

39. Draw a hypothetical diagram for two elements, showing partial solubility in the solid state in which the effect of either element is to raise the freezing point of the other. Would you expect such a system to exist in a simple unqualified form? Why?

40. When a liquid solution of two components and a solid solution of the same components are in equilibrium, (a) How can the compositions of the solutions be changed? (b) What, if anything, happens to the compositions of the solutions if the temperature is changed? Why?

41. What happens if pure water freezes in the cooling system of an automobile? Does a water-alcohol mixture freeze at constant temperature? Why? What effect will some impurity or alcohol in the water have even though the combination freezes?

42. Why is a peritectic reaction often incomplete? Which reaction would you expect to take place more rapidly, a peritectic or a peritectoid? Why? How would you interpret the term "reaction rim"?

43. In what respects does a ternary alloy system differ from a binary system? In what respects are binary and ternary alloy systems the same?

44. What is the composition in weight percent of uranium hydride (UH_2)? In atomic percent?

45. The phase diagram for the A-B system is given. At temperature T_1 liquid containing 50% B and solid containing 15% B coexist to the left of the eutectic point, and liquid containing 65% B and solid containing 95% B coexist to the right of the eutectic point. At temperature T_2 solid α containing 15% B and solid β containing 5% A coexist. The eutectic composition is 40% A and 60% B.
Consider Alloy I, which is 65% A and 35% B.
(a) Which is this, a hypoeutectic or hypereutectic alloy?
(b) Sketch the type of cooling curve expected for Alloy I under equilibrium conditions.

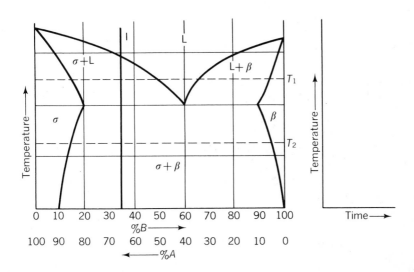

(c) At temperature T_1 the relative amounts of solid and liquid are: solid _____ %; liquid _____ %.

(d) At temperature T_2 the relative amounts of α and β are: α _____ %; β _____ %.

(e) At temperature T_2 the relative amounts of α and β in the eutectic are: α _____ %; β _____ %.

(f) At temperature T_2 the relative amounts of α and eutectic are: α _____ %; eutectic _____ %.

(g) At temperature T_2 the relative amounts of β and eutectic are: β _____ %; eutectic _____ %.

46. Consider Alloy I in problem 45. (a) Determine the relative fractions of the phases present and the composition of each phase at a temperature very slightly above the eutectic temperature. (b) Repeat for a temperature very slightly below the eutectic temperature.

47. BCC iron is stable up to 910°C. Above 910°C, FCC iron is stable. If W, Mo, or Cr is added to iron, this allotropic transformation temperature is increased. If Ni is added, however, the temperature is lowered. Explain.

8

RATE PROCESSES

8-1 INTRODUCTION

Chapter 7 discussed equilibrium and relationships between phases in equilibrium, although we seldom attain true equilibrium outside the laboratory. In addition, we often prefer to retain a metastable state or to exercise control over the change of a given system to some state intermediate between the starting state and equilibrium. This implies a need for determining the time factor involved in a process, i.e., the rate at which the process occurs, since this may be more important than knowledge of the equilibrium state. Nucleation, growth, solid-state reaction or transformation, precipitation from solutions (or glasses), decomposition of phases (e.g., hydrates or carbonates), polymerization, oxidation, corrosion, creep, and diffusion rates are some of the more important rates that we should know.

Silica provides a good example of metastability in comparison with equilibrium. Figure 7-2 shows a schematic equilibrium diagram for SiO_2. The α-quartz-β-quartz polymorphic transformation at 573°C is reversible and rapid. The other transformations shown in Fig. 7-2 are sluggish and require long times to achieve equilibrium. The diagram can be extended to show metastable forms of SiO_2 which are possible (Fig. 8-1). The phase with the lowest vapor pressure (solid lines) is the most stable at any temperature, i.e., it is the equilibrium phase. Once a high temperature phase is formed, its transformation is so sluggish that it will commonly transform into a metastable phase rather than the equilibrium form. For example, β_2-tridymite will normally transform into α-tridymite and β-tridymite rather than β-quartz. Liquid SiO_2, upon cooling, can form silica glass which remains in that form indefinitely at room temperature. At a constant temperature, however, there is a tendency for a metastable phase to change to a phase of

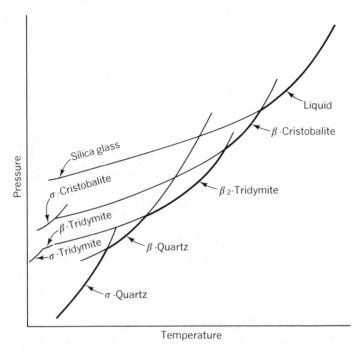

Fig. 8-1. Schematic temperature-pressure diagram for SiO_2 including metastable phases.

lower free energy although this change is not necessarily to the lowest energy form (i.e., equilibrium). At 1100°C, for example, silica glass can transform into β-crystobalite, β_2-tridymite, or β-quartz. Which, if any, of these transformations will occur depends on the kinetics of the system in the existing state.

8-2 METASTABILITY

We can gain an idea of metastability by using the mechanical model (Fig. 8-2) of a solid block in three different positions on an "infinite" plane, i.e., a model of three different energy states. States A and B are metastable while state C is stable. If we rotate the block about the proper axes, we can change A to B and both A and B to C. These changes are accompanied by a net decrease in free energy (potential energy only in this model). State C is the condition of minimum free energy, and this is in the condition of equilibrium (Chapter 7). In order to make these changes, it is necessary to initially add some energy even though there is a net release of energy when the model reaches a more stable state. The energy which must be supplied to initiate the

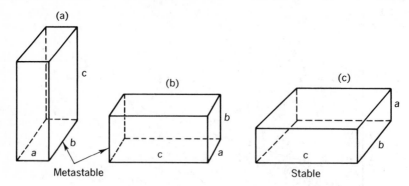

Fig. 8-2. A mechanical analogy of stability and metastability.

change [the energy necessary to raise the center of gravity from the meta-stable state (B) to the unstable state (D)] is the *activation energy*. The energy released by the change is the difference in energy between the two states under consideration. The energy relationships for our simple model are shown in Fig. 8-3.

Chemical and physical changes can be activated in an analogous manner to transform a system from a metastable condition to a more stable one. In general, in order to transform a system to a more stable state, the atoms of

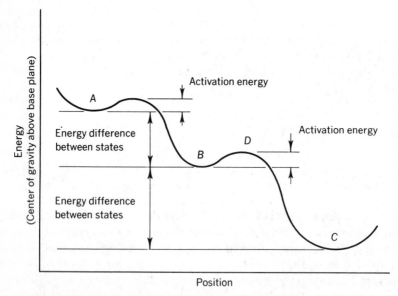

Fig. 8-3. Schematic energy relationships for model in Fig. 8-2. States A and B are metastable; state C is stable.

the system, or at least the atoms of the part of the system where the reaction starts, have to pass through a condition in which their energy is increased. If we let Fig. 8-3 represent the potential energy of an atom as a function of its position, it is obvious that for an atom to move from its metastable position B to the stable one at C, it must pass through the unstable position at D. If it cannot temporarily gain the extra energy necessary to pass over this potential energy barrier, it must remain in position B. The smallest energy which allows the atom to pass over the barrier is the activation energy. This energy may be supplied by a thermal, mechanical, electrical, or any other convenient source, although thermal sources are most common.

The rate at which the reaction takes place depends on: (1) the number, n, of atoms in the metastable position; (2) the frequency, v, with which an atom in this position attempts to scale the barrier; and (3) the probability that the atom has the necessary activation energy, q, during this attempt. This probability is generally small (particularly in solids) and the flux, f, of reacting atoms, i.e., the number which cross the barrier in unit time, is

$$f = nve^{-q/kT} \tag{8-1}$$

where q is the activation energy per atom, k is Boltzmann's constant and T is the absolute temperature. The rate depends on the relative stabilities of the two phases involved, since this often affects the magnitude of the activation barrier.

8-3 ARRHENIUS EQUATION

It is common practice to express the activation energy in terms of calories per mole and to replace Boltzmann's constant by the universal gas constant, R. The rate equation is then written as the Arrhenius equation, i.e.,

$$\text{Rate} = Ae^{-Q/RT} \tag{8-2}$$

where A is a constant containing nv and Q is the activation energy per mole. This formula is obeyed in many physical and chemical changes and provides a good first approximation in other cases. It is semiempirical, yet it fits very well a large number of reactions, both homogeneous and heterogeneous, particularly gaseous and liquid reactions.

We can physically interpret the Arrhenius equation as the existence of equilibrium between normal and activated atoms, the latter being the ones that actually react. A temperature increase has the effect of increasing the number of active atoms and thus increasing the rate. Q is the energy needed to activate a gram-mole of atoms.

The Boltzmann equation, which is similar to the Arrhenius equation, can be derived from statistical mechanics. Application to particular problems gives equations of various forms, but the exponential term is common to all.

EXAMPLE 8-1

(a) What is the change in the rate of a given process having an activation
    energy of 30 Kcal/mole in going from room temperature (27°C) to
    127°C? (b) In going from 227°C to 337°C? (c) In going from 227°C
    to 337°C if the activation energy is 60 Kcal/mole?

Solution:

Using Eq. (8-2),

$$\text{Rate} = Ae^{-Q/RT}$$

(a) $\dfrac{\text{rate}_2}{\text{rate}_1} = \dfrac{Ae^{-30,000/(2 \times 400)}}{Ae^{-30,000/(2 \times 300)}} = e^{+12.5} = 270 \times 10^3$

(b) $\dfrac{\text{rate}_2}{\text{rate}_1} = \dfrac{Ae^{-30,000/(2 \times 600)}}{Ae^{-30,000/(2 \times 500)}} = e^{+5} \simeq 150$

(c) $\dfrac{\text{rate}_2}{\text{rate}_1} = \dfrac{Ae^{-60,000/(2 \times 600)}}{Ae^{-60,000/(2 \times 500)}} = e^{+10} = 22,000$

8-4 TRANSFORMATIONS

Processing of materials often involves transformation from one phase to
another. Solidification of liquid metal in a casting and change from one solid
phase to another during heat treatment are examples. Allotropic changes,
changes in solubility, suppressed eutectic or eutectoid transformations are
other examples. In general, the rate of change from one metastable state to
another or from a metastable state to a stable state is of great practical interest. These phenomena involve three important, and somewhat inseparable
phenomena, i.e., nucleation, growth, and diffusion. Although the nature of
a given phase transformation depends on the phases involved, every transformation depends on: (1) a thermodynamic factor that determines whether
the transformation is possible; and (2) a kinetic factor that determines
whether the transformation will proceed at a practical rate. These principles
are applicable in metallic, ceramic, and polymeric systems.

8-4-1 NUCLEATION

The process by which a new phase appears in an existing phase is called
nucleation. Nucleation is homogeneous when it takes place at random within
a homogeneous phase; it is heterogeneous when it occurs at interfaces, impurities, surfaces, etc. The distinction is important since we believe that
solid-state nucleation is heterogeneous (which makes analysis difficult).
Homogeneous nucleation, however, can take place in transformations from
gas to liquid, and the theory for this is well developed.

Nucleation occurs because of variations in compositions and energies due to the statistical behavior of atoms in a system. In addition, thermodynamically speaking, one phase is stable relative to another if it has a lower free energy. If the system is metastable, we must supply some activation energy although not all the atoms in the system must be activated simultaneously. We can rewrite the free energy equation (7-2b) for a phase change as

$$\Delta F = \Delta H - T\Delta S \qquad (8\text{-}3)$$

in which ΔF is a useful measure of the tendency for a reaction to occur. This change in Gibbs free energy must be negative for the reaction to take place.

8-4-2   GAS-LIQUID TRANSFORMATION

Consider the transformation of a gas or vapor to a liquid in a unary system with an equilibrium temperature of T_0. At temperatures above T_0, the gas is stable, i.e., $F_g < F_l$ for $T > T_0$. At temperatures below T_0, the reverse is true, i.e., $F_g > F_l$ for $T < T_0$. At T_0, the two phases are in equilibrium, i.e., $F_g = F_l$. At T_0, we thus find (for unit volume):

$$\left. \begin{array}{c} F_g - F_l = \Delta F_v = \Delta H_v - T_0\Delta S_v = 0 \\[2mm] \Delta S_v = \dfrac{\Delta H_v}{T_0} \end{array} \right\} \qquad (8\text{-}4)$$

where the subscript v indicates quantity per cm^3 of the system in question. Assuming that ΔH_v and ΔS_v are temperature independent, the free energy change accompanying liquefaction of gas at any temperature T is:

$$\Delta F_v = \Delta H_v - T\Delta S_v = \Delta H_v\left(\frac{T_0 - T}{T_0}\right) \qquad (8\text{-}5)$$

Heat is normally released during liquefaction (ΔH_v is negative), and ΔF_v is negative, i.e., the reaction is possible, provided $T < T_0$. It is well known that these transformations do not occur at temperatures just infinitesimally below the equilibrium temperature but require appreciable supercooling (Fig. 4-33) before a practical reaction rate is achieved. This behavior is a direct result of the mechanism by which phase transformations take place.

The most satisfactory theory of nucleation of liquid from vapor proposes that clusters of atoms exist in the gas in equilibrium with the gas and that these clusters are potential nuclei for the liquid phase. In order to form a drop of liquid of radius r (cm), an interface must form between the liquid and the surrounding gas. If σ is the energy required to create one cm^2 of interfacial area, and ΔF_v is the free energy change per cm^3 for liquefaction, we find the overall free energy change for formation of a spherical drop of liquid within the gas is:

$$\Delta f = 4\pi r^2\sigma + \tfrac{4}{3}\pi r^3\Delta F_v \qquad (8\text{-}6)$$

The first term on the right of this equation is the energy needed to create the interface and is always positive. The second term is the energy released by the condensing phase and is negative when ΔF_v is negative. These terms contain r^2 and r^3, respectively, and vary with radius as shown in Fig. 8-4.

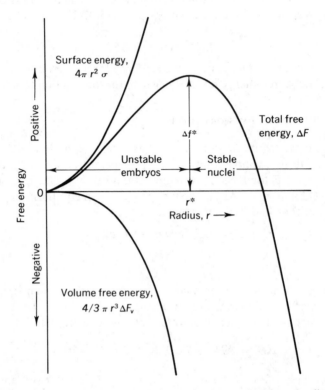

Fig. 8-4. Free energy change in a drop of liquid as a function of radius.

Initially, the r^3 term is smaller but becomes greater for larger values of r. The sum Δf thus has a maximum at a critical radius r^*, whose value is:

$$r^* = -\frac{2\sigma}{\Delta F_v} \tag{8-7}$$

The corresponding value of Δf is:

$$\Delta f^* = \frac{16\pi\sigma^3}{3(\Delta F_v)^2} \tag{8-8}$$

The drops of liquid described by the net free energy curve in Fig. 8-4 are formed by random collision of two atoms to form a pair, addition of a third atom from a second collision, and so on. At any instant, there will be an equilibrium distribution of all sizes, and it can be shown that the number of drops of radius r^* per cm³ of gas is:

$$n^* = n \cdot e^{-\Delta f^*/kT} \tag{8-9}$$

where n is the number of atoms per cm³ and k is Boltzmann's constant. Drops with a radius equal to, or greater than r^* differ from smaller drops in that the addition of another atom will decrease the free energy. A drop of critical radius, or larger, will tend to grow spontaneously and is therefore called a *nucleus*. Smaller drops will tend to decrease in size and will grow only if subjected to the proper combination of chance collisions. These smaller drops are *embryos*.

If the number of critical nuclei n^* is given by Eq. 8-9, the rate of formation of stable nuclei can be determined from the rate at which critical nuclei acquire one additional atom. If z is the number of atoms striking unit area per second, the rate of nucleation per cm³ per second, N_v, is:

$$N_v = z \times n^* \cdot 4\pi(r^*)^2 \tag{8-10}$$

The kinetic theory of gases states the collision frequency is $z = P/\sqrt{2\pi mkT}$ where P is the pressure of the gas and m is the mass of an atom. The nucleation rate then is:

$$N_v = \frac{nP4\pi(r^*)^2}{\sqrt{2\pi mkT}} e^{-\Delta f^*/kT} \tag{8-11}$$

If this is rewritten on a molar basis:

$$N_v = \frac{nP4\pi(r^*)^2}{\sqrt{2\pi mkT}} e^{-\Delta F^*/RT} \tag{8-11a}$$

The similarity to the Arrhenius equation (Eq. 8-2) is obvious if the total free energy change is effectively the activation energy.

If we substitute appropriate numbers in Eq. 8-11, we find that very large amounts of supercooling are necessary to obtain practical rates of nucleation of drops in the volume of the gas, i.e., homogeneous nucleation. It appears that, in practice, drops of liquid form on nucleation catalysts, such as dust particles in the gas or on the container walls, and we have heterogeneous nucleation.

8-4-3   LIQUID-SOLID TRANSFORMATION

The argument used to develop the homogeneous nucleation rate equation (8-11) for gas-liquid transformation applies, in large measure, to homogeneous nucleation of a solid from a liquid phase. The kinetic theory of gases does not apply, however, and modification is required. The equation corresponding to Eq. 8-10 is:

$$N = n^* v \tag{8-12}$$

The frequency v with which a liquid atom will cross the liquid-solid interface and become part of the solid nucleus is given (from theory of absolute reaction rates) by:

$$v = \left(\frac{kT}{h}\right)e^{-\Delta f_A/kT} \tag{8-13}$$

where h is Planck's constant and Δf_A is the energy required to cross the barrier of the interface (also known as the diffusion activation energy). Eq. 8-12 then becomes

$$N = n\frac{kT}{h}e^{-[\Delta f^*/kT + \Delta f_A/kT]} \tag{8-14}$$

or

$$N = n\frac{kT}{h}e^{-[\Delta F^* + \Delta F_A]/RT} \tag{8-14a}$$

where n is the number of atoms per cm³ of liquid.

Use of appropriate values for metals in Eq. 8-14 indicates 100° of supercooling is required before homogeneous nucleation occurs. This type of nucleation can be produced in the laboratory, but it is clear that solidification of most metals occurs by heterogeneous nucleation, since observed supercooling is normally only a few degrees. As in the gaseous phase, heterogeneous nucleation in a liquid occurs on impurities floating in the liquid or on the container walls. Differences in interfacial tensions account for the relative ease of heterogeneous nucleation in comparison with the homogeneous form.

If we assume embryos of solid (α) form as spherical caps on a plane nucleating surface, equilibrium among the interfacial tensions is given by:

$$\sigma_{Sl} = \sigma_{\alpha S} + \sigma_{\alpha i} \cos \theta \tag{8-15}$$

where S refers to the surface, l to the liquid, α to the solid being formed, and θ is the contact angle of the spherical cap. As the embryo develops and extends the area covered on the nucleating surface, the S-l interface is replaced by the α-S interface with an energy change equal to the difference of the two interfacial energies, i.e., $\sigma_{\alpha S} - \sigma_{Sl}$. This free energy change can be evaluated from Eq. 8-15, namely,

$$\sigma_{\alpha S} - \sigma_{Sl} = -\sigma_{\alpha l} \cos \theta \tag{8-16}$$

Accounting for this additional term in Eq. 8-6 gives:

$$\Delta f_S = \left(\pi r^2 \sigma_{\alpha l} + \frac{\pi r^3 \Delta F_v}{3}\right)(2 - 3\cos \theta + \cos^3 \theta) \tag{8-17}$$

The critical values of r and Δf become:

$$r_S^* = \frac{2\sigma_{\alpha l}}{\Delta F_v}\sin \theta = r^* \sin \theta \tag{8-18}$$

$$\Delta f_S^* = \Delta f^* \frac{(2 + \cos \theta)(1 - \cos \theta)^2}{4} \tag{8-19}$$

where r^* and Δf^* are the critical values for homogeneous nucleation. By reasoning analogous to that used to obtain Eq. 8-14, we find that the equation for the rate of heterogeneous nucleation is:

$$N_S = n_s \frac{kT}{h} e^{-[\Delta f^*_s + \Delta f_A]/kT} \tag{8-20}$$

or

$$N_S = n_s \frac{kT}{h} e^{-[\Delta F^*_s + \Delta F_A]/RT} \tag{8-20a}$$

where n_s is the number of atoms of liquid absorbed on one cm² of nucleating surface.

Additional understanding of the nucleation process can be had by analyzing the free energy terms in Eqs. 8-14 or 8-20. The Δf^* (as in Eq. 8-8) will be proportional to the cube of the surface energy and inversely proportional to the square of the volume free energy. The volume free energy is zero at the (equilibrium) transformation point, and the rate of nucleation is zero. As temperature decreases, the volume free energy term increases (negatively), and ΔF^* will decrease with decreasing temperature (for approximately constant surface energy). This has the effect of increasing the nucleation rate. As the degree of supercooling increases, ΔF^* becomes smaller relative to ΔF_A, and at some point the diffusion activation energy becomes controlling. The required activation energy for diffusion, in turn, increases with decreasing temperature, since it becomes more difficult to move atoms from one place to another. Eventually the energy requirement for diffusion becomes so great that nucleation stops. This process is shown schematically in Fig. 8-5.

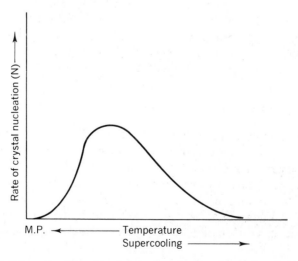

Fig. 8-5. Schematic effect of temperature on the rate of crystal nucleation (N).

8-4-4   SOLID-SOLID TRANSFORMATIONS

Nucleation in the solid state, as in allotropic transformations, precipitation, recrystallization, etc., follows substantially the theories for gas-liquid and liquid-solid transformations. There are additional features, such as the consideration of strain energy, the orientation relationships between the crystal structures of the matrix and the new phase, and the effect of grain boundaries.

If the new phase forms at relatively high temperatures, i.e., with little supercooling, the transformation is similar to formation of a solid from a liquid. The interfacial energy is high, and nucleation is heterogeneous with grain boundaries usually serving as nucleation catalysts. If the matrix phase is supercooled far below the equilibrium temperature at which the second phase becomes stable, platelike particles may form under conditions requiring a relatively small amount of surface energy. The new phase normally has a volume differing from the matrix, and thus solid-solid phase changes tend to have an increase in strain energy. At high temperatures, the strain energy is generally low since the matrix can yield. As temperature decreases, however, strain energy can become a major factor. The amount of strain energy developed by a given volume of second phase depends on the shape of the particles formed. Spherical particles of a specific volume have a minimum surface energy but produce a maximum strain energy. Particles in the form of disks (platelets) minimize the strain energy but increase the total surface area.

If we include the effects of all the additional features of the solid-solid transformations under the heading of strain energy, we can express the rate of nucleation in a form similar to Eq. 8-14 or Eq. 8-20, i.e.,

$$N = A e^{-[\Delta F^* + \Delta F_A + \Delta F_E]/RT} \tag{8-21}$$

where ΔF_E is the strain energy term. Since this strain energy term is positive, much like the surface energy term, we could include it within ΔF^*, if we so desired. Experimental proofs of this equation are lacking because of difficulties in making observations. In general, however, the rate of nucleation appears to follow the pattern of Fig. 8-5.

8-4-5   GROWTH

Once nucleation has occurred, whether in recrystallization, neocrystallization, phase precipitation, or some other process, we find that the new crystals normally increase in size at the expense of the matrix (often with concurrent nucleation) until, in many cases, the entire matrix has been consumed by the new phase. In addition, when a polycrystalline solid is heated, larger grains normally increase in size at the expense of smaller ones. The rate at which a characteristic linear dimension of the crystal changes with time is the growth rate, G.

Growth rate can be determined experimentally, but it is relatively difficult to make accurate predictions of growth rate. In recrystallization, growth occurs because of a decrease in strain energy or a difference in surface energies of neighboring grains. In neocrystallization (allotropic change), it may occur because of a difference in free energy of formation of the phases involved. In precipitation, growth occurs thanks to a release of chemical energy. In any case, grains grow because their growth tends to minimize the free energy of the system. All crystals can grow, because any polycrystalline material would achieve minimum free energy if it existed as a single crystal of minimum surface area.

Available evidence indicates that a curved boundary migrates toward its center of curvature. The atoms on the concave side of the boundary are more stable than those on the convex side since they are more completely surrounded by atoms. The radius of curvature is inversely proportional to grain size, and the rate of growth would be greater for greater curvatures. This has been confirmed by observation in many cases. It follows that:

$$G = \frac{dD}{dt} = \frac{k}{D} \qquad (8\text{-}22)$$

where D is the diameter, t is time, and k is a proportionality "constant" which is a function of many variables, including temperature and boundary-surface energy. Integration of Eq. 8-22 and evaluation of the integration constant at $t = 0$, gives:

$$D^2 - D_0^2 = kt \qquad (8\text{-}23)$$

If D_0 is small in comparison with D, or if grain growth follows an allotropic change,

$$D = (kt)^{1/2} \qquad (8\text{-}24)$$

In practice, however, the exponent is only rarely 0.5 but is rather 0.1–0.3 for most materials. Grain growth studies during recrystallization generally indicate a constant growth rate. Once recrystallization is complete, or following a phase transformation, the growth rate normally decreases with time.

Grain growth involves movement or rearrangement of atoms, and this movement is a function of temperature. By analogy to other rate processes, growth rate is normally related to temperature through the Arrhenius equation,

$$G = G_0 e^{-Q/RT} \qquad (8\text{-}25)$$

Although the equation is valid in many cases, its validity appears to be primarily fortuitous. Attempts to relate growth activation energy to diffusion activation energy have not been noticeably successful, since the activation energy required for grain boundary movement is generally appreciably higher than the activation energy for self-diffusion. This has been explained in some cases by concentrations of impurities at the grain boundaries where the impurities act as anchors and restrict growth.

8-5   DIFFUSION

Many reactions in materials, e.g., nucleation and grain growth, phase changes, precipitation, etc., involve redistribution of the atoms present. Reaction rates are necessarily controlled by rates of migration of participating atoms. This migration or movement is known as diffusion. Movements in diffusion may be relatively short-range, as in allotropy, recrystallization, and in precipitation.. Diffusion is fundamental to phase changes and is important in heat treatments. It is also basic to many processes, such as case hardening of steel, production of strong bodies by sintering powders, homogenization of castings etc. Diffusion is basic to many reactions and processes in metals, ceramics, and polymers. It is *not* limited to any one of these.

Diffusion is basically statistical in nature, and the term applies to macroscopic flow (not individual movements) resulting from innumerable random movements of individual atoms. The path of an individual atom is random, zigzag, and unpredictable. Nonetheless, when large numbers of atoms make such movements, they can produce a systematic flow.

The relationship between individual atomic movements and diffusion is illustrated in Fig. 8-6 for movement of radioactive atoms among normal atoms of the same kind (self-diffusion). Fig. 8-6(a) has a central region containing a uniform concentration, c_1, of radioactive atoms and two adjoining regions initially containing no radioactive atoms. By mechanisms to be discussed later (and known to be essentially the same for both normal and radioactive atoms), each atom will tend to move (or jump) from its position to a neighboring position. After an average of one jump per atom, the distribution of atoms might be that shown in Fig. 8-6(b). As random movements of individual atoms continue, the atoms will continue to migrate until a uniform distribution [Fig. 8-6(c)] is produced. The atoms continue to move randomly, as before, but the concentration is constant, and we can no longer observe diffusion.

8-5-1   VOLUMETRIC DIFFUSION

The above implies that the condition necessary for diffusion to take place between two points is existence of a difference in concentration between the points. We assume:

1. Concentration c (atoms/cm³) varies only in the x-direction (Fig. 8-7).
2. Concentration gradient $(\delta c/\delta x)$ is small enough for the difference between neighboring atomic planes to be considered infinitesimal.
3. Each atom migrates randomly by a series of jumps from site to site.

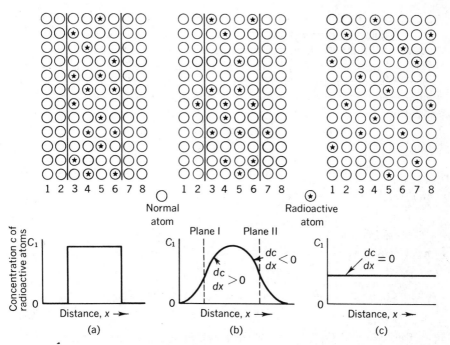

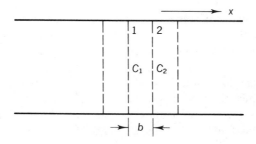

Fig. 8-6. The process of self-diffusion: Atomic distributions (above) and the corresponding concentration curves (below) for three stages in the diffusion process; (a) Condition before diffusion; (b) condition after brief diffusion (one jump per atom); (c) uniform condition produced by prolonged diffusion.

Fig. 8-7. Atomic planes normal to concentration gradiant.

4. The average frequency, f, of jumps by a single atom is constant for constant composition.

5. The jump length, b, is constant.

Consider a sequence of atomic planes, normal to the concentration gradient, spaced a jump distance (b) apart (Fig. 8-7). If c_1 and c_2 ($c_1 < c_2$)

are the concentrations in planes 1 and 2, respectively, the numbers of atoms in the planes are $n_1 = c_1 b$ and $n_2 = c_2 b$. In a small time increment (dt), the number of atoms which jump from plane 1 is $n_1 f\, dt$. On the average, however, only half these atoms jump from plane 1 to plane 2, i.e., $\frac{1}{2}(n_1 f\, dt)$. Likewise, the number of atoms in plane 2 that jump to plane 1 is $\frac{1}{2}(n_2 f\, dt)$. Thus the net number (n_d) of atoms which pass through a plane of unit area in time dt is given by:

$$n_d = \tfrac{1}{2}(n_1 - n_2)f\, dt = \tfrac{1}{2}b(c_1 - c_2)f\, dt$$

But

$$c_1 - c_2 = -b\frac{\delta c}{\delta x}$$

and therefore

$$\left. \begin{aligned} n_d &= \tfrac{1}{2}b^2 f \frac{\delta c}{\delta x}\, dt \\[2mm] n_d &= -D\frac{\delta c}{\delta x}\, dt \end{aligned} \right\} \tag{8-26}$$

or

where D, the diffusion coefficient, is measured in units of area/time and is:

$$D = \tfrac{1}{2}b^2 f \tag{8-27}$$

Derivation of Eq. 8-26, known as Fick's First Law, indicates that the rate of flow is proportional to concentration gradient and is directed down the gradient. It is equally possible to derive the equation in terms of grams of solute rather than number of atoms. In this case, Fick's first law becomes:

$$dM = -D\frac{\delta c}{\delta x}\, dt; \quad \text{i.e.,} \quad \frac{dM}{dt} = -D\frac{\delta c}{\delta x} \tag{8-26a}$$

where dM is the grams of solute that pass through one cm^2 of a plane normal to the direction of diffusion, D is the diffusion coefficient (cm^2/sec), $\delta c/\delta x$ is the concentration gradient at the plane (c in grams/cm^3) and t is time in seconds.

Fick's first law has limited application since it assumes a steady state of diffusion not often obtained in solids. A typical steady-state example is diffusion of a gas through a permeable diaphragm. If a constant pressure differential is maintained on opposite sides of the diaphragm, a steady state is reached, and dM/dt can be measured directly. An approximation to this is found in one type of oxidation of metal in which diffusion of only the metal atoms need be considered. Since the difference, Δc, between the metal concentration, c_0, at the oxide surface (Fig. 8-8) and the concentration, c_m, at the metal surface is constant, the rate of transfer of metal atoms per unit area through any oxide thickness, x, is:

$$dM = -D\frac{\Delta c}{x}\, dt$$

At the same time, dM is directly proportional to the change in thickness, dx, i.e., $dM = K' \, dx$. Upon rewriting,

$$dx = \frac{-D}{K'} \frac{\Delta c}{x} \, dt$$

Integration of this equation gives:

$$x^2 = Kt \qquad\qquad (8\text{-}28)$$

which indicates that the thickness of the oxide layer increases as the square root of the period of oxidation as shown in Fig. 8-8(b).

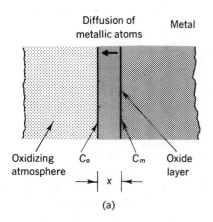

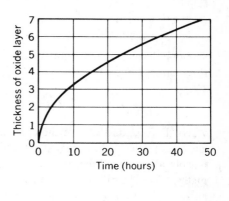

(a)   (b)

Fig. 8-8. The oxidation of metals, a diffusion process: (a) The process of oxidation of a metal; (b) a typical parabolic relation btween the thickness of the oxide layer and time of oxidation.

In general, concentration changes with time. Returning to Fig. 8-7, we assume the two planes are not neighbors but are a distance h apart (greater than b but small enough to regard concentration difference along x as infinitesimal). The concentration at plane 1 is c and at plane 2 is $c + h \, (\delta c / \delta x)$.

From Eq. 8-26a, flow across plane 1 is:

$$\frac{dM}{dt} = -D \frac{\delta c}{\delta x}$$

and flow across plane 2 is:

$$\frac{dM}{dt} + h \frac{\delta \left(\dfrac{dM}{dt} \right)}{\delta x} = -D \frac{\delta c}{\delta x} - h \frac{\delta \left[D \dfrac{\delta c}{\delta x} \right]}{\delta x}$$

The accumulation rate in the element is the difference, i.e.,

$$-h \frac{\delta \left(\dfrac{dM}{dt} \right)}{\delta x} = h \frac{\delta \left[D \dfrac{\delta c}{\delta x} \right]}{\delta x}$$

Since no atoms are gained or lost during the process,

$$\frac{\delta\left(\frac{dM}{dt}\right)}{\delta x} = \frac{-\delta c}{\delta t}$$

Equating, we obtain Fick's second law:

$$\frac{\delta c}{\delta t} = \frac{\delta}{\delta x}\left(D\frac{\delta c}{\delta x}\right) \tag{8-29}$$

For the special case in which D is constant this becomes:

$$\frac{\delta c}{\delta t} = D\frac{\delta^2 c}{\delta x^2} \tag{8-29a}$$

The one dimensional formulae may be generalized to Fick's second law in three dimensions:

$$\frac{\delta c}{\delta t} = \frac{\delta}{\delta x}\left(D_x\frac{\delta c}{\delta x}\right) + \frac{\delta}{\delta y}\left(D_y\frac{\delta c}{\delta y}\right) + \frac{\delta}{\delta z}\left(D_z\frac{\delta c}{\delta z}\right) \tag{8-30}$$

where D_x, D_y, and D_z are the coefficients of diffusion along the x, y, and z axes, respectively. In symmetrical crystals such as cubes, the diffusion coefficients along the three axes are generally equal. In crystals of lower symmetry, however, diffusion can be anisotropic. Tin and zinc, for example, have anisotropic diffusion. Bismuth (rhombohedral structure), near its melting point, has a rate of diffusion perpendicular to the rhombohedral axis approximately 10^6 times greater than the rate parallel to the axis.

Many solutions (mathematical or graphical) of Eq. 8-29 assume concentration is a function of x and t, i.e., $c = f(x/\sqrt{t})$. This is justified only on an experimental basis.

The dimensionless quantity, $x/\sqrt{Dt}$, has considerable practical importance. For example, a given point will have a concentration halfway between the initial one and the final one when $x/\sqrt{Dt} = 0.95$, or when t is approximately x^2/D. We can use this information to make a rapid but approximate calculation to determine if an appreciable change in composition has occurred by diffusion under a given set of conditions.

8-5-2   EFFECT OF CHEMICAL POTENTIAL
ON DIFFUSION

Both of Fick's laws are used rather widely for diffusion work, especially in metals, but do not explain diffusion which increases rather than decreases concentration. Actually, *atoms will migrate preferentially in directions which cause lowering of the free energy*, i.e., directions which cause the entire system to approach equilibrium. Thus the essential condition for diffusion to occur is a difference in chemical potential. Rate of flow is assumed proportional to the gradient of chemical potential. Equilibrium is reached when the chemical potential of each species present is the same throughout the system. An example of "uphill diffusion," with an accompanying decrease in internal energy, occurs in a homogeneous solution in which solute atoms concentrate to form nuclei of a precipitating second phase. Diffusion is just as necessary for this case as for homogenization of castings. Thus, Eq. 8-26 is better written as:

$$n_d = -L \frac{\delta \mu}{\delta x} dt \qquad (8\text{-}31)$$

where L is a coefficient having significance similar to D and μ is the chemical potential. We find that

$$D = \frac{L}{c} \frac{\delta \mu}{\delta \log c} \qquad (8\text{-}32)$$

8-5-3   EFFECT OF TEMPERATURE ON DIFFUSION

Experience tells us that diffusion speeds up with increasing temperature, and it is common practice to express the relationship by the Arrhenius equation,

$$D = D_0 e^{-\Delta H/RT} \qquad (8\text{-}33)$$

where D_0 is the frequency factor, and ΔH, the enthalpy change, is the effective activation energy. Both are considered constant over the range of measurement (for a given composition). Use of this equation is within experimental error in most sets of data but probably should be considered only as a good approximation.

Earlier (Eq. 8-27), we found $D = \frac{1}{2}(b^2 f)$ in which f is the frequency of "jumping" along the x-axis. This frequency is the product of the number of vibrations per unit time (in the correct direction) and the probability that the atom can surmount the energy barrier, i.e.,

$$f \simeq \frac{1}{3} v e^{-\Delta H/RT}$$

where v is the vibration frequency. (The factor of $\frac{1}{3}$ is required because there is an equal probability of the atom jumping in the y- and z-directions as well as in the x-direction.) The diffusion coefficient thus becomes:

$$D = \frac{1}{6} b^2 v e^{-\Delta H/RT}$$

and the frequency factor would appear to be:

$$D_0 = \frac{b^2 v}{6}$$

Measurements of D_0 are found to vary widely (by orders of magnitude in many cases) from calculated values. Much of the difference comes from the fact that entropy in the system has to increase while the atom is jumping, since the surrounding atoms are temporarily under severe elastic strain. Thus a better value for D_0 is:

$$D_0 = \frac{b^2 v}{6} e^{\Delta S/R} \tag{8-34}$$

where ΔS is the entropy of activation and can be estimated from:

$$\Delta S \simeq \beta \left(\frac{\Delta H}{T_m} \right) \tag{8-35}$$

where T_m is the melting point (absolute temperature), and β is a parameter which is a function of the elastic constants, and varies between 0.25 and 0.45 for most metals. These modifications give predicted values which are in good agreement with experimental values.

Approximate values of D_0 and ΔH for a number of binary systems are given in Table 8-1.

Table 8-1. Approximate values of D_0 and ΔH for typical systems

Diffusing component	Matrix metal	D_0 (cm²/sec)	ΔH (Kcal/gram-mol)
carbon	γ-iron	0.21	33.8
carbon	α-iron	0.0079	18.1
nitrogen	α-iron	0.007	18
iron	α-iron	5.8	59.7
iron	γ-iron	0.58	67.9
nickel	γ-iron	0.5	66.0
manganese	γ-iron	0.35	67.5
copper	aluminum	2.0	33.9
zinc	copper	0.033	38.0
silver	silver (volume diffusion)	0.72	45.0
silver	silver (grain boundary diffusion)	0.14	21.5

8-5-4 EFFECT OF CONCENTRATION ON DIFFUSION

We rarely find symmetrical concentration-depth curves among the alloys. Even in as simple a system as Cu-Ni, the curve deviates from the ideal since diffusion occurs more readily in a Cu-rich solution than in a Ni-rich one. Similar effects are common in concentrated solid solutions and lead to the conclusion that the diffusion coefficient (better termed an interdiffusion coefficient) is not constant, even at constant temperatures, but varies with alloy composition. Thus we must use the general form of Fick's second law

(Eq. 8-29). Techniques for determining D as a function of composition exist and have been applied to many primary solid solutions. We find that D may increase by a factor of 10 as concentration increases from zero to the solubility limit.

This dependence on concentration rests on two factors:

1. Cohesive forces generally vary with composition. In the Cu-Ni system, interatomic binding in Ni is greater than in Cu. It is reasonable to expect the activation energy for diffusion to be at least roughly proportional to the binding energy. Consequently, the diffusion coefficient at constant temperature should decrease as Ni content increases.
2. Migration of atoms in a nonideal solution is not completely random, but takes place in a direction which lowers the free energy.

8-5-5 KIRKENDALL EFFECT

A second, independent phenomenon in alloys is often confused with the dependence of the interdiffusion coefficient on composition. This is the fact that the intrinsic diffusion coefficients for the two components in a given binary alloy may have different values, and thus one migrates faster than the other with no relationship to the concentration. As a result, when two metals with different diffusion coefficients interdiffuse, there is a net transport of material across the plane of the original interface between the two specimens. This is known as the Kirkendall effect.

The effect is revealed by an experiment shown schematically in Fig. 8-9 for the Au-Ni system. A specimen is made by joining Au and Ni bars so that diffusion will take place across an interface which has inert markers (very fine W or Mo wires). After many hours at a suitably elevated temperature (e.g., 900°C), there is interdiffusion of Au and Ni and a change of concentration distribution as shown. D_{Au} is greater than D_{Ni}, which means more Au atoms than Ni atoms diffuse past the inert markers. These markers in turn will move toward the Au end of the specimen. This movement is known as the Kirkendall shift.

The marker shift increases with temperature for a constant annealing time and increases with increasing difference in composition between the two halves of the specimen. In all cases, marker shift is toward the side of the specimen with the lower melting point, i.e., toward Cu in Cu-Ni specimens but toward Au in Au-Cu specimens.

The plane at the position of the original interface (the Matano interface) has the property that the two areas A_1 and A_2 are equal (Fig. 8-9), indicating that equal amounts of Ni and Au, respectively, have crossed this interface. If the Matano interface is used to describe diffusion behavior, only the overall (inter)diffusion coefficient D is needed. This is related to the intrinsic dif-

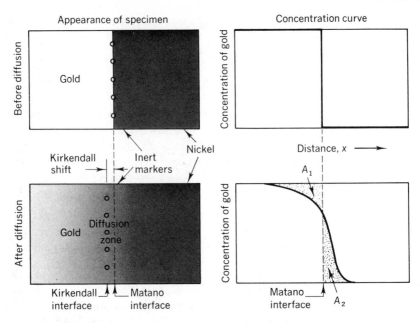

Fig. 8-9. The Kirkendall effect during interdiffusion of gold and nickel.

fusion coefficients through the equation:

$$D = X_{Ni}D_{Au} + X_{Au}D_{Ni} \tag{8-36}$$

where X indicates the atomic fraction of the element.

8-5-6   MECHANISM OF DIFFUSION

In interstitial solutions, such as carbon in γ-iron, there is limited solubility, and a large number of interstices remain unfilled. We believe diffusion consists simply of a solute atom jumping from one interstitial site to a neighboring one. Experimental evidence indicates that this belief is reasonable.

For diffusion in substitutional solutions, however, the nature of the process is not as obvious. The variety of ideas proposed for this process has been gradually reduced to three main theories based on:

1. Migration of vacancies.
2. Migration of interstitialcies.
3. Place exchange.

In our discussion of crystal structure (Chapter 4), we found that a crystal contains an appreciable number of defects under equilibrium conditions. Vacancies [Schottky defects, Fig. 4-38(b)] can (and do) wander continually through the crystal by interchanging positions with individual atoms. Each interchange displaces an atom from one site to the next. Enough of these, over a period of time, produces concentration changes which we observe as diffusion. Interstitialcies [Frenkel defects, Fig. 4-38(a)], wandering through the crystal, can produce diffusion in much the same way. The place exchange mechanisms are different since they involve direct interchange of two or more neighboring atoms in their sites. Figure 8-10 illustrates schematic examples of such mechanisms.

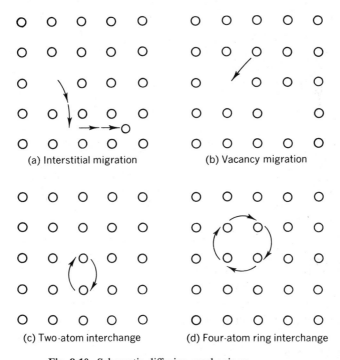

(a) Interstitial migration

(b) Vacancy migration

(c) Two-atom interchange

(d) Four-atom ring interchange

Fig. 8-10. Schematic diffusion mechanisms.

With three theories, we ask what indications are given by experimental evidence. Calculated values for activation energy for Cu, for example, are (in Kcal/mol) 64 for vacancy migration, 230 for interstitialcy migration, 240 for the two-atom interchange, and 90 for the four-atom ring. The experimental value of 50 favors the vacancy mechanism. A second consideration is the frequency factor, D_0. Calculations, based on statistical procedures, have been made for the various processes with the conclusion that observations are consistent with vacancy migration in FCC metals. Correlation of D_0 and ΔH is possible in BCC metals, however, only on the basis of the four-atom ring. The third consideration is the Kirkendall effect which has been generally established in FCC alloys. It clearly cannot be produced by place exchange mechanisms. In vacancy migration or interstitialcy migration, however, it is possible for atomic defects to exchange places more often with atoms of one kind than with the other. Evidence in favor of the vacancy or interstitialcy mechanisms is given by other experiments which show that the rate of diffusion is greater in a quenched alloy than in a slowly cooled one. This phenomenon is due to a greater number of vacancies, since some of those existing under equilibrium at elevated temperature will be retained upon quenching.

The four diffusion mechanisms shown in Fig. 8-10 are considered possible for ceramics as well as for metals. Direct exchange [Fig. 8-10(c)] is not probable, however, because of high strain energy developed when two atoms squeeze past each other, especially in ionic solids which would require interchange of anions and cations. The ring interchange, Fig. 8-10(d), is possible but has not been demonstrated in any real system. The vacancy mechanism, Fig. 8-10(b), is the most common mechanism of diffusion in ceramics.

In polymers, diffusion of very small molecules (diffusate) takes place as a result of random Brownian movements of the diffusate. Molecules which are about the same size as, or larger than, the chain segments which contribute most to heat motion in polymers diffuse by more complicated mechanisms which depend on motions of both polymer and diffusate molecules.

If organic vapors are the diffusate, there are some features of diffusion unique to polymers. The molar volumes of the penetrants are relatively large and cause swelling of the polymer leading to three considerations: (1) The plane of reference for diffusion must be a plane across which there is no net flow of mass. (2) The penetrant acts as a plasticizer, lowering the glass transition temperature, and loosening segmental motions at any given temperature. Since diffusion depends on frequency of molecular jumps, the diffusion coefficient will increase as the concentration of penetrant increases. (3) As the polymer swells, chain segments must assume a new average distribution in space which involves alternate stressing and relaxing of contorted chains as segments move through the entangled mass. The rate of dimensional changes may sometimes be too great for the most probable distribution of segments to be maintained during swelling. Therefore, the diffusion coefficient may vary with time while the chains are relaxing.

8-5-7 GRAIN BOUNDARY AND SURFACE DIFFUSION

The above discussion was in terms of volume (or lattice) diffusion, i.e., migrating atoms diffusing through the body of the crystal. Many times, however, the observation has been made that D_0 is much smaller (sometimes by orders of magnitude) than the predicted value. This small D_0 is always accompanied by a much lower value of ΔH than expected. These anomalously low values indicate that diffusion occurs along a number of "short circuit" paths. Study has established that atoms can diffuse along grain boundaries and surfaces as well as through crystal lattices.

There is much quantitative data available on lattice diffusion but only a limited amount on grain boundary and surface diffusion. Qualitative evidence, however, indicates that the grain boundary frequency factor is much greater than the lattice frequency factor for penetration of solute atoms into a solvent metal. This is not surprising, since grain boundaries are regions of marked crystal imperfection. Various experiments have also qualitatively demonstrated that atoms have exceptionally high mobility in free surface layers. This is reasonable since these atoms are less tightly bound to a given position than an atom within the lattice. This is also consistent with the fact that diffusion in a noncrystalline glass with a random structure is more rapid than diffusion in a crystalline glass.

There is evidence, in some systems, that diffusion is easiest along surfaces, intermediate along grain boundaries and most difficult through the volume. Activation energies for surface and grain boundary diffusion are less than for lattice diffusion. Thus surface and grain boundary diffusion tend to be important at relatively low temperatures while volume diffusion is more important at higher temperatures. Grain boundary diffusion is probably more important in ceramics than in metals.

A polymer sometimes behaves as more than one phase. Diffusion may also take place through an interconnecting capillary or crack system. A partially crystalline polymer has almost impermeable crystallites mixed with relatively permeable amorphous regions. The interfaces between these may also act much like a network of cracks.

8-5-8 DIFFUSION COUPLES

In the Au-Ni diffusion couple (Fig. 8-9), we find a nonsymmetrical concentration distribution and a Kirkendall shift. Upon examination, we would find a solid solution in the region of the interface shaded area, Fig. 8-9. This solution would range from pure Au to pure Ni, would extend over an appreciable distance in the specimen, and would provide a continuous bond between the two pure metals.

It is possible to form a continuous bond by diffusion between two metals under a wide variety of conditions, but in every case some diffusion must occur. Since diffusion occurs only in solid solutions, some solid solubility is required for diffusion bonding. This requirement can be desirable or undesirable. Consider Pb, which is essentially insoluble in Fe, thus making it possible to heat iron or steel in molten Pb to provide protection from the atmosphere. On the other hand, if we wish to coat Fe with an adherent layer of Pb, we must add a small quantity of a third element which alloys with Fe, e.g., Sn, to obtain adequate bonding.

A complicating factor arises when the two metals being joined form intermediate phases. Two primary solid solutions are formed, but the intermediate phases which exist at the diffusion temperature are also formed. This is illustrated by the Cu-Zn system (Fig. 7-24), in which the intermediate phases are the β, γ and ϵ "electron compounds." Examination of the diffusion couple after some time at temperature shows not only the two pieces with primary solutions merging into the original material but also shows three distinct bands or zones between the primary solutions. These bands occur in the vicinity of the original interface, and each zone is one of the intermediate phases. This is a concentration gradient across each zone with an apparent interface between adjoining zones. Each of the diffusion zones is a single phase region of varying composition. The two-phase regions in the phase diagram do not appear as regions in the diffusion couple but constitute the interface between adjoining one phase regions.

If brittle intermediate phases are formed, the bond between the two materials will be brittle. In addition, many intermediate phases have specific volumes which differ from those of the components. Extended diffusion, with resultant accumulation of these phases, imposes large strains on the materials being bonded.

This problem occurred in the fuel element in the early X-10 nuclear reactor. Metallic uranium was sheathed in an aluminum alloy tube. Under operating conditions, there was a diffusion couple between the two metals with formation of three intermetallic compounds at the interface between U and Al. This resulted in rupture of the Al tube and release of fission products into the coolant stream, a condition which cannot be tolerated. The problem was solved by placing a diffusion barrier (an Al-Si alloy) between the U and Al.

EXAMPLE 8-2

If the time required to reach maximum hardness in a precipitation hardening alloy (see Sec. 9-5-1) of 4% copper in aluminum is 10 hours at 150°C, how long will it take to reach maximum hardness at 100°C?

Solution:

This involves diffusion with x^2/Dt, a constant. Since maximum hardness is the same at both temperatures, x is constant. Therefore,

$$(Dt)_{150°C} = (Dt)_{100°C}$$

Using data from Table 8-1,

$$\frac{t_{100°C}}{t_{150°C}} = \frac{D_{150°C}}{D_{100°C}} = \frac{2e^{-33,900/2 \times 423}}{2e^{-33,900/2 \times 373}} = e^{+5.43}$$

$$t_{100°C} = (10)(230) = 2300 \text{ hours}$$

QUESTIONS AND PROBLEMS

1. If the activation energy for a given process is 50 Kcal/mol, how much must the temperature be raised above 0°C for the reaction rate to be increased by a factor of three?

2. If we assume that a given reaction occurs in 1 second at 1000°K with an activation energy of 40 Kcal/mol, how long will it take for the reaction to occur at 300°K?

3. Verify Eqs. 8-7 and 8-8.

4. A reaction $A \longrightarrow B$ is investigated as a function of temperature. The data are:

Time to initiate	Temperature
77 min, 50 sec	327°C
13.8 sec	427°C
0.316 sec	527°C
1 millisec	627°C

What is the activation energy?

5. Indicate the procedures that should be used (by controlling solidification) to produce the following structures in a casting:
   (a) coarse grain throughout.
   (b) fine grain throughout.
   (c) fine grain on the surface with a coarse grain in the interior.
   (d) coarse grain on the surface with a fine grain in the interior.

6. Consider pure iron heated to 2000°F. At 1670°F, the alpha iron transforms to gamma iron (neocrystallization) and the gamma grains begin to grow. The data given show the relationship between grain size and time. Determine the isothermal growth rate equation for this case.

Time (hrs)	Diameter (mm)
0	0
20	.16
40	.18
80	.205
120	.215
200	.24

7. Many liquids may be supercooled; that is, under special condictions, freezing may be avoided at temperatures well below the equilibrium freezing point. Explain such behavior and indicate the conditions and procedures which promote supercooling.

8. (a) A bottle of a soft drink is placed in a freezer and left for a day or two. What happens? Why? (Is there a difference in your answer if the bottle is returnable or nonreturnable?) (b) An identical bottle is placed in the freezer and left for an hour, removed from the freezer and opened. What happens? Why?

9. What is diffusion? What part does it play in metallurgical reactions? Is it microscopic or macroscopic in nature? What role is played by statistics in diffusion? What is the distinction between diffusion and self-diffusion?

10. What is the difference between Fick's first and second laws? What is meant by a diffusion coefficient? Is it constant with concentration? Is it isotropic? What effect do the two previous answers have on the application of Fick's laws?

11. If the value of D for the diffusion of carbon in α-iron is 7.2×10^{-7} cm^2/sec at 700°C and 12.5×10^{-7} cm^2/sec at 761°C, what is the value of the activation energy in Kcal/gram-atom?

12. The diffusion rates of carbon in α-iron (ferrite) and γ-iron (austenite) are given by

$$\text{Rate } \alpha = 0.0079 \text{ cm}^2/\text{sec } e^{-18,100/RT}$$

$$\text{Rate } \gamma = 0.21 \quad \text{cm}^2/\text{sec } e^{-33,800/RT}$$

(a) Calculate the two diffusion coefficients at 800°C and 1000°C.

(b) Explain the relative magnitudes of the two diffusion rates in terms of structure.

13. If the vibration frequency in aluminum is 8×10^{12}/sec and the activation energy for vacancy motion is 14 Kcal/mole, determine the jump frequency (in all directions) (a) at room temperature, (b) at the melting point (660°C).

14. Pure iron is to be nitrided to produce a concentration of 0.10 a/o at 0.05 cm below the surface. Gaseous diffusion will be performed at 800°C with a N concentration at the surface of 0.5 a/o. How long will be required to obtain the desired concentration? The solution of Fick's second law (Eq. 8-29) for this special situation is

$$\frac{C_s - C}{C_s - C_o} = \text{erf}\left(\frac{x}{2\sqrt{Dt}}\right)$$

where C_s = surface concentration,
$\quad\quad C_o$ = initial concentration in matrix,
$\quad\quad C$ = concentration at depth x after time t,
$\quad\quad$ erf = error function found in mathematical tables.

15. What is the "driving force" for self-diffusion?

16. Is it possible for diffusion to take place *up* the concentration gradient? If so, how can this be explained? If so, of what significance is it?

17. What is the Kirkendall effect? What is its significance? How can it be explained?

18. What are the mechanisms which are considered the major possible explanations for diffusion? What are the arguments and experimental evidence for each? What appears to you to be the most probable mechanism?

19. Discuss the conditions, if any, under which a metastable phase could be found in a binary diffusion couple.

20. Constructively criticize this statement: "The diffusion of atoms along the grain boundaries and surface of a piece of metal must be slow, since these types of diffusion are neglected in most commercial alloys."

21. Carburization of steels parts at 1600°F is preferred rather than at 1700°F because finer grain size can be obtained.
    (a) Calculate the diffusion coefficient of carbon in gamma iron at 1600°F using the data in Table 8-1.
    (b) What time will be required at 1600°F to give the same results as 10 hours at 1700°F? (Neglect any changes in solubility of carbon in iron.)

22. Would cold working a cast alloy prior to an elevated temperature soaking increase or decrease the rate of homogenization of alloy content? Why? Compare the relative effects of cold working and hot working in this situation.

23. The diffusion coefficient of a metal just below the melting point is 10^{-7} cm²/sec. If $D_o = 1$ cm²/sec, determine the diffusion coefficient at one half the melting point.

24. Verify Eq. 8-28.

25. Given the following data:

Oxide thickness, A	Time, minutes
0	0
500	10
1000	40
2000	160

Test for "fit" to the linear, parabolic, or cubic rate laws.

9

NONEQUILIBRIUM
RELATIONSHIPS IN ALLOYS

9-1 INTRODUCTION

Chapter 7 applies solely to equilibrium conditions. Once equilibrium is reached, a system does not change in any manner unless acted on by some external force. Phase diagrams indicate phase relationships which exist only when cooling has been slow enough to assure equilibrium or when equilibrium has been established by holding the material long enough at some temperature. Either method has the potential difficulty that the temperature at which equilibrium is desired will be too low to permit sufficient diffusion to establish equilibrium within a reasonable time. In any system, the ease with which equilibrium is established depends on the system and the temperature range under consideration, as we found in Chapter 8. We expect (and find) faster diffusion rates and greater ease in establishing equilibrium at higher temperatures, but even an apparently high temperature is not always sufficient. For instance, in the Fe-Mn and Fe-Ni systems, extremely slow cooling and heating over several hundred hours fail to place the temperature range of the $\alpha \rightleftharpoons \gamma$ transition at the same temperature for both cooling and heating.

9-2 HEAT TREATMENT

Heating and cooling rates commonly encountered in practice are highly conducive to producing nonequilibrium relationships. These nonequilibrium conditions may, or may not, be produced intentionally, and the results may, or may not, be desirable. If done deliberately, we have *heat treatment*, i.e., *a combination of heating and cooling operations timed and applied in sequence to produce desired properties.*

9-3   COLD-WORK-ANNEAL CYCLE

We found (Chapter 6) that as cold deformation continues (Fig. 9-1), the resistance to further deformation eventually becomes equal to the resistance to fracture, and fracture occurs upon further loading. Even if deformation is stopped prior to fracture, it is possible that the desired final shape has not been obtained, e.g., a drawn wire may be larger than the desired final size. To continue working and obtain the required dimensions and/or shape, we must restore the metal to a condition approximating that prior to deformation.

Fortunately we can regain the original condition, with little change in shape, by a relatively simple procedure. The strain-hardened condition of a metal is metastable. If temperature is increased, the system more closely approaches equilibrium, and the initial properties of the metal are gradually recovered. This process of heating is known as *annealing*. This term, however, is used in so many different ways that it should have a specific adjective. In this case, we call it *process annealing*, one important type of heat treatment.

The effect of process annealing on the properties of cold-worked metal can be divided into three general stages: (1) recovery or stress relief; (2) recrystallization; and (3) grain growth. These stages and their general effects on properties and microstructure are shown schematically in Fig. 9-1.

Recovery (or stress relief) is a low temperature phenomenon which removes most internal stresses with little change in mechanical properties and no discernible change in microstructure. Although slight increases may occur in tensile and yield strengths, the major property change is a noticeable increase in electrical conductivity. Stress relief is commercially important. For example, removal of internal stress will reduce susceptibility to corrosion without loss of increased mechanical strength due to cold working.

Recrystallization is a process by which distorted grains of cold-worked metal are replaced by new, strain-free grains during heating above a specific minimum temperature. This minimum temperature is known as the recrystallization temperature and is defined as the temperature at which the first tiny new grain appears. This temperature is a function of the particular metal, its purity, the amount of prior deformation, and the annealing time.

Recrystallization (formation of new grains from old) has a marked effect on properties and microstructure. It is recognized by: (1) a relatively abrupt change in mechanical properties; (2) disappearance of distorted and elongated cold-worked grains; (3) disappearance of slip bands; (4) further relief of internal stress; and (5) grain refinement (usually). There is no change in crystal structure during recrystallization, i.e., FCC metal remains FCC. Recrystallization should not be confused with the formation of new grains in allotropic phase transformation (neocrystallization, Sec. 9-5-5), in which

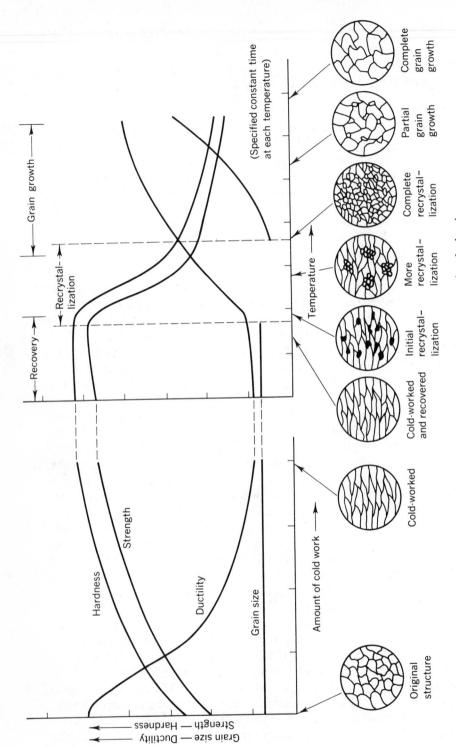

Fig. 9-1. Schematic representation of the cold-work-anneal cycle showing the effects on properties and microstructure.

there is a change of crystal structure. Following recrystallization, properties are approximately the same as those before cold deformation.

Grain Growth is an increase in grain size. When the material is held for longer times at a given temperature (above recrystallization temperature), or when it is heated to a higher temperature, the grain size increases and there is an accompanying secondary loss in strength and a gain in ductility. Small grains become still smaller and ultimately disappear upon absorption by the larger grains. Final grain size is more dependent on temperature than on time since grain growth normally continues (time permitting) to a grain size characteristic of the temperature. If time is too short to obtain this characteristic size, grain size depends on the time of heating.

9-3-1 RECRYSTALLIZATION TEMPERATURE

We define recrystallization temperature as the temperature at which the first tiny new grain appears during process annealing of cold-worked metal. Since this temperature is influenced by amount of prior deformation and annealing time, conditions must be specified.

Recrystallization temperature is sometimes helpful in evaluating high-temperature behavior of a metal, since metals cannot be stressed in service for long periods at temperatures approaching the recrystallization temperature. A very crude approximation is that recrystallization temperature is about 40% of the melting point (in degrees absolute). At best, this is an approximation and may be seriously in error for metals which have polymorphic transformations. Typical examples of recrystallization temperatures are 150°C for aluminum (m.p., 660°C), 450°C for iron (m.p., 1539°C) and 900°C for molybdenum (m.p., 2625°C). Recrystallization actually extends over a considerable temperature range for any given metal.

Recrystallization starts only when there is sufficient energy available to overcome resistance to movement of atoms. Part of this energy is supplied by distortion energy in the grains. The balance must be supplied by heat. Obviously, less heat energy is required for greater deformation, which implies that recrystallization temperature varies inversely with extent of cold working. In addition, a given amount of heat energy can be supplied in a short time at a high temperature or in a long time at a low temperature, which implies that lower recrystallization temperatures require longer times of annealing. The relationship between recrystallization temperature, amount of cold working, and time of annealing is indicated in Fig. 9-2.

We should emphasize that a minimum amount of cold deformation is required before any recrystallization takes place. Although this minimum amount varies from one metal to another, it is on the order of 1%. Recrystallization starts in the most severely deformed regions (strain centers), but it cannot start unless there is at least one strain center. As the amount of cold

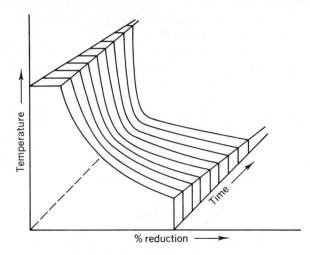

Fig. 9-2. Schematic diagram of relationship between recrystallization temperature, time, and amount of cold deformation.

work increases above the minimum, there are additional strain centers and finer grain size after recrystallization.

Recrystallization temperature is also affected by purity (or alloy content), initial grain size, and temperature at which deformation occurs. Pure metals have low recrystallization temperatures which are increased by the presence of impurities or alloy additions, especially those which enter into solid solution.

For equal amounts of overall deformation, more strain hardening is introduced into initially fine-grained metal than into initially coarse-grained metal. As a result, recrystallization temperature is lower for a given annealing time for fine-grained metals. Likewise, with lower temperatures of cold working, greater amounts of strain are introduced, effectively decreasing recrystallization temperature for a given annealing time.

9-3-2 RECRYSTALLIZATION TEXTURE

Cold working of metal introduces preferred orientation known as deformation texture. If the metal recrystallizes during process annealing, we find that the new grains, although strain-free, sometimes have preferred orientation, known as *annealing* or *recrystallization texture*, which may differ from deformation texture. For example, aluminum wire has a deformation texture of [111]. For recrystallization at temperatures below 500°C, this texture is maintained. The texture becomes increasingly random with recrystallization at temperatures above 500°C, especially in lower-purity metal. Aluminum of 99.95% purity develops a new texture, [112], when recrystallized at 600°C.

Annealing texture introduces potentially serious problems in design, especially in HCP materials.

EXAMPLE 9-1

A given cold-rolled aluminum sheet must be annealed for 8 hours at 200°C to achieve essentially complete recrystallization.
(a) How long will be required at 300°C for the same effect, if the activation energy for recrystallization is 25 Kcal/mole?
(b) What will be the comparative grain sizes from recrystallization at the two temperatures?

Solution:

(a) rate $= \dfrac{1}{\text{time}}$

$$\frac{t_{300°C}}{t_{200°C}} = \frac{\text{rate}_{200°C}}{\text{rate}_{300°C}} = \frac{Ae^{-(25,000/2\times473)}}{Ae^{-(25,000/2\times573)}} = e^{-5.3}$$

$$t_{300°C} = (480)(0.00505) = 2.4 \text{ min}$$

(b) Sec. 9-3–1 gives the recrystallization temperature for aluminum as about 150°C. The new grains from recrystallization at 200°C would be relatively small, i.e., the sheet would be fine-grained. The sheet recrystallized at 300°C would probably have somewhat larger grains. If the sheet were held at 300°C for some minutes after the 2.4 min calculated as necessary for recrystallization, the sheet would be rather coarse-grained.

9-4 NONEQUILIBRIUM COOLING

Conditions produced by nonequilibrium cooling rates can drastically alter properties. This may be desired or detrimental depending on the purpose intended and the cooling rate. Some of these effects are considered below.

9-4-1 CORING

A solid-solution portion of a phase diagram is shown in Fig. 9-3 for a solution which has a cooling rate faster than that of equilibrium. In alloy X, we expect that cooling liquid will first produce solid of composition α_1, at temperature T_1. If temperature is lowered to T_2, we expect solid of α_2 composition to freeze out of the melt of L_2 composition. Under equilibrium, we would have solid of entirely α_2 composition coexisting with L_2 liquid. Since the cooling rate is faster than that of equilibrium, diffusion is incomplete, and we find the center of a grain is essentially α_1 while the outside is α_2. The average com-

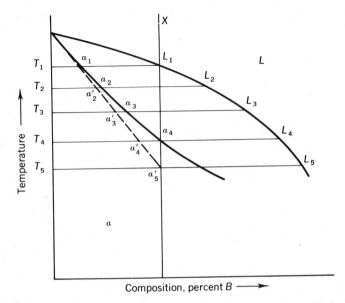

Fig. 9-3. Schematic solidification of a solid solution under nonequilibrium cooling conditions (producing coring).

position of the grain is intermediate, say α_2'. The melt, on the other hand, has essentially equilibrium composition since diffusion is relatively rapid in liquid.

Under equilibrium, solid α_2' cannot coexist with liquid L_2. The solid must have a higher percentage of B, requiring B to diffuse from liquid to solid. This would, in turn, require the liquid composition to become richer in A. Since this is not permissible (according to the diagram), additional amounts of α_2 are deposited simultaneously with diffusion of B. If sufficient time is allowed, the process continues until all the solid is α_2.

Under more-rapid-than-equilibrium cooling, the concentration gradient between the center and the exterior of each grain continues throughout solidification with the average composition of solid following the dotted line in Fig. 9-3. Application of the lever rule to this nonequilibrium situation shows more liquid present than under equilibrium and less solid. Under equilibrium, we expect complete solidification at T_4. In this case, however, we find some liquid remaining, and it is not until we reach T_5 that the average composition of the solid reaches the overall composition, and solidification is complete. We find increasingly greater depression of T_5 below T_4 as the cooling rate increases.

This manner of solidification results in solid α having a continuous series of layers beginning with the highest melting α_1 (richest in A) at the "center" and ending with the lowest melting α'_s (richest in B) at the outside. A photomicrograph (Fig. 9-4) shows a typical *cored* structure indicating crystal segregation as differentiated from ingot segregation (Chapter 12). The particular appearance of this single-phase structure is based on two factors: (1) solidification occurs in a dendritic pattern; and (2) etching agents generally attack alloys of different compositions at varying rates. Due to coring, axes of the dendrites are much richer in A than the surrounding or later-solidifying material.

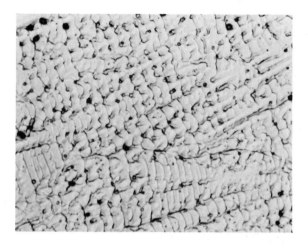

Fig. 9-4. Cored structure obtained in chill-cast Monel (approx. 68% Ni, 30% Cu).

Cored structures of this type are commonly found in as-cast metals. Since coring creates grain-boundary layers which are distinctly different in composition from the alloy as a whole, the results are frequently weakness and brittleness as well as a serious lack of uniformity in mechanical and physical properties. At least equally serious, in many cases, is the increased susceptibility to corrosion which exists in the cored structure because of preferential attack by a corrosive medium.

Coring can be avoided by slow freezing from the melt, but this results in large grain size and requires excessive time. A common practice is to chill-cast the material (producing a fine grain with highly nonequilibrium conditions and coring) followed by hot working or annealing to eliminate coring.

9-4-2 EUTECTIC SHIFT

In general, undercooling of liquids is rather limited in metallic systems. There are, however, a number of intermetallic compounds and some elements (principally silicon, tin, bismuth, and antimony) in which appreciable undercooling can take place. In a eutectic alloy, undercooling tends to shift the eutectic to a lower temperature and toward a higher content of the component which undercools more readily. If the solid phases are solutions, undercooling may also induce solubility limits appreciably beyond maximum solubilities indicated by the phase diagram.

If alloy X in Fig. 9-5 is cooled under equilibrium conditions, the first solid appears at T_1, followed by formation of the eutectic at T_E. If the liquidus is

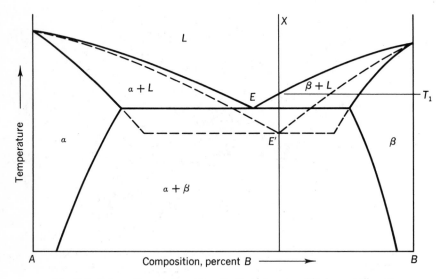

Fig. 9-5. Schematic representation of effect of nonequilibrium cooling on a simple eutectic.

depressed, however, solid may not appear until point E' is reached. In this case, the liquid is saturated with respect to both solid phases and will solidify as a eutectic. If the liquidus were depressed even further, alloy X (expected to precipitate primary β plus eutectic) would actually precipitate primary α followed by eutectic. Depression of the liquidus, accompanied by eutectic shift, has been demonstrated. For instance, the binary Al-Si system has an equilibrium eutectic containing 11.6% Si at 577°C. It is possible to cool Al-Si alloys rapidly enough for the eutectic to occur at 14% Si at 564°C.

9-4-3 METASTABLE EUTECTIC

Consider an alloy such as the one shown in Fig. 9-6. Under equilibrium cooling of alloy X, solid of α_1 composition will first appear at T_1. Continued equilibrium cooling allows composition of the melt to shift along the liquidus

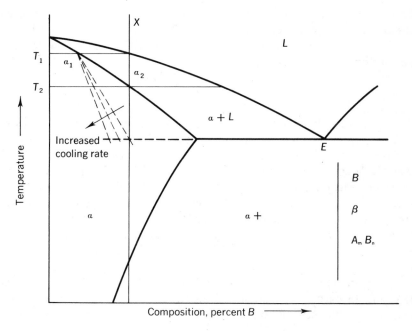

Fig. 9-6. Schematic representation of formation of metastable eutectic in an alloy of solid solution composition.

and composition of the solid to shift along the solidus until the melt entirely disappears at T_2. On the other hand, at a more rapid cooling rate, the first solid of approximately α_1 composition again appears at approximately temperature T_1. At T_2, time has been too short for all the solid to reach α_2 composition, and the same process occurs as in Fig. 9-3. In alloy X (Fig. 9-6), melt composition shifts along the liquidus and may reach eutectic composition before solidification is complete. If this happens, the remaining liquid solidifies as a eutectic, and the microstructure has cored primary α plus eutetic in an alloy which, under equilibrium, will be a homogeneous solid solution. As cooling rate increases, we expect more eutectic.

In general, we expect a relatively small amount of metastable eutectic. This eutectic has a definite tendency to become divorced and thus appear as small islands and interdendritic layers of the second phase in the cored primary α crystals. In Al-Cu, for example, this second phase is an intermetallic compound, $CuAl_2$, which is hard and brittle. Since this phase appears as a more or less continuous network, the divorced metastable eutectic tends to make the whole structure brittle. In addition, upon heating, the metastable eutectic starts melting at the eutectic temperature even though it may be completely divorced from the solution and does not belong there in the first place.

Since slow cooling rates necessary to avoid (or minimize) coring and formation of metastable eutectics are not generally practical, the usual practice is to allow the eutectic to form. This is followed by annealing the alloy just below the eutectic temperature (so that diffusion is relatively rapid, but so that the eutectic will not melt).

9-4-4 COMPOUND SUPPRESSION

Formation of a compound can be entirely suppressed in some systems by sufficiently rapid cooling. This is fairly common if the compound forms from a peritectic reaction, although it does occur in other compounds. As an example, a schematic alloy system is shown in Fig. 9-7. In equilibrium, a peritectic reaction occurs at T_1 and converts B_3A_2 and liquid into BA. If cooling is sufficiently rapid, the peritectic reaction does not occur nor does BA precipitate from the melt at temperatures below T_1. Instead, B_3A_2 continues to form at temperatures down to about T_2, ending with final solidification of liquid as a metastable eutectic (A-B_3A_2) as indicated by dashed lines in Fig. 9-7.

Complete suppression is possible, but partial suppression is far more common. In partial suppression the melt reacts peritectically with existing solid to form the new and different solid. Obviously, this reaction begins at

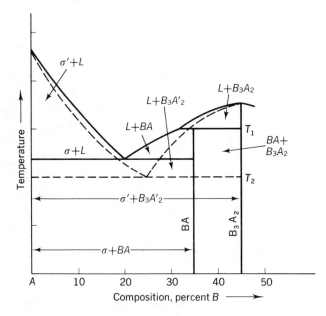

Fig. 9-7. Portion of the A-B alloy system showing effect of nonequilibrium cooling in suppressing formation of BA and establishment of metastable A-B_3A_2 eutectic. Prime marks indicate metastable situation.

the surface of existing crystals and continues by diffusion through the solid reacted layer that separates the remaining melt from the unreacted core. Diffusion and reaction become more difficult as the reacted layer thickens. With rapid cooling and relatively large original crystals, diffusion will not proceed to the center of the crystals. At room temperature, we find grains with an unreacted center of metastable phase surrounded by a peritectic wall of stable phase produced by the peritectic reaction. This wall, in turn, is surrounded by another phase, e.g., a solid solution or a eutectic.

In either complete or partial suppression, unstable phases can be avoided by very slow cooling during freezing. In both cases, unstable phases can be also eliminated by annealing somewhat below peritectic equilibrium temperature.

9-5   SOLID-STATE REACTIONS

Alloying elements are added to base metal to modify various properties and/ or gain desirable characteristics. If alloying elements enter into a solid solution, there is an increase in mechanical strength and hardness (solid-solution hardening) with a decrease in ductility and electrical conductivity, as well as changes in other properties (see Sec. 7-2–6). We can also appreciably increase strength and hardness in a metal or alloy by cold working (see Sec. 6-8–6 and 9-3). If a solid solution is cold-worked, we obtain a "summation" of both effects.

In many alloys it is possible to obtain even more drastic modifications through solid-state reactions. These are especially significant because: (1) they increase hardness far above that possible by solution hardening; (2) no plastic deformation is required; and (3) the reaction can be activated at any desired point in the fabricating process. There are some restrictions: (1) not every alloy can undergo a given solid-state reaction; (2) occurrence of a solid-state reaction under equilibrium conditions does not give large changes (to obtain these, nonequilibrium conditions must exist); and (3) the degree of change varies with alloy systems and may be negligibly small in some alloys.

The reactions of major interest are:

1. Eutectoid decomposition [Fig. 9-8(a), basis for Chapter 10].
2. Precipitation from solid solution [Fig. 9-8(b)].
3. Ordering of disordered solid solution [Fig. 9-8(c)].
4. Diffusion reaction [Fig. 9-8(d)].

9-5-1   PRECIPITATION HARDENING

For precipitation hardening, it is necessary, but not sufficient, for the alloy to have decreasing solid solubility with decreasing temperature. The aluminum-

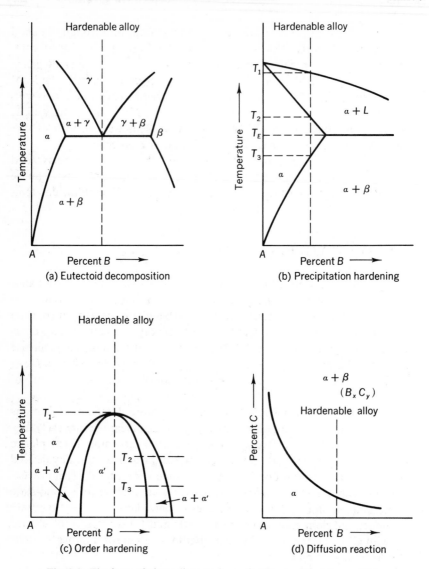

Fig. 9-8. The form of phase diagram for each of four solid-state reactions.

rich end of the aluminum-copper system (Fig. 7-25) is typical, and aluminum base alloys containing 4% and 4.5% copper are important commercial alloys. Similar behavior is found in several other systems, such as Al-Ag, Al-Mg, and Cu-Be. In the latter, an alloy containing 2% Be can be precipitation-hardened and is often used for tools (e.g., wrenches and chisels) which must be non-sparking. In fact, there are some 200 precipitation-hardening alloys in commercial use.

To generalize, consider the schematic diagram in Fig. 9-8(b). If liquid alloy X is cooled under equilibrium, we know solidification starts at T_1 and finishes at T_2. There is no further change under equilibrium until T_3 is reached. On continued cooling, a second phase precipitates along grain boundaries and on slip planes since the limit of solubility has been exceeded. This precipitating phase may be an essentially pure metal, a solution, or an intermediate phase.

On the other hand, if the cooling rate is more rapid than that for equilibrium, we find incomplete precipitation. In fact, quenching in water from temperatures between T_2 and T_3 normally gives sufficiently rapid cooling to permit retention of all the second phase in solution in most alloys, i.e., the alloy is *supersaturated*. Since more of B is in solution than possible under equilibrium, there is a tendency for precipitation to take place. Whether or not precipitation actually occurs in practical times depends on the diffusion rate. If diffusion is sufficiently rapid, precipitation can occur at room temperature. In this case, the alloy "ages" naturally, as in the two Al-Cu alloys mentioned above. In most alloys, however, room temperature diffusion is too slow. Heating to temperatures less than T_3 increases diffusion and allows precipitation. In this case, the alloy "ages" artificially.

The strength of many primary solid solutions which exhibit decreasing solubility with decreasing temperature can be increased by precipitation hardening treatment in three steps: (1) solutionizing, (2) quenching, and (3) precipitation treatment.

Solutionizing (*solution heat treatment*) consists of heating the alloy above T_3 [Fig. 9-8(b)] but below T_E and holding to homogenize. T_3 is the lowest temperature allowing complete solution, and T_E is the maximum temperature used in order to avoid melting of a possible metastable eutectic (Fig. 9-6). In practice, the highest temperature possible is used to obtain most rapid diffusion. In Al-Cu, alloy 2024 (4.5 w/o Cu) has a eutectic temperature of 932°F. Solution heat treatment is normally carried out between 910° and 930°F.

Quenching (*solution quenching*) is cooling of the alloy from solutionizing at a rate great enough to retain all the second phase in a supersaturated solid solution. Water quenching is normally used for alloy 2024.

The *precipitation treatment* (*aging*) requires holding the material at the proper temperature until the desired increase in hardness and strength occurs. Since the supersaturated solution is unstable, there is a definite tendency for the second phase to precipitate. At treatment temperature, the excess B atoms, which α solution wants to reject, tend to diffuse to, and concentrate along, certain crystallographic planes, i.e., those on which they create minimum distortion. The slip planes are preferred since they have maximum interplanar spacing. As this process continues, distortion of the planes increases and, in effect, inhibits slip along those planes, thus resulting in a stronger alloy. Any mechanical working, e,g., bending, stamping, riveting, etc., is usually done between quenching and precipitation treatments. If precipitation takes place at constant temperature, we find that room temperature hardness changes with time as indicated in Fig. 9-9. This also shows the effect of various treatment temperatures. Obviously, precipitation and resulting properties are functions of time and temperature. To obtain optimum properties (or desired properties) we must control both time and temperature with the proper combination depending on the alloy. For instance, properties of aluminum alloy 2024 are optimum after about four days at room temperature. Aluminum alloy 7075, however, requires months at room temperature to approach the properties developed at 250°F in 24 hours.

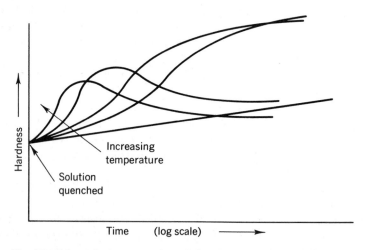

Fig. 9-9. Schematic representation of the effect of various precipitation treatment temperatures on the room temperature hardness of a precipitation-hardenable alloy.

9-5-2 THEORY OF PRECIPITATION HARDENING

In precipitation hardening, increase in hardness can result from coherency, solid-solution hardening, and fine dispersion of a second phase (mechanical interference with slip). At the same time, hardness may be decreased by relief of coherency strains, solid-solution depletion, coalescence of the dispersed phase, and recrystallization. Obviously, these reactions compete, and the "aging" curves (Fig. 9-9) are the sum of their effects. The effect of time at constant temperature for each reaction is shown in Fig. 9-10.

After solutionizing and quenching, the primary α solution is supersaturated and would prefer to reach the equilibrium state. The excess B atoms in solution tend to diffuse to, and concentrate along, certain crystallographic planes. These concentrations of B atoms tend to be platelets rather than spheroids, since the platelet produces a lower energy configuration.

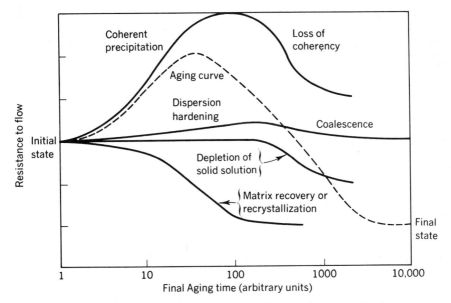

Fig. 9-10. Schematic representation of the effects on hardness of various processes that occur during aging.

There is a strong tendency for these crystal embryos of the second phase to remain in registry with the lattice of the matrix (the primary solution). This is known as *coherency*. Because the second phase has lattice parameters which differ from those of the matrix, these attempts to maintain registry cause distortion. Since the crystal embryo (incipient precipitate) is small in volume compared with the matrix, it is distorted much more than the matrix. At the same time, distortion in the matrix sets up strains over a considerably larger volume than if the second phase were a discrete particle. This distortion or straining of the matrix is a major source of hardening during precipitation treatment.

This relatively large amount of distortion prevents the incipient precipitate from forming a structure with equilibrium parameters. The precipitate thus goes through a transition or intermediate stage [Fig. 9-11(b)] between the time it starts coming out of the supersaturated solid solution [Fig. 9-11(a)] and the time that it forms an equilibrium structure [Fig. 9-11(c)]. During precipitation treatment, there is a continuous process at any given point which ranges from incipient precipitation through a transition lattice to an equilibrium structure of the precipitating phase. When equilibrium structure is obtained, the alloy no longer has maximum hardness and strength but is "over aged." In this condition, we normally expect to see a precipitated phase under a microscope, but we do not normally expect discernible precipitation at maximum hardness after precipitation treatment.

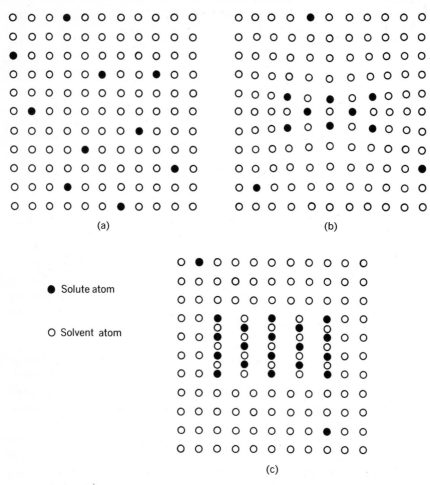

Fig. 9-11. Stages in formation of an equilibrium precipitate: (a) Supersaturated solid solution; (b) transition lattice coherent with solid solution; (c) equilibrium precipitate essentially independent of solid solution.

The above discussion is general and applies broadly, but it is based on the behavior of an aluminum-base alloy such as 2024. For Al-Cu, there is evidence that the incipient precipitate is in the form of platelets (Guinier-Preston zones) which are rich in copper, are about 2 atom layers thick, and extend about 50 to 800 angstroms. In addition, the lattice spacing of the matrix in the vicinity of these zones or platelets is slightly smaller than elsewhere. As time at temperature increases, these zones gradually transform into an intermediate or transition phase (called θ') which subsequently becomes θ or $CuAl_2$. $CuAl_2$ forms a tetragonal structure while the transition structure is, apparently, severely distorted FCC, and the α solution is FCC.

Our discussion implies a continuous process. This is true at any given point. Considering the alloy as a whole, however, the process will be at different stages as we go from point to point. In addition, coherency effects may exist up to maximum hardness (and beyond), since not all particles will break away at once. In fact, even a given particle may not make a complete break with the matrix all at once.

In certain other alloys, hardening is apparently due to direct, simple precipitation of a stable phase more or less uniformly dispersed throughout the matrix. Still others appear to form a metastable phase which precipitates and later gives way to a more stable phase. In other words, no one pattern appears to be followed by all alloys capable of precipitation hardening, but, rather, a number of patterns are possible.

A decrease in solid solubility with decreasing temperature is a necessary (but not sufficient) condition for precipitation hardening. We find that appreciable hardening occurs only when the precipitating phase is relatively complex (an intermetallic compound in most alloys). The transition structure must be capable of registry with the parent matrix yet there must be some relatively small amount of misfit. For example, in Al-Cu, Al-Ag, and Al-Mg, strain required to maintain registry is relatively small, and platelets grow to a thickness of about 300 angstroms before coherency is lost. These alloys can precipitation harden to an appreciable degree. Annealed aluminum alloy 2024 has a tensile strength of about 33,000 psi which increases to over 66,000 psi after the usual treatment. Ductility is reduced, however, from about 16 to about 11 (% in two inches). Al-Zn, Mg-Pb, and Mg-Zn, on the other hand, require about 10 times as much strain to maintain registry. Platelets in these systems grow to only about 10-angstroms thickness before precipitating. These alloys have little capacity for precipitation hardening.

EXAMPLE 9-2

An aluminum base alloy with 7 w/o copper is heated to 540°C, water quenched, and then reheated to develop hardness.
(a) What fraction of the alloy will harden by precipitation?
(b) Is it possible to precipitation harden a copper base alloy with 7 w/o aluminum?

Solution:

From Fig. 7-25, at 540°C the alloy is composed of κ (about 94.3 w/o Al — 5.7 w/o Cu) and θ (about 47.5 w/o Al — 52.5 w/o Cu). Using the lever rule,

$$\kappa = \frac{93.0 - 47.5}{94.3 - 47.5} = 97\%$$

This is the fraction of the alloy which precipitation hardens.
(b) Fig. 7-25 shows a copper base alloy with 7 w/o aluminum is a single

phase (α) throughout the entire temperature range for solid material of this composition. Therefore, it cannot be precipitation hardened.

9-5-3 ORDER HARDENING

When atoms in a disordered solid solution at elevated temperature rearrange themselves in an ordered array at lower temperature (Sec. 7-2-4), two structures are possible. The first is an isostructural ordered phase [Fig. 9-12(b)], in which there is little change in atomic position from the disordered phase or random solution [Fig. 9-12(a)]. There is also little hardening possible. If, on the other hand, a neostructural ordered phase is formed [Fig. 9-12(c)],

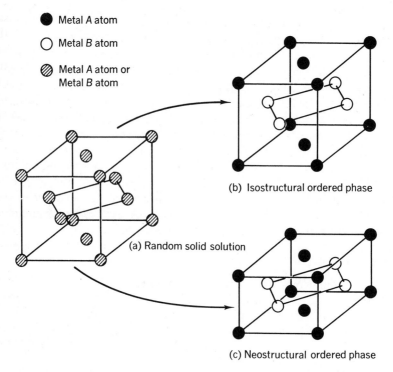

Fig. 9-12. Unit cells of typical ordered phases in alloy AB in order-disorder reactions.

which differs significantly in structure from the random solution, appreciable hardness is possible. This hardness can be obtained by heating to somewhat above T_1 [Fig. 9-8(c)], homogenizing, cooling rapidly to room temperature, and reheating to between T_2 and T_3. During reheating, nuclei of the ordered phase form and grow at the expense of the disordered solution. Hardening continues until a maximum is reached. With continued heating, softening will occur, much as in precipitation hardening.

9-5-4 DIFFUSION REACTION HARDENING

Diffusion reaction hardening depends on a change in alloy composition. If A dissolves B but has limited solubility for B_xC_y, hardness can sometimes be obtained by diffusing C into a solid solution of B in A [Fig. 9-8(d)]. The overall composition shifts from the ternary solution region into the α plus B_xC_y region with B_xC_y tending to precipitate. Hardening appears to be a result of lattice coherence existing in the early stages of B_xC_y precipitation. Only the surface is hardened. Nitriding for surface hardening in steels is an example of this reaction.

9-5-5 POLYMORPHIC TRANSFORMATION

We recall (Chapter 4) that, upon heating or cooling, certain metals undergo a change of crystal structure known as polymorphic transformation. In alloy systems based on these metals, we find the transformation will persist in at least the terminal (i.e., primary) solid solution, although transformation temperature is usually altered by the presence of solute atoms.

This polymorphic transformation is sometimes called *neocrystallization* because of its similarity to *recrystallization* in the cold-work-anneal cycle. There are differences, however. In *recrystallization*, the initial phase is made unstable as a result of cold working and tends to revert to the stable, unworked state at all temperatures. Recrystallization temperature is simply the temperature at which this process occurs at a fairly rapid rate. *Neocrystallization*, however, requires no cold working since one phase is stable above a certain equilibrium temperature and the other phase is stable below, and the transformation occurs only when the alloy passes through this temperature. These two processes are compared in Fig. 9-13. If strain-free α is heated above the transformation temperature, grains of γ form from α by neocrystallization. On the other hand, if the same initial α is cold-worked, the plastically deformed grains will recrystallize at a temperature below the neocrystallization temperature and produce a new set of strain-free grains of α. If these recrystallized grains are heated into the γ region, they will change to γ by neocrystallization. Likewise, γ will change to α by neocrystallization upon cooling below the transformation temperature.

In the alloy, if there is an appreciable difference between solid solubilities in the two phases, the transformation often occurs by a eutectoid or peritectoid (occasionally a monotectoid) reaction. Regardless of the specific reaction involved, this is a true phase change and shows a thermal arrest in the cooling curve under equilibrium. Since the transformation occurs in the solid state,

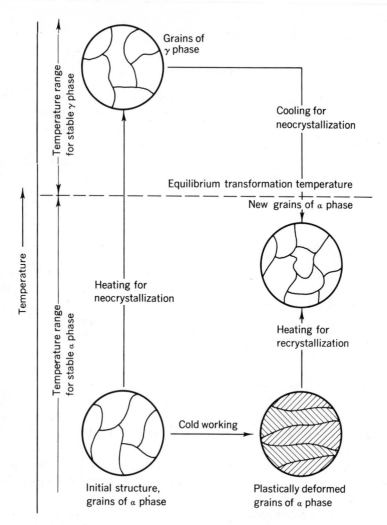

Fig. 9-13. Schematic representation of neocrystallization and recrystallization in a metal (e.g., iron) that has a polymorphic transformation at elevated temperature.

however, it is usually susceptible to drastic undercooling or superheating. Drastic undercooling, for example, can produce a metastable state appreciably different from the equilibrium state with drastic changes in alloy properties.

Heat treatments based on the polymorphic transformation exist for a large number of alloys. The most prominent and best understood is the iron-carbon system which is most important commercially and will be discussed in Chapter 10. Similar reactions, however, are found in many alloys.

QUESTIONS AND PROBLEMS

1. Define heat treatment. Is it practically and commercially important and why? Is heat treatment limited to a few possible treatments?

2. What effect does heating have on the properties of a strain-hardened metal? What effect does it have on the shape?

3. What is meant by recovery? Why is it of practical importance?

4. What is recrystallization? What practical significance does it have? How does recrystallization take place?

5. Where do crystal nuclei form first during recrystallization? Why?

6. What changes occur because of recrystallization? Why?

7. What is the recrystallization temperature? What factors determine the recrystallization temperature? Indicate the effect of each.

8. High purity lead can be worked by rolling almost indefinitely at room temperature without any appreciable hardening.
   (a) Why is this so?
   (b) What difference in behavior would you expect in this situation if it were rolled at the boiling point of liquid nitrogen ($77°K$)?

9. What is coring? What is its practical significance? Is it objectionable and why? How can it be eliminated? How does hot working remove coring?

10. Using the lever rule, estimate the proportion of liquid in (a) the equilibrium cooled alloy and (b) the cored alloy at temperatures T_1, T_2, T_3, T_4, and T_5 in Fig. 9-3. In the rapidly cooled alloy will there be more or less than equilibrium amounts of solid at any given temperature when liquid and solid coexist?

11. Discuss and compare some of the general effects on phase diagrams of cooling more rapidly than equilibrium rates. How do these affect microstructure, freezing point, and solubility limits?

12. What effects, other than hardening, may be achieved by heat treatment?

13. How are the properties of a solid-solution alloy affected by grain size, by cold work, and by heat treatment? What heat treatments can be applied?

14. What requirements must be met for an alloy to be precipitation hardenable?

15. What steps are required in shop fabrication for precipitation hardening?

16. What is the precipitation-hardening mechanism? How does it work?

17. Using a sketch of the Al-Cu diagram (Al-rich end), show the series of treatments required to transform a cored casting into a wrought precipitation-hardened part.

18. What is meant by "aging" and "over aging"? In what kind(s) of alloys do they occur? What effect does each have on the properties of an alloy?

19. Discuss some possible uses for precipitation-hardening alloys which harden but do not overage (in any reasonable time) at room temperature.

20. An alloy of 96 w/o Al-4 w/o Cu at 540°C is cooled very rapidly (quenched) to room temperature. (a) What is the resulting structure? (b) What would the structure be if the alloy were cooled very slowly (quasiequilibrium cooling)?

21. One source of electrical wire scrap is a mixture of 65 w/o Cu and 35 w/o Al. It can be purchased at about half the price of the two metals separately. How much of this scrap can be used per ton in melting metal to produce an aluminum base precipitation hardening alloy like 2014 (96 Al-4 Cu)?

22. A proposal has been made to add 5% Sn to Pb to make a precipitation hardenable alloy. Discuss the prospects of this proposal.

23. A box of Duralumin rivets (Al base, Cu principal alloying element) is found in a stock room with an unknown date of acquisition. The Production Department desires to use them. Should they be given to Production directly or should they be given some treatment first? Why? What treatment, if any?

24. Why are aluminum alloys often refrigerated after solutionizing?

25. Why is it preferable to precipitation-harden alloys (such as Al-4.5% Cu) by first quenching to a low temperature and then reheating rather than by quenching directly to the precipitation-hardening temperature?

26. Should precipitation hardenable alloys present welding difficulties? How and why? Can this be eliminated or minimized? How?

27. Discuss the similarities and differences between precipitation hardening and order hardening.

28. Discuss the similarities and differences between precipitation hardening and diffusion reaction hardening.

29. Upon heating, pure iron has an allotropic transformation from BCC to FCC at 910°C. Does iron expand, contract, or retain its original volume during this transformation?

30. The grain size of a copper sample can be refined (made smaller) only by heating it after it has been cold-worked. The grain size of a large-grained iron sample can be refined by heating it without any prior cold-working. Explain this difference.

31. Discuss the differences and similarities of recrystallization and neocrystallization. Will both always, never, or sometimes, occur in the same alloy? Explain your answer.

32. An alloy of 90 Al-10 Cu is heated to 540°C (1004°F) for solutionizing, followed by quenching and precipitation treatment. What fraction, if any, of the alloy will precipitation harden?

10

NONEQUILIBRIUM
RELATIONSHIPS IN STEELS

10-1 INTRODUCTION

The iron-carbon system provides the most prominent example of heat treatment and property alteration based on polymorphic transformation and eutectoid decomposition. Because of its outstanding commercial importance, the iron-carbon system has been studied in more detail than most alloy systems. Many reactions were first found in the iron-carbon system and later in other alloys. Discussion in this chapter is specifically pertinent to steels, but the reactions occur in many alloy systems in identical or modified form.

10-2 IRON-CARBON ALLOYS

Pure iron has two polymorphic changes, as indicated in Fig. 10-1. Addition of carbon (a metalloid) to iron produces effects which sometimes seem out of proportion to the amount of carbon added. Alloys of iron and carbon (with other elements added for special purposes) comprise the commercially important ferrous-base alloys known as steels and cast irons. These alloys, particularly the steels, are susceptible to heat treatment, and a wide range of properties can be obtained by proper variation and timing of heating and cooling cycles. Of the various effects of carbon on iron, the effect on the $\gamma \rightleftharpoons \alpha$ transformation at 910°C in pure iron is perhaps the most important.

We found (Chapter 7) that carbon atoms are small compared to iron atoms, and have a radius ratio (carbon to iron) of 0.63. Consequently, any solute carbon forms an interstitial solution. Since the biggest interstices in gamma (FCC) iron are larger (0.52 A radius) than the largest in alpha (BCC)

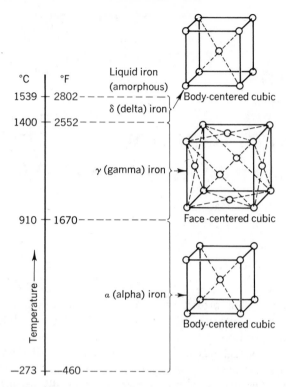

Fig. 10-1. Temperature ranges in which allotropic forms of iron exist under equilibrium conditions.

iron (0.36 A radius), we expect greater solubility of carbon in γ than in α. This does occur, as indicated in Fig. 10-2, the so-called Iron-Carbon Diagram. Fig. 10-2 indicates that γ can dissolve a maximum of 2.0 w/o carbon at 1130°C, while α can dissolve a maximum of only 0.025 w/o carbon at 723°C. This difference in solubility, combined with control of cooling rate and diffusion, is basic to most of the possible variations in steel properties. Fig. 10-2 contains several terms which may need definition.

Ferrite is a primary solid solution based on α or BCC iron. In immediate context, ferrite is a solution of carbon in α-iron, but the term is also applicable when additional elements are added to an alloy which still retains the BCC structure. Ferrite is relatively soft (Brinell hardness number: 50 to 100) and ductile unless it has been cold-worked. It can undergo extensive cold working.

Austenite is a primary solid solution based on γ or FCC iron. As with ferrite, the term has wider connotation than a binary solution of carbon. Austenite is generally soft and ductile unless cold-worked, although it is generally stronger and less ductile than ferrite. Austenite is denser, has a

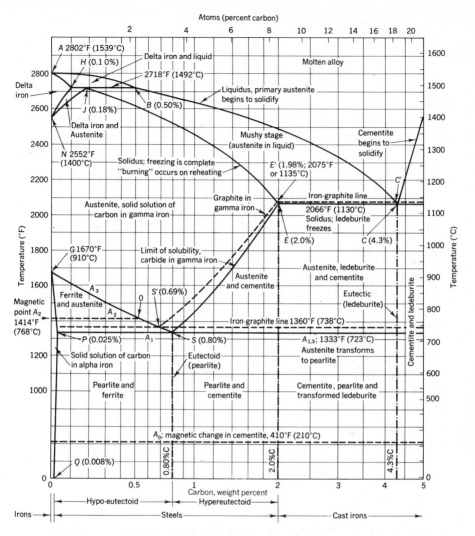

Figure 10-2. Iron-carbon equilibrium diagram. Dash lines show true equilibrium of iron and graphite. Solid lines show a metastable phase diagram of iron and iron carbide (Fe₃C). The metastable diagram is used in the same manner as a true equilibrium diagram. The distinction between the two is negligible above 2100°C.

higher electrical resistance and a higher thermal coefficient of expansion than ferrite, and is essentially nonmagnetic (paramagnetic).

Cementite is the hard (about 1400 Brinell), brittle intermetallic compound of iron with 6.67 w/o carbon (iron carbide). The amount and distribution of this phase in a ferritic or austenitic matrix greatly influence the steel's properties. Iron carbide (Fe₃C) has an orthorhombic structure with lattice parameters (Table 4-1) of $a = 4.5235$, $b = 5.0888$, and $c = 6.7431$ A. The unit cell has 12 iron atoms and 4 carbon atoms which are located interstitially in an almost close-packed structure of iron atoms where each carbon atom is surrounded by six iron atoms. The lattice parameters vary somewhat with the temperature at which iron carbide is brought into equilibrium with austenite. Cementite becomes ferromagnetic below 210°C (Fig. 10-2).

The Iron-Carbon Diagram shown (Fig. 10-2) is actually only a part (up to 5 w/o C) of the Iron-Iron Carbide Diagram. *Plain carbon steels* are generally considered to contain up to 2 w/o carbon whereas *cast irons* range from 2 to 5 w/o carbon. Fig. 10-2 has been called an equilibrium diagram, but it is actually a metastable phase diagram since (with sufficient annealing) cementite decomposes into iron plus graphite. This is shown in Fig. 10-2 by dashed lines with labels indicating graphite rather than cementite. In effect, then, the solid-line diagram is a pseudoequilibrium diagram. It is useful, nevertheless, in interpreting iron-carbon alloy reactions and is a most important tool of the ferrous metallurgist. At the same time, this diagram gives no indication of structures which can be formed by treatments which do not approach pseudoequilibrium.

The plain carbon steels (0 to 2 w/o carbon) are iron-base alloys with carbon as principal alloying element, small amounts (about 0.5 w/o) of silicon and manganese, and even smaller amounts (preferably less than 0.5 w/o) of sulfur and phosphorus. We will neglect these impurities and treat the alloys as iron-carbon binaries.

EXAMPLE 10-1

AISI 1030 steel is in equilibrium at room temperature.
(a) What is the structure? (b) If this steel is in equilibrium at 900°C, what is the structure?

Solution:

(a) In Fig. 10-2, the region containing 0.30 w/o carbon at room temperature shows pearlite and ferrite. Neglecting the solubility of 0.008 w/o carbon and applying the lever rule, then

$$\text{ferrite} = \frac{0.80 - 0.30}{0.80 - 0} = 62.5\%$$

$$\text{pearlite} = \frac{0.30 - 0}{0.80 - 0} = 37.5\%$$

But pearlite, in turn, is a eutectoid structure of ferrite and cementite (Fe_3C) in the proportions

$$\text{ferrite} = \frac{6.67 - 0.80}{6.67 - 0} = 88\%$$

$$\text{cementite} = \frac{0.80 - 0}{6.67 - 0} = 12\%$$

(b) Fig. 10-2 shows this is 100% austenite, a solid solution of carbon in FCC iron.

The development of various properties in steels can be obtained by control of the eutectoid decomposition from FCC to BCC. In other words, we can control the transformation of austenite into other phases and thereby obtain a variety of structures and properties. We are specifically interested in five transformations, namely:

1. Austenite ————————→ Ferrite
2. Austenite ————————→ Cementite
3. Austenite ————————→ Pearlite
4. Austenite ————————→ Martensite
5. Austenite ————————→ Bainite

Which transformation will occur depends principally on alloy content (amount of carbon, kind and amount of other alloying elements) and cooling rate.

10-3   THE PEARLITE TRANSFORMATION

Figure 10-2 shows a peritectic at 1492°C and 0.18 w/o carbon and a eutectoid at 723°C and 0.80 w/o carbon. The peritectic has little practical significance, but the eutectoid reaction is most important. In steels, austenite decomposes (under isothermal equilibrium) into ferrite and cementite. In eutectoid austenite (0.80 w/o C), pearlite forms on equilibrium cooling below the critical (i.e., three-phase equilibrium) temperature.

Pearlite is a lamellar aggregate of cementite and ferrite. The name derives from its lustrous appearance (similar to mother-of-pearl) when viewed in white light under a microscope. Pearlite is readily recognized by its lustrous appearance and its structure of alternate plates of ferrite and cementite.

Transformation of austenite to pearlite starts by formation of cementite crystal nuclei at austenite grain boundaries. Carbon diffuses from the surrounding austenite to the cementite, and the growth of carbide begins. As carbon diffuses, the adjacent austenite is depleted in carbon and transforms to ferrite. These two steps are shown schematically in parts (a) and (b), respectively, of Fig. 10-3. With formation of ferrite, there is rejection of carbon from the ferrite region, i.e., effective enrichment of the adjacent austenite. This results in the formation of additional nuclei of cementite [Fig. 10-3(c)]. Because of the alternate formation of cementite and ferrite, cementite can only grow away from the boundary of the original austenite grain as a platelet, somewhat like the filling in a sandwich [Fig. 10-3(d)].

Nucleation and growth of alternate plates of cementite and ferrite occur

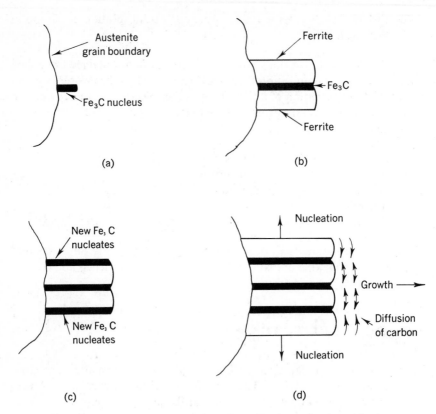

Fig. 10-3. Schematic representation of pearlite formation by nucleation and growth; (a) through (d) indicate successive steps in time sequence.

at several points along the austenite grain boundaries. This forms pearlite colonies which are approximately hemispherical regions of alternate parallel plates of cementite and ferrite. Usually several such colonies form within any one austenite grain. These pearlite colonies grow until the entire austenite grain has been consumed and has become a pearlite nodule. The process of pearlite formation is sometimes referred to as "sidewise nucleation and edgewise growth."

In "edgewise growth," the common interface between the lamellae (cementite and ferrite) and austenite advances into the latter. The growth process attains a steady state in which both the velocity of motion of the interface and the spacing between the lamellae remain constant for isothermal transformation of homogeneous austenite. This growth involves two distinct steps: (1) redistribution of carbon; and (2) FCC austenite must change to orthorhombic cementite and BCC ferrite.

The rate of growth of pearlite colonies into austenite grains is controlled by the rate of diffusion in the vicinity of the interface. Spacing between plates, i.e., platelet thickness, is determined by a balance of factors. For large spacing, little energy is involved at the ferrite/cementite interfaces, but the carbon atoms must diffuse over relatively large distances. The rate of carbon redistribution is slow since the only "driving force" is the energy released by the austenite $\rightarrow$ pearlite transformation. If spacing is small, there is a large amount of ferrite/cementite interfacial energy with little "driving force" left for diffusion. A balance, with a resulting specific spacing, is obtained when the ratio of "driving force" to diffusion distance is maximized. The amount of energy released by the austenite $\rightarrow$ pearlite transformation increases with increased undercooling, resulting in decreased interlamellar spacing as the transformation temperature is decreased.

Many characteristics of this eutectoid reaction are independent of the iron-carbon system. In fact, a major difference between the austenite $\rightarrow$ (cementite + ferrite) transformation and other eutectoid transformations is that carbon dissolves interstitially rather than substitutionally. Decomposition of β eutectoid in Cu-Al and Al-Zn systems, for example, follows the same pattern as decomposition of austenite. Formation of most eutectics and eutectoids of the lamellar type can be explained by the above mechanism.

EXAMPLE 10-2

(a) AISI 1080 steel is in equilibrium at 750°C. What is the structure?

(b) If the same steel is in equilibrium at 700°C, what is the structure?

Solution:

(a) Fig. 10-2 shows that an iron base alloy with 0.80 w/o carbon at 750°C is 100% austenite.

(b) Fig. 10-2 shows the structure is pearlite, i.e., alternating platelets of ferrite (α iron) and cementite (Fe_3C) with about 88% ferrite and 12% Fe_3C (see Example 10-1).

10-4 FERRITE AND CEMENTITE
FORMATION

Discussion of pearlite formation was in terms of eutectoid steel (0.80 w/o C). If we have a hypoeutectoid steel, i.e., less than 0.80 w/o C, the situation is only slightly different. In this case, upon slow cooling of homogeneous austenite, there is no noticeable change until the temperature reaches that of the point on the line G-S (Fig. 10-2) which corresponds to the given composition. At this temperature, the austenite starts to precipitate ferrite at grain boundaries. The composition of this ferrite is indicated by the intersection of an isothermal tie-line with the line G-P (Fig. 10-2). Since this ferrite is much lower in carbon content than the austenite, the remaining austenite is enriched with carbon. As temperature decreases, the composition of ferrite follows line *G-P* while that of austenite follows line *G-S*.

If carbon content is 0.025 w/o (or less), austenite will be converted entirely into ferrite. If carbon content is between 0.008 w/o and 0.025 w/o, there will be a precipitation of a small amount of secondary cementite as room temperature is approached. This small quantity is normally not visible except at very high magnifications. If, on the other hand, the carbon content is between 0.025 w/o and 0.80 w/o C, ferrite cannot dissolve all the carbon. On reaching 723°C, any austenite which has not yet transformed to proeutectoid ferrite undergoes the eutectoid reaction to form pearlite. Proeutectoid ferrite and ferrite in the pearlite have different orientations relative to the austenite. In a hypoeutectoid steel, the microstructure contains "islands" of pearlite in a ferrite matrix.

A hypereutectoid steel, i.e., one containing between 0.80 w/o C and 2.0 w/o C, upon slow cooling of homogeneous austenite, precipitates proeutectoid cementite when the temperature reaches that given by line *E-S* (Fig. 10-2). This cementite forms at austenite grain boundaries as does proeutectoid ferrite in hypoeutectoid steel. As cooling continues, austenite is saturated with respect to carbon, and the solubility limit follows line *E-S*. When the temperature reaches 723°C, the remaining austenite transforms into pearlite. Cementite in the pearlite grows directly from the proeutectoid cementite. The microstructure has "islands" of pearlite in a matrix of cementite.

The microstructure of slowly cooled hypereutectoid steel resembles that of slowly cooled hypoeutectoid steel, but there are relatively simple tests which help determine the proeutectoid constitutent. The bases of these tests are:

1. *Shape of grains.* Proeutectoid cementite tends to be thin and acicular (needle-like), whereas proeutectoid ferrite tends to be relatively thick and rounded.
2. *Thickness of the proeutectoid layer.* In any steel, the proportion of proeutectoid cementite will never be more than 15% by weight (it is rarely over 5%) while the proportion of proeutectoid ferrite may be any amount up to 100%. Consequently, if the proeutectoid layer is thick it is necessarily ferrite.
3. *Relative hardness.* Cementite is very hard, and ferrite is relatively soft. If, after mechanical polishing, there is detectable relief on the surface of the specimen, proeutectoid cementite stands above while proeutectoid ferrite stands below the plane of the surface. If there is no detectable relief, a simple scratch test or a microhardness test will readily indicate the nature of the proeutectoid layer.
4. *Special etchants.* Most common etchants used for steels attack neither

cementite nor ferrite but only the grain boundaries of interfaces between them. A few etchants, however, show preferential attack on one
or the other of these phases and thus allow a clear-cut distinction
between them. A common one is hot sodium picrate, which darkens
cementite while leaving ferrite bright.

In low-alloy steels and plain carbon steels which have been cooled relatively slowly through the critical range (between the lines G-S-E and P-S,
Fig. 10-2), it is possible to obtain a reasonable estimate of carbon content by
examining microstructure. This estimate is somewhat more accurate for
hypoeutectoid steels than for hypereutectoid steels.

EXAMPLE 10-3

(a) AISI 1045 steel is in equilibrium at room temperature. What are the
amounts of proeutectoid and eutectoid ferrite?

(b) A steel with 1.2 w/o carbon is in equilibrium at room temperature.
What are the amounts of proeutectoid and eutectoid cementite?

Solution:

(a) From Fig. 10-2 and the lever rule,

$$\text{proeutectoid ferrite} = \frac{0.80 - 0.45}{0.80 - 0} = 44\%$$

$$\text{eutectoid (pearlite)} = \frac{0.45 - 0}{0.80 - 0} = 56\%$$

But pearlite is 88% ferrite and 12% Fe_3C, therefore

ferrite:	
proeutectoid	44.0%
in eutectoid	49.5%
	93.5%
cementite (Fe_3C)	
(all in pearlite)	6.5%

(b)

$$\text{proeutectoid } Fe_3C = \frac{1.20 - 0.80}{6.67 - 0.80} = 6.8\%$$

$$\text{eutectoid (pearlite)} = \frac{6.67 - 1.20}{6.67 - 0.80} = 93.2\%$$

But pearlite is 88% ferrite and 12% Fe_3C, therefore

cementite:	
proeutectoid	6.8%
in eutectoid	11.2%
	18.0%
ferrite	
(all in pearlite)	82.0%

10-5   THE MARTENSITE TRANSFORMATION

Austenite transforms into proeutectoid ferrite, proeutectoid cementite, and/or pearlite, under relatively slow cooling. A very rapid cooling rate, however, transforms austenite into a metastable phase known as martensite. The cooling rate must exceed a critical value which depends on composition and metallurgical history. Martensite will not form unless austenite is cooled below a certain critical temperature (M_s) which depends on composition. Formation continues only if temperature continues to decrease. There is a lower temperature (M_f) at which transformation to martensite is complete. Martensite formation is essentially independent of time, i.e., formation of martensite has little relationship to cooling rate provided it equals or exceeds a critical value. The martensite reaction cannot be suppressed or prevented in plain carbon steel by even the most rapid possible cooling rate, although the reaction can be suppressed in many alloy steels.

Although interlamellar spacing of pearlite is influenced by cooling rate, the mechanism of formation is not. Martensite, however, is the product of a different mechanism of transformation with no precipitation of carbon. Martensite is a single-phase, supersaturated solution of carbon in ferrite with carbon atoms located interstitially in a body-centered tetragonal lattice; i.e., the excessive supersaturation distorts the normal BCC structure to body-centered tetragonal. The lattice distortion is reflected in mechanical properties of high tensile strength and hardness and low ductility. In fact, martensite is brittle and, by itself, has limited application. The major interest lies in the desirable combinations of properties which can be obtained by heat treating (tempering) this structure (Sec. 10-10).

Formation of martensite is similar to slip and twinning since it is a shear transformation, but it differs since there is a change in crystal structure. In slip, twinning, and other shear transformations (such as martensite formation), the composition of the sheared region is the same as that of the parent material. Such transformations are diffusionless, and the sheared region is coherently joined to the matrix. Shear propagates at about the speed of sound through the crystal.

Martensite forms from austenite as individual platelets, as temperature decreases. Each platelet is formed in a short time interval (perhaps less than a microsecond). Additional transformation on continued cooling is by formation of additional plates rather than by growth of existing plates. The plane of the plate has a characteristic orientation relative to the axes of the parent austenite (known as the habit plane). There is also a characteristic relative orientation between the axis of the martensite cell and the axes of the parent austenite. There is a macroscopic distortion of austenite as it changes to martensite consisting of a shear on the austenite-martensite interface and a small expansion normal to the interface. This is consistent with the fact that martensite has a specific volume about 104% that of austenite. This macroscopic distortion differs from microscopic distortion since application of the macroscopic distortion to austenite does not generate martensite.

The mechanism of martensite formation is not completely determined, but we assume nucleation and shear, which imply a progressive shear wave rather than a homogeneous shear. The concept of nucleation in martensite formation rationalizes the lack of isothermal transformation in the presence of untransformed austenite, the shape of the transformation-versus-temperature curve, and dependence of M_s upon composition. The proposed nucleation and growth mechanisms for martensite formation are diffusionless and therefore differ from those discussed in Chapter 8. A conclusion often reached is that the strain energy of misfit between the martensite and the untransformed austenite matrix is minimized by formation of platelets (as observed) upon which a nonhomogeneous shear is imposed. This shear cancels the macroscopic changes in dimensions in the plane of the plate.

Many nonferrous systems have phase changes, called martensitic, with regular movements of atoms causing the change in the transforming region. The phase changes are also called "athermal" to distinguish them from nucleation and growth transformations where transfer of atoms is clearly due to thermal activation. They are also called "diffusionless" to emphasize that they occur with no change in composition in the transforming region. Several nonferrous alloys change from FCC to face-centered tetragonal by a martensitic mechanism during cooling. In one (indium with 20.75 a/o thallium), martensite formation apparently occurs by two consecutive ⟨110⟩ shears on different {110} planes at 60° to each other. "Pure" lithium, zirconium, and cobalt exhibit a martensitic change from BCC to HCP. The β phase of the Cu-Al system exhibits a martensitic reaction, changing from BCC to distorted HCP.

10-6 TIME-TEMPERATURE-TRANSFORMATION RELATIONS

When a phase has been brought into an unstable region, e.g., austenite (0.80 w/o C) cooled below 723°C, the equilibrium diagram no longer applies nor does it tell us anything about the structure of transformation products or the mechanism of their formation.

We know very slow cooling produces pearlite, and very rapid cooling produces martensite. This leaves a middle ground for isothermal transformation, continuous cooling transformation, or some combination of both. Isothermal transformation of austenite at subcritical temperatures (less than

723°C for 0.80 w/o C) is shown on Time-Temperature-Transformation (TTT) diagrams, whereas continuous cooling is shown on Continuous-Cooling-Transformation (CCT) diagrams. One diagram is required for each alloy composition in contrast to an equilibrium diagram which covers an entire alloy system.

10-6-1 TTT DIAGRAMS

TTT diagrams are graphical summaries of isothermal transformation data obtained as follows: a large number of relatively small specimens, at equilibrium above critical temperature, are quenched in a liquid bath maintained at constant subcritical temperature. Specimens are withdrawn at regular intervals and quenched in iced brine to prevent further transformation. Microstructures are examined to determine the beginning (usually 0.5% or 1%) and the end (99.5% or 99%, respectively) of transformation at the given temperature. A series of these tests at several temperatures allows construction of diagrams such as Fig. 10-4 (for eutectoid carbon steel). Because of their shape, these curves are sometimes called S curves or C curves although more common practice is to call them TTT diagrams.

Transformation of eutectoid steel (Fig. 10-4) just below critical temperature is to pearlite at a very slow rate. As transformation temperature is decreased, the rate increases rapidly to a maximum at about 540°C (1000°F). Transformation of austenite to pearlite is a nucleation and growth process, dependent on diffusion and therefore on temperature. Nucleation rate increases, and growth rate (directly controlled by carbon diffusion) decreases with decreasing temperature (see Chapter 8). In a somewhat different aspect, the "driving force" is the difference in free energy between unstable austenite and pearlite. This free energy difference increases with decreasing temperature, i.e., increased undercooling. At the same time, the energy required for diffusion increases with increased undercooling. The relative size of the two effects determines transformation rate at a given temperature.

Transformation temperature influences both reaction rate and resulting microstructure. Pearlite which forms just below 723°C has relatively thick lamellae which are essentially parallel. Since the nucleation rate increases as temperature decreases to that of the pearlite "nose," we expect an increasing number of pearlite colonies growing from austenite grain boundaries. This has been observed. We also find that the lamellae are thinner and tend to be arranged in a radial or fan-shaped pattern rather than remain parallel. The pearlite structure thus becomes progressively finer until transformation temperature is about 500–600°C (930–1110°F), i.e., at the "nose" in Fig. 10-4. Below this temperature, austenite transforms to bainite.

The change in the spacing between alternate lamellae in pearlite has a

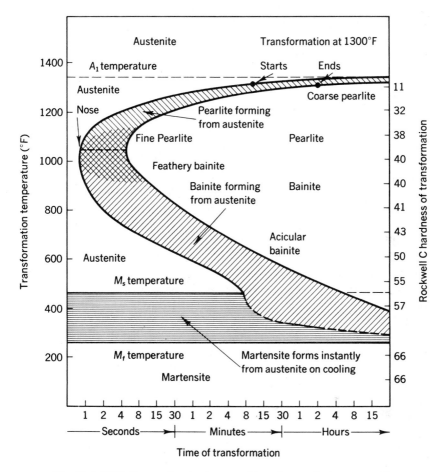

Fig. 10-4. TTT diagram for the decomposition of austenite in a eutectoid carbon steel (AISI 1080).

pronounced effect on mechanical properties. This is indicated by the hardness numbers given at the right side of Fig. 10-4 where we see an increase from about 15 R_c at the critical temperature to about 40 R_c at the "nose."

It might be emphasized that Fig. 10-4 is the TTT diagram for plain carbon steel of eutectoid content (AISI 1080). If the Fe-C binary alloy is either hypoeutectoid or hypereutectoid, there is a further complication in isothermal transformation. This is somewhat analogous to the proeutectoid formation discussed in Sec. 10-4. For example, if we have a hypoeutectoid alloy such as AISI 8620 (Fig. 15-2) undergoing isothermal transformation above about 550°C, ferrite forms first, followed by transformation to pearlite.

10-6-2   THE BAINITE TRANSFORMATION

Bainite is the name given to the structures that form on isothermal transformation at temperatures below the nose of the TTT curves. *Bainite is an isothermal transformation product and cannot be produced by continuous cooling.*

Bainite is an intimate mixture of ferrite and cementite, as is pearlite. Pearlite has alternate plates of ferrite and cementite. In bainite, however (when the structure can be resolved), cementite apparently exists as tiny spheroids uniformly distributed throughout a ferrite matrix. In upper bainite (formed at temperatures just below the nose of the TTT curve), there is evidence of some pattern in the cementite arrangement since the microstructure has a feathery appearance. In lower bainite (formed at temperatures approaching M_s) the cementite becomes too fine for resolution, and an acicular (needlelike) pattern is found.

Due to differences in size, shape, and distribution of cementite, bainite is normally harder, stronger, and tougher than fine pearlite of the same chemical composition. In addition, bainite exhibits considerable variation in properties depending on the temperature at which it is formed. For eutectoid carbon steel, hardness increases from about 40 R_c to about 55 R_c (Fig. 10-4).

The mechanism of the austenite $\rightarrow$ bainite transformation and its difference from the austenite $\rightarrow$ pearlite and the austenite $\rightarrow$ martensite transformations are not fully understood. One difference between pearlite and bainite formation is that pearlite is initiated by cementite precipitation whereas bainite is initiated by ferrite precipitation. This, however, provides little explanation for observed difference in microstructure. One theory suggests that bainite forms initially as supersaturated ferrite by a lattice shearing process (somewhat similar to martensite). Another leading theory suggests separate precipitation of ferrite and cementite from austenite. In any event, the slow rate of propagation indicates that diffusion-controlled processes are involved in bainite formation.

10-6-3   MARTENSITE IN THE TTT DIAGRAM

Although most of the TTT diagram is based on isothermal transformation of austenite, it is common to show the transformation to martensite which occurs only on continuous cooling, e.g., Fig. 10-4 shows the M_s and M_f temperatures. The critical cooling rate for martensite formation is one which allows us to just get through the "gate," i.e., past the "nose," without any transformation of austenite.

At temperatures below M_s, part of the austenite will transform to martensite during cooling, making impossible a strictly isothermal transformation of all austenite below M_s. As temperature is decreased, a larger fraction of austenite is transformed to martensite. If a specimen is held isothermally at a temperature between M_s and M_f, the remaining fraction of austenite will eventually transform to bainite.

10-6-4 CONTINUOUS COOLING TRANSFORMATION

TTT diagrams are available for many steels. These diagrams are used to predict the cooling rate required to produce a given structure or the structure produced by a given cooling rate. This is done by superimposing a cooling curve on the TTT diagram, but caution must be used. TTT curves are obtained by quenching, holding for isothermal transformation, and quenching to room temperature. Continuous cooling generally displaces the pearlite and bainite transformation lines toward (slightly) longer times and (appreciably) lower temperatures than shown on the TTT diagram. The martensite transformation lines, however, are unaffected since they are determined by continuous cooling. Construction of an actual continuous-cooling-transformation (CCT) diagram, analogous to a TTT diagram, is difficult, and few exist. An approximation to such a diagram for eutectoid carbon steel is given in Fig. 10-5.

Keeping in mind the effective shift of the pearlite and bainite lines, we can interpret a continuous cooling curve, superposed on TTT curves, as a series of brief isothermal transformations, each at a successively lower temperature. Intersection of the cooling curve with the beginning TTT curve indicates the start of transformation. Time at temperature, in proportion to the time required for complete transformation, indicates the degree to which the transformation has progressed.

Interpretation of effect of cooling rate in terms of TTT curves is shown in Fig. 10-6. Cooling curve *A* indicates formation of relatively coarse pearlite (see Fig. 10-7). Curve *B* has considerably faster cooling (e.g., oil quench) with formation of very fine pearlite (Fig. 10-7). Curve *C* is a cooling rate rapid enough to avoid formation of ferrite, cementite, pearlite, or bainite. This might be a water quench to produce martensite (Fig. 10-7). A cooling rate which lets transformation start but which recrosses the beginning line before transformation of the austenite is complete (Curve *D*) gives a mixed microstructure, in this case very fine pearlite and martensite.

With continuous cooling of steels, we can produce only pearlite or martensite (or a mixture of the two). There is no mechanism by which pearlite can be transformed into martensite or martensite into pearlite without first reheating to obtain austenite.

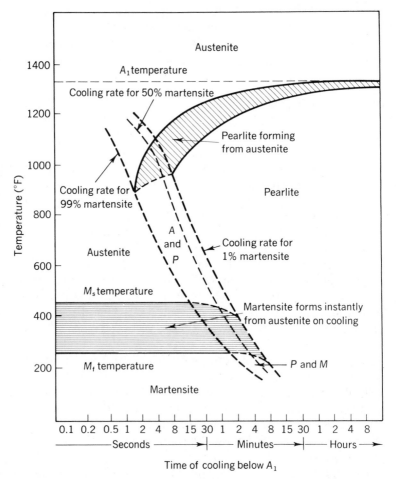

Fig. 10-5. An approximate CCT (continuous cooling transformation) diagram for eutectoid carbon steel (AISI 1080).

EXAMPLE 10-4

Consider AISI 1080 steel which is homogeneous at a temperature of 1400°F. Determine the structure at the end of the treatment cycle indicated.

(a) No further treatment.

(b) Cooling below 900°F in less than 2 sec and reaching room temperature in less than 15 sec.

(c) Cooling to 600°F in 1 sec, holding for 30 min, and cooling rapidly to room temperature.

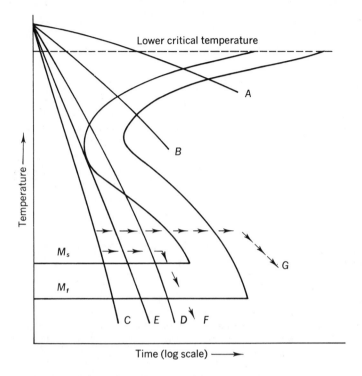

Fig. 10-6. Schematic TTT curves with superposed cooling curves.

(d) Continuous cooling to 800°F in 30 sec, further cooling to room temperature in 5 to 10 sec.

(e) Continuous cooling to 800°F in 1 sec, further continuous cooling to 500°F at the end of 30 sec, with further continuous cooling to room temperature at the end of 1 min.

(f) Continuous cooling to 600°F in 1 sec, holding at 600°F for 4 min, followed by continuous cooling to room temperature at the end of 5 min.

(g) Continuous cooling to 1200°F in 20 min, rapid cooling to 500°F, holding at 500°F for 15 min, with further continuous cooling to room temperature in 20 sec.

(h) Continuous cooling to 500°F in less than 8 sec, holding at 500°F for 30 sec, followed by cooling to room temperature in another 5 to 10 sec.

(i) Continuous cooling to 1200°F in 3 sec, holding for 45 sec, further cooling to room temperature in another 5 to 10 sec.

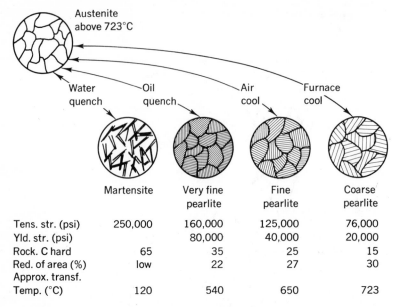

	Martensite	Very fine pearlite	Fine pearlite	Coarse pearlite
Tens. str. (psi)	250,000	160,000	125,000	76,000
Yld. str. (psi)		80,000	40,000	20,000
Rock. C hard	65	35	25	15
Red. of area (%)	low	22	27	30
Approx. transf. Temp. (°C)	120	540	650	723

Fig. 10-7. Schematic interpretation of the effect of cooling rate on approximate transformation temperature, microstructure, and properties of a eutectoid carbon steel.

Solution:

Using Figs. 10-4 and 10-5, as appropriate:

(a) austenite
(b) martensite
(c) bainite with a Rockwell C hardness of about 50
(d) fine pearlite
(e) martensite
(f) martensite plus bainite
(g) coarse pearlite
(h) martensite
(i) pearlite with a Rockwell C hardness of about 32

10-7 STEEL HEAT TREATMENTS

Heat treatments are commercially important and require careful control, since losses due to careless heat treatment may far exceed the cost of the best possible laboratory and control methods. Poor or careless heat treatment may result in failure of parts, either immediately or (often) after expensive machining, fitting, and assembly. Parts often crack during rapid heating or cooling, particularly during quenching. Properly selected steels, quenching media, and techniques are of definite importance, but design for heat treatment is equally important. Temperature gradients should be minimized during heating and quenching, and, therefore, sharp corners, re-entrant angles, and severe and abrupt changes in thickness should be avoided whenever possible. If design cannot be changed, selection of alloy steel may help since some alloys allow slower cooling rates with less drastic temperature gradients. Generous fillets or padding help to equalize temperature and minimize (or prevent) distortion or cracking during quenching. The particular heat treatments used depend on composition and intended application.

Annealing is so general a term that it should not be used without a modifying adjective. Annealing always implies a softening with the extent indicated by the adjective. The purpose of annealing may be to: (1) relieve or remove stresses, (2) induce softness, (3) alter ductility, toughness, electrical, magnetic, or other properties, (4) refine grain structure, (5) remove gases, or (6) produce a definite microstructure. In all cases, three steps are involved: (1) heating to a suitable temperature; (2) holding for a definite time at temperature; and (3) slow cooling.

Stress-relief annealing (or stress-relieving) relieves (essentially eliminates) stresses induced by casting, quenching, machining, cold working, welding, etc. This was discussed in Sec. 9-3 and applies equally well to ferrous and nonferrous metals.

Process annealing is applied to soften metal for further cold working. This was also discussed in Sec. 9-3 and applies to both ferrous and nonferrous metals. For ferrous-base alloys (plain carbon steel and low-alloy steels), this normally means heating to between 1000°F and 1200°F, holding, and cooling at the desired rate (usually by air cooling).

Full annealing for ferrous metals involves heating to above the transformation range, holding at this temperature until austenization is complete, and slow cooling. The steel is usually allowed to cool in the furnace after the heat source is shut off, although any arrangement which allows equally slow cooling is acceptable. This treatment is used for grain refining and for improving machinability. Full annealing for any nonpolymorphic metal or alloy requires heating, holding long enough for homogenization, and slow cooling. Full annealing implies the closest practical approach to equilibrium.

Full annealing is commonly used for hypoeutectoid steels by heating above line *G-S* (Fig. 10-2), holding for one half hour to one hour per inch of thickness, and furnace cooling. The resulting microstructure is ferrite plus pearlite. For hypereutectoid steels, heating is to above 723°C but not normally above *S-E* (Fig. 10-2). This does not completely austenitize the steel but leaves excess cementite mixed with austenite. After cooling, the microstructure is cementite and pearlite.

Normalizing is similar to annealing, but cooling is accomplished in still air rather than in the furnace. For hypoeutectoid steel, the same temperature is used as for full annealing. This treatment gives somewhat greater strength than full annealing plus considerable ductility and a microstructure of ferrite and fine pearlite.

In hypereutectoid steels (especially with 1 w/o or more C), heat treatments which involve slow cooling normally leave the excess cementite as a network around pearlite grains. This network is stable and makes further heat treatment difficult since it is unaffected by ordinary annealing treatments and does not entirely dissolve at the normal hardening temperatures. Normalizing can be used to break up the network and keep it from reforming by heating to above the A_{cm} line (line S-E, Fig. 10-2) to completely austenitize the alloy. The following air cooling is sufficiently rapid to prevent excess cementite from reforming the network or forming large plates. The resulting microstructure is a mixture of fine cementite and fine pearlite. This structure gives somewhat greater hardness and strength but less ductility than fully annealed steel.

Normalizing is also often used to homogenize structures in "mild" (low) carbon steels, particularly heavy forgings. It is common to find nonuniform structure and grain size variation due to unequal amounts of working and possible thermal gradients. Normalizing develops uniformity.

Spheroidizing produces a rounded or globular form of carbide in steel. For example, if a normalized high-carbon steel is reheated to just below 723°C and held, the carbide particles coalesce into spheroids. The microstructure then consists of rounded carbide particles in a ferrite matrix, and the steel is relatively soft and machinable. In addition, this structure is well suited to absorption of carbide during subsequent hardening treatments, thus increasing the potential hardness. Spheroidizing greatly reduces the danger of cracking during hardening, but it does so at the expense of considerably increased heat-treating time.

Quenching is accelerated cooling. This can be accomplished by contact with a quenching medium which may be a gas, liquid, or solid. Since rapid heat removal is desired, fluids, especially liquids, are generally used as quenching media. The choice of quenching medium is based on the desired rate of heat removal and the required temperature interval. Other factors such as boiling point, viscosity, flash point (if combustible), stability under repeated use, possible reactions with the material being quenched, and relative cost are of interest in selection of the quenching medium. Metal-to-metal quenches are also possible, for example, with local heating as in welding, brazing, induction hardening, etc. This "self-quench" can have serious consequences.

Hardening, applied to steels, normally implies heat-treating operations which produce microstructures which are entirely or predominantly martensitic. The procedure involves heating a steel above 723°C or lines G-S

and *S-E* in Fig. 10-2 to partially or completely convert it to austenite, soaking (holding) at temperature long enough to ensure the desired degree of austenization, and cooling at a rate equal to or greater than the critical cooling rate. The critical cooling rate is the slowest rate which permits no austenite transformation above M_s, e.g., curve *E* in Fig. 10-6.

The pseudostable phases below the critical temperature are ferrite and cementite. *There is no method by which these phases can be caused to transform directly into martensite.* Since martensite forms only by quenching austenite, quenching from below the critical range is completely ineffective as a hardening treatment.

Tempering is any process of reheating quench-hardened steel to below the critical range, holding, and cooling at any desired rate. Martensite is extremely hard and strong but has low ductility and toughness. There is ample evidence that steels with martensitic microstructures cannot undergo appreciable plastic deformation or resist sudden applications of load. Tempering imparts ductility and toughness to the strong and hard martensitic steel.

Martensite is a supersaturated solid solution of carbon in alpha iron and, like all supersaturated solutions, is unstable. The crystallographic distortion which determines its properties is perhaps somewhat analogous to the distortion found during the early stages of precipitation hardening. Martensite, like precipitation-hardening alloys, is susceptible to a process similar to overaging. During this process, Fe_3C is precipitated from solution, the unstable body-centered tetragonal matrix structure reverts to stable BCC, and hardness, strength, and brittleness of the steel are markedly reduced. Precipitation of Fe_3C is temperature dependent and probably occurs by normal nucleation and growth. Some indication of the degree of dependence is given by the fact that Fe_3C precipitation takes place in the order of seconds above 200°C, in the order of minutes at 100°C, and in the order of many years at atmospheric temperatures. In fact, undecomposed martensite has been identified in prehistoric steels.

In the early stages of tempering, it is presumed that an unresolvably fine precipitate of cementite appears along certain crystallographic planes. Precipitation of cementite is accompanied by changes in the martensite lattice which slowly approaches that of ferrite. With sufficient time, martensite will entirely disappear. In effect, tempering softens martensite by precipitation of carbide particles from supersaturated solid solution followed by growth of these particles and of ferrite having a normal α lattice. As tempering temperature is increased, the degree of growth of these phases increases, resulting in structures having less hardness but more ductility than martensite. Tempering at 200°C–500°C gives a structure consisting of submicroscopic particles of carbide in ferrite. As tempering temperature is increased, the structure gradually changes (by indistinguishable degrees) into one in which the carbides have a distinctly granular appearance. At still higher tempering temperatures, the final stage of agglomeration of the carbide is spheroidite.

A qualitative indication of the effects of heating and cooling on steel structures is indicated in Fig. 10-8. Representative microstructures of these

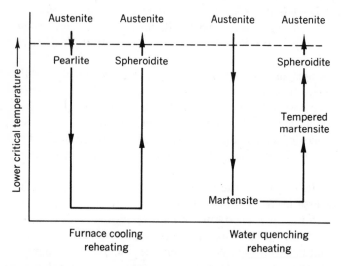

Fig. 10-8. Qualitative representation of the relative effects of cooling rate and reheating of a plain carbon eutectoid steel.

hardened and tempered steels are given in Fig. 10-9. As tempering temperature increases, the microstructures merge gradually and transform continuously from one to the next. Thus, we call all of them "tempered martensite," and we specify the degree of hardness. Tempered martensite structures appear very similar to structures obtained by isothermal transformation of austenite.

When carbide particles in tempered martensite can be observed, they are spheroidal and interrupt the soft, tough, ferrite matrix to a lesser degree than cementite platelets in pearlite. A given steel, at a specific hardness level, is stronger and less brittle as tempered martensite than as pearlite. Tempered martensite generally has one of the best attainable combinations of strength and toughness for a given steel.

10-7-1 INTERRUPTED QUENCHING

In plain carbon steel and in some low-alloy steels there is a narrow gate (Fig. 10-4) which requires a rather drastic cooling rate (water or iced-brine quench) for full hardening. Such rapid cooling rates are often hazardous to the pieces

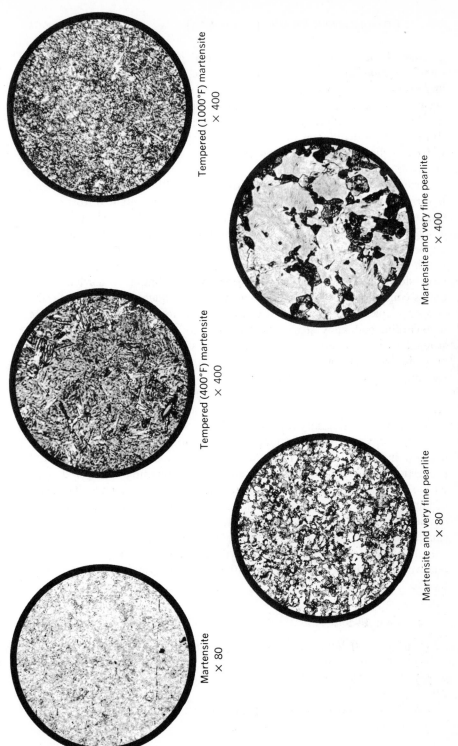

Martensite
× 80

Tempered (400°F) martensite
× 400

Tempered (1000°F) martensite
× 400

Martensite and very fine pearlite
× 80

Martensite and very fine pearlite
× 400

Fig. 10-9. Structures of hardened steels.

because of the stresses developed by thermal gradients and by the $\gamma \rightleftharpoons \alpha$ transformation. The change from FCC to BCC is accompanied by a decrease in density. The temperature at which transformation and volume changes occur decreases as cooling rate increases. Thermal gradients introduced by rapid cooling also affect volume changes since different portions of the piece will transform at different times. The net effect may be to crack or at least warp and distort the piece. Attempts to mechanically straighten a warped piece are apt to cause cracking.

Proper design can mitigate these stresses but will seldom eliminate them. *Interrupted quenching* can also reduce stresses and resultant distortion. In this two-stage process the piece is first quenched in a medium which provides a sufficiently rapid cooling rate to pass through the gate into the "bay" opposite the bainite region. The piece is further quenched in a second medium which cools it rapidly enough to avoid the bainite transformation but much more slowly than the first medium. Since transformation to martensite requires continuous cooling but is essentially independent of cooling rate, the two-stage quenching develops stresses of much lower magnitude. In some high-carbon and alloy steels, the bainite "bay" allows sufficient time for straightening or otherwise mechanically working the piece while it is still austenite, without appreciably affecting its capacity to transform to martensite upon further cooling. Curve *F*, Fig. 10-6, is a typical cooling curve. This process is generally limited to steels of high hardenability which, being alloy steels, are relatively expensive. It can be used for some low-alloy steels, but a drastic quench is necessary for plain carbon steels.

A different type of interrupted quenching to form bainite is shown by curve *G* in Fig. 10-6. This is not a hardening treatment. In structure and properties, however, the bainite formed closely resembles tempered martensite. In general, steels treated thus are tougher and more ductile than steels of tempered martensite having equal hardness and tensile strength. A major limitation is that size is restricted to relatively thin sections so the entire piece can quickly reach the temperature of the quenching bath. For steels of higher hardenability, larger sections can be used. An additional disadvantage is the relatively long time required for the isothermal transformation of austenite to bainite.

EXAMPLE 10-5

Consider AISI 1080 steel. Determine the structure at the end of the treatment cycle indicated.

(a) If the structure formed in Example 10-4b is reheated to 500°F for 30 min.

(b) If the structure formed in Example 10-4b is reheated to 1200°F for 30 min.

(c) If the structure formed in Example 10-4b is reheated to 1350°F for 30 min.

(d) If the structure formed in Example 10-4i is reheated to 500°F for 15 min.

(e) If the structure formed in Example 10-4i is reheated to 1200°F for 30 min.

(f) If the structure formed in Example 10-4i is reheated to 1350°F for 15 min.

(g) If the structure formed in Example 10-4b is reheated to 500°F for 15 min.

(h) If the structure formed in Example 10-4b is reheated to 400°F for 15 min.

Solution:

(a) tempered martensite
(b) spheroidite
(c) austentite
(d) remains pearlite
(e) spheroidite
(f) austenite
(g) tempered martensite which is somewhat harder than that of part (a) above
(h) tempered martensite which is somewhat harder than that of part (g) above

10-8 RETAINED AUSTENITE

The degree of completeness of the transformation of austenite to martensite is not determined by cooling rate or holding time but by the temperature to which the steel is cooled. At any given temperature, quasiequilibrium is reached when a definite proportion of martensite has formed. Since the coefficient of thermal expansion of martensite is considerably greater than that of austenite, further cooling decreases the stress and "equilibrium" is upset. Additional austenite transforms to martensite until a new "equilibrium" is reached. With each decrease in temperature there is an incremental transformation of austenite to martensite until M_f is reached, i.e., the transformation is complete.

Austenite, unless extensively cold-worked, is relatively soft, ductile, and tough but has only moderately high strength. We might expect retained austenite in a hardened steel to reduce strength and hardness and increase plasticity and toughness. The expected effect on strength and hardness is found to some extent. It is possible, with 30–40% retained austenite, to perform light cold-working operations on the quenched alloy, but smaller amounts of retained austenite contribute nothing to plasticity or toughness. In fact, smaller proportions often appear to make hardened steel more brittle than if it were 100% martensite. One possible explanation is that retained austenite is inherently unstable, and deformation makes transformation to martensite possible by providing the required activation energy or by temporarily and locally reducing the stress field.

Retained austenite transforms very slowly at room temperature even without mechanical deformation. This probably occurs by transformation to exceedingly fine bainite with a small increase in total volume. This change is unimportant for many applications but can give trouble in many precisely machined parts and is disastrous in precision gages or test blocks.

Retained austenite is undesirable in many hardened steels, particularly tool steels, since retained austenite prevents attainment of the best possible combination of strength, hardness, toughness, and dimensional stability. Two practical methods exist for eliminating the retained austenite: tempering (previously discussed) and cold treatment, which simply requires cooling the steel below M_f.

10-9 GRAIN SIZE TREATMENTS

Heat treatments to adjust grain size in steel are common and important since additional strength, toughness, and impact resistance are generally associated with fine grain size. Grain size alteration is also a by-product of many steel heat treatments. *Grain size* in steels refers to austenitic grain size, i.e., the size of the austenite grain prior to its transformation (unless ferritic grain size is specified).

Since there is little or no austenite in steel (except at high temperatures or when highly alloyed), the emphasis on austenitic grain size may seem strange. It is the nature and distribution of austenite transformation products that are directly important, but these are strongly influenced by the grain size of the pre-existing austenite. For example, martensite orientation is directly and intimately related to orientation of the parent austenite. As another example, if a hypereutectoid steel is slowly cooled, we find a proeutectoid cementite network around pearlite "nodules." In large austenite grains, this proeutectoid layer is relatively thick and seriously embrittles the steel. If, however, austenite grains are small, the cementite layer is thinner, and the steel is somewhat tougher.

In plain carbon steel or in low-alloy steel, room-temperature microstructure is principally proeutectoid ferrite (or cementite) plus pearlite or martensite. The amount of retained austenite, if any, is small and unstable.

Reheating transforms it to ferrite plus cementite whereas cooling transforms it to martensite. To alter grain size in this transformed structure, we must make austenite grains reappear. The only way to do this is to heat within (or above) the critical range.

Upon slow reheating, the changes produced by cooling reverse. At the lower critical temperature, austenite nucleates at interfaces between ferrite and cementite. Regardless of the specific arrangement of ferrite and cementite there are a large number of interfaces. Consequently, many austenite grains are nucleated just above the lower critical temperature. At this temperature, the average austenitic grain size is about a minimum. Unless the steel is nearly eutectoid composition, however, there will be significant proportions of ferrite or cementite associated with these new austenite grains. Ferrite and cementite can be transformed to austenite only by heating above the upper critical temperature. During this heating the new austenite grains grow. Most heat treatments desire complete austenization, and this coarsening is accepted for hypoeutectoid steels. With hypereutectoid steels, however, temperature required for complete solution of cementite coarsens grain size to an undesirable degree. For these steels, most treatments involve heating only above the lower critical temperature where grain size is an approximate minimum.

In general, steel is held at austenizing temperature just long enough to ensure complete austenization and then cooled enough for further growth to be insignificant. A significant period is required to reach equilibrium grain size ("characteristic grain size") which, at the usual austenizing temperature, is considerably coarser than normally desired. Close control of soaking times minimizes growth. Since characteristic size is not attained during soaking, it is possible (in principle) for grains to continue to grow during cooling. In practice, little growth occurs. No mechanism exists whereby grains grow smaller during cooling.

The basic principle of grain size control is: *In steel, grain refinement occurs only upon heating within or above the critical temperature range and never occurs during cooling.* Once an austenitic grain size has been established during heating, it cannot be refined during cooling, regardless of the cooling rate. It can be refined, however, by subsequent reheating.

10-10 HARDENABILITY

Hardenability is the property of a steel which determines the depth and distribution of hardness induced by quenching. It is *not* an indication of the *hardness*. Hardenability is usually interpreted as the ability to become uniformly hard or to harden in depth. Hardenability can be evaluated by determining the minimum cooling rate which transforms austenized steel to a predominantly or entirely martensitic structure, or by determining the greatest thickness that can be transformed to this structure by a specific quenching treatment.

The principal variables which affect hardenability are alloying elements (see below), grain size (greater hardenability with larger size), and homogeneity. Lack of homogeneity affects hardenability through premature transformation according to local composition. Segregation and incomplete austenization effectively reduce hardenability.

10-11 HARDENING POWER

Although increased carbon content increases hardenability, the effect is small compared to that of alloy content. This must not be confused with the effect of carbon on *hardening power*, which is a measure of the "potential" maximum hardness attainable in a given steel. Maximum potential hardness of steel is determined by carbon content and is essentially independent of other alloy content. Strength and hardness of martensite come from distortion produced by supersaturation of the solution of carbon in α-iron. A low-carbon martensite has little distortion and differs little from ferrite in structure and properties. As carbon content increases, the maximum hardness possible also increases up to about 0.7 w/o C, as indicated in Fig. 10-10.

10-12 ALLOY STEELS

Alloy elements are added to base metal to obtain a desired effect. Addition of a given element does not necessarily make the alloy superior to base metal in every respect, since each element tends to improve some characteristics at the expense of others. Consequently, the amounts of each element added represent a compromise to achieve a desired combination of properties. Ferrous alloys provide excellent examples.

Properties of plain carbon steels depend upon carbon content and treatment. Alloying elements may merely modify the reactions in plain carbon steels as discussed above. In sufficient quantity, however, they can cause extensive alteration of reactions and properties.

Alloying elements commonly used in steels include Mn, Ni, Cr, W, V, Mo, Si, Cu, Ti, Co, Al, and Zr. These elements may be divided into two groups: those like Mn, Cr, W, V, and Mo, which form carbides, are generally BCC and are alpha stabilizers; those like Ni, Cu, and Si, which do not form carbides, are generally FCC and are gamma stabilizers.

In general, alloy additions which are retained in solid solution should

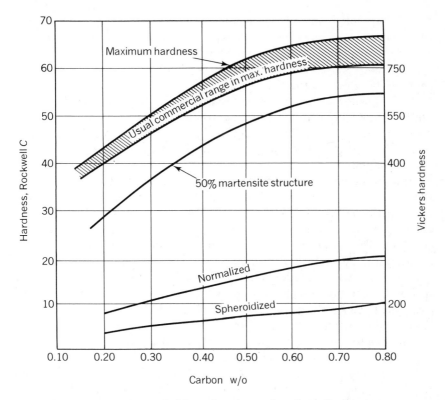

Fig. 10-10. Hardness of plain carbon, low and medium alloy steels as a function of carbon content. Data are approximate since hardness depends on spacing of lamellae in pearlite and on grain size.

increase strength, elastic limit, and hardness while decreasing ductility. In the case of nickel, for example, decrease in ductility is relatively slight whereas improvement in strength and hardness is appreciable. Alloy additions that form carbides which separate out on slow cooling are seldom used when free carbides will exist because of their relative brittleness. The value of carbide-forming additions rests upon: (1) the readiness with which carbides can be absorbed at high temperatures; (2) the readiness with which carbides can be retained in solid solution upon cooling at a suitable rate; and (3) the alloy's resistance to tempering after absorption of carbides.

The carbides are of three types: (1) simple carbides resulting from direct chemical combination of carbon and alloy element; (2) isomorphous mixtures of iron carbide and the carbide of the alloy element; and (3) double (complex) carbides of the alloy element and iron. At sufficiently high temperature these carbides dissolve in γ-iron and may be retained (at least partly) in solution at room temperature in γ- or in α-iron. Actual behavior depends on amount of carbon, particular alloy element, amount of alloy, austenization temperature, and cooling rate. With a relatively small amount of alloy addition, carbides may be "completely" rejected on slow cooling and will be embedded in a pearlitic matrix. If carbides are retained in solution in γ-iron at room temperature, the steel is austenitic.

In many alloy steels, phases may be produced during slow cooling which can be produced in carbon steels only by very rapid cooling. Carbon steels, even with very drastic cooling, cannot be retained in a wholly austenitic condition, although many high-alloy steels remain completely austenitic after slow cooling.

Martensitic or austenitic structures occur on slow cooling because some alloy additions depress the critical range. As a general rule: (1) if the lower critical temperature remains above 600°C, the steel is pearlitic; (2) if the lower critical temperature is depressed below about 300°C, the steel is martensitic; and (3) if the lower critical temperature is depressed below atmospheric temperature, the steel is austenitic.

There is a wide variation among alloy elements in effective shifting of critical temperature range with some elements forming martensitic and austenitic steels much more readily than others. Some elements never cause sufficient depression of the critical range to produce martensitic steel, and, in fact, they raise, rather than lower, the critical temperature range.

Metals tend to be more soluble in each other if they have the same crystal structure. Consequently we expect BCC elements to be more soluble in α-iron and FCC elements more soluble in γ-iron. Each class of elements retards allotropic transformation to the form in which they are less soluble; i.e., it stabilizes the form in which they are more soluble. Generally, addition of any alloying element tends to decrease the carbon content of the eutectoid; i.e., alloying elements tend to decrease the amount of free ferrite.

The M_s and M_f lines in the TTT diagram are always isothermal for a given alloy. Both temperatures are decreased by increased content of carbon (Fig. 10-11) and all other elements, except Al and Co. This can produce appreciable quantities of retained austenite at room temperature.

Each alloy addition is made for some specific purpose. Choice of elements depends principally upon intended use of the alloy. Choice and amount of each element depend, to some degree, on other elements in the alloy. The principal alloying elements for steel and their general effects are indicated in Table 10-1.

10-12-1 ALLOY CLASSIFICATIONS

Steels are generally considered in three classes: (1) plain carbon steels; (2) low-alloy steels; and (3) high-alloy steels. The first includes all iron-base alloys

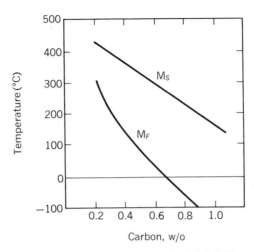

Fig. 10-11. Martensite start (M_s) and martensite finish (M_f) temperatures as a function of carbon content in plain carbon steel.

with carbon additions (plus impurities) and has been discussed in considerable detail.

The low-alloy steels are subdivided into two broad groups, the AISI steels and the high-strength low-alloy structural steels. The AISI (American Iron and Steel Institute) steels are generally used in machine construction and normally have less than about 5% total addition of elements such as Cr, Ni, Cu, Mn, Mo, V, etc. Although machinability, wear resistance, weldability, formability, etc., are of significance, the property of major interest in this group is generally hardenability. The high-strength low-alloy structural steels are essentially ordinary structural steels with 1 to 2% alloy addition. These have the advantages over low-carbon structural steel of: (1) increased yield strength by solid solution hardening with little change in cold formability; (2) balancing the character of austenite decomposition between the desirability of (a) increasing strength by forming fine pearlite (with proeutectoid ferrite) on cooling after hot working and (b) avoiding hardening and embrittlement after welding; and (3) an increase in resistance to atmospheric corrosion.

The high-alloy steels usually have several percent of one or more alloy additions and include several groups. Tool steels are high-quality alloys used for tools, dies, etc., which require special characteristics of hardenability, wear resistance, resistance to softening on heating, and the like. High-temperature steels require good mechanical properties, particularly strength, at elevated temperatures. Stainless steels are used for improved corrosion resistance.

Table 10.1. Specific effects of alloys in steel*

Element	Solid solubility		Influence upon ferrite	Influence upon austenite (hardenability)	Influence exerted through carbide		Principal functions
	In gamma iron	In alpha iron			Carbide forming tendency	Action during tempering	
Aluminum Al	1.1% (Increased by C)	30% ±	Hardens considerably by solid solution	Increases hardenability mildly, if dissolved in austenite	Less than Fe (Graphitizes)	—	1. Deoxidizes efficiently 2. Restricts grain growth (by forming dispersed oxides or nitrides) 3. Alloying element in nitriding steel
Chromium Cr	12.8% (20% with 0.5% C)	Unlimited	Hardens slightly; increases corrosion resistance	Increases hardenability moderately, similarly to manganese	Greater than Mn; Less than W	Mildly resists softening	1. Increases corrosion and oxidation resistance 2. Increases hardenability 3. Adds some strength at high temperatures 4. Resists abrasion and wear (with high carbon)
Cobalt Co	Unlimited	80% ±	Hardens considerably by solid solution	Decreases hardenability as dissolved	Similar to Fe	Sustains hardness by solid solution	1. Contributes to red hardness by hardening ferrite
Manganese Mn	Unlimited	15 to 18%	Hardens markedly; reduces plasticity somewhat	Increases hardenability moderately, similarly to chromium	Greater than Fe; Less than Cr	Very little, in usual percentages	1. Counteracts brittleness from the sulphur 2. Increases hardenability inexpensively 3. Forms batter resistant steel (high Mn, high C)
Molybdenum Mo	3% ± (8% with 0.3% C)	32% (Less with lowered temperature)	Provides age-hardening system in high Mo-Fe alloys	Increases hardenability strongly (Mo > Cr)	Strong; Greater than Cr	Opposes softening, by secondary hardening	1. Deepens hardening 2. Raises coarsening temperature of austenite 3. Raises hot and creep strength, red hardness 4. Enhances corrosion resistance in stainless 5. Forms abrasion-resisting particles
Nickel Ni	Unlimited	25% ± (Irrespective of carbon content)	Strengthens and toughens by solid solution	Increases hardenability mildly, but tends to retain austenite with higher carbon	Less than Fe (Graphitizes)	Very little	1. Strengthens unquenched or annealed steels 2. Toughens pearlitic-ferritic steels (especially at low temperature) 3. Renders high chromium-iron alloys austenitic

*Courtesy of Metal Progress.

Table 10-1. Specific effects of alloys in steel (*cont.*)

Element	Solid solubility		Influence upon ferrite	Influence upon austenite (hardenability)	Influence exerted through carbide		Principal functions
	In gamma iron	In alpha iron			Carbide forming tendency	Action during tempering	
Phosphorus P	0.5%	2.5% (Irrespective of carbon content)	Hardens strongly by solid solution	Increases hardenability, similarly to manganese	Nil	—	1. Strengthens low carbon steel 2. Increases resistance to corrosion 3. Improves machinability in free-cutting steels
Silicon Si	2% ± (9% with 0.35% C)	18.5% (Not much changed by carbon)	Hardens with loss in plasticity (Mn < Si < P)	Increases hardenability more than nickel (Ni < Si < Mn)	Less than Fe (Graphitizes)	Sustains hardness by solid solution	1. Used as general purpose deoxidizer 2. Alloy for electrical and magnetic sheet 3. Improves oxidation resistance 4. Increases hardenability of steels carrying nongraphitizing elements 5. Strengthens low alloy steels

Titanium Ti	0.75% (1% ± with 0.20% C)	6% ±, (Less with lowered temperature)	Provides age-hardening system in high Ti-Fe alloys	Probably increases hardenability very strongly, *as dissolved.* Its carbide effects reduce hardenability	Greatest known (2% Ti renders 0.50% carbon steel unhardenable)	Persistent carbides probably unaffected. Some secondary hardening	1. Fixes carbon in inert particles
							  (a) Reduces martensitic hardness and hardenability in medium chromium steels
							  (b) Prevents formation of austenite in high chromium steels
							  (c) Prevents localized depletion of chromium in stainless during long heating

Tungsten W	6% (11% with 0.25% C)	32%, (Less with lowered temperature)	Provides age-hardening system in high W-Fe alloys	Increases hardenability moderately in small amounts	Strong	Opposes softening by secondary hardening	1. Forms hard, abrasion resistant particles in tool steels 2. Promotes red hardness and hot strength
Vanadium V	1.5% ± (4% with 0.20% C)	Unlimited	Hardens moderately by solid solution	Increases hardenability very strongly, *as dissolved*	Very strong (V < Ti or Nb)	Maximum for secondary hardening	1. Elevates coarsening temperature of austenite (promotes fine grain) 2. Increases hardenability (when dissolved) 3. Resists tempering and causes marked secondary hardening

327

Many possibilities exist for structural modification (and thus property alteration or development) in alloy steels. As an example, one type of martensitic reaction, known as *maraging*, does not depend on carbon for developing strength. In one alloy containing 18 w/o Ni, 0.4 w/o Ti, 8 w/o Co, and 5 w/o Mo, heating martensite to 500°C causes precipitation of fine particles of Ni_3Ti and Fe_2Mo to produce appreciable strength with adequate ductility.

QUESTIONS AND PROBLEMS

1. What is steel? What is meant by eutectoid, hypoeutectoid, hypereutectoid steels? What is the relationship between them?

2. Define: ferrite, cementite, austenite, pearlite, martensite, bainite. What are the general characteristics of each?

3. Of what significance is the existence of the $\gamma \rightleftharpoons \alpha$ allotropic transformation in steels?

4. What is meant by "critical range" or "critical temperature range"? How is this related to heating and cooling? What practical significance does it have?

5. What is the distinction between plain carbon steels and alloy steels? Between low-alloy and high-alloy steels?

6. Describe qualitatively the changes that occur upon cooling and heating plain carbon steels with 0.20, 0.80, and 1.20 w/o carbon through the critical range. Sketch and label the microstructures found in these alloys at room temperature after reasonably slow cooling.

7. What are the failings of the Fe-C "equilibrium" diagram (Fig. 10-2) from the viewpoint of practical heat treatment?

8. Define: annealing, quenching, normalizing, tempering.

9. Describe the mechanism of the transformation of austenite to (a) pearlite; (b) ferrite; (c) cementite; (d) bainite; (e) martensite.

10. Discuss the differences and similarities between pairs of the transformations discussed in question 9.

11. Compare the structures and properties of austenite, ferrite, cementite, pearlite, martensite, and bainite in plain carbon steels.

12. Explain the characteristic "S" shape of the TTT curves, i.e., very slow at just below the transformation temperature and just above the M_s but very rapid at 550°C (1030°F). What practical significance does the "knee" on a TTT curve have?

13. What information is supplied by TTT curves? By CCT curves? What relation, if any, does this information have to question 7?

14. Discuss the relative usefulness of the equilibrium diagram, the TTT diagram, and the CCT diagram in predicting the effect of (a) normalizing; (b) water quenching a $\frac{3}{4}$ inch bar of eutectoid carbon steel.

15. What feature of an alloy or a pure metal allows it to undergo a martensitic transformation?

16. (a) A plain carbon steel has 0.40 w/o C (AISI 1040). What is the equilibrium structure and relative fraction of constituents at 724°C (1335°F)? Sketch the microstructure. (b) This specimen is now cooled to 700°C (1292°F) within 10 seconds and allowed to transform isothermally. What is the structure at the end of 10 seconds? What are the relative fractions of the constituents? What is the structure after transformation is complete? What are the relative fractions of the constituents? Sketch microstructures. (c) If the specimen is cooled from 724°C (1335°F) to 400°C (752°F) in 1 sec what is the structure at that time? What are the relative fractions of the constituents? If isothermal transformation then goes to completion, what is the structure? What are the relative fractions of the constituents? Sketch microstructures. (d) If the specimen is cooled from 724°C (1335°F) to 600°C (1112°F) in 2 sec and then to 430°C (806°F) at the end of 8 sec and then to 100°C (212°F) at the end of 4 min and then to room temperature, what is the structure? What are the relative fractions of the constituents? Sketch microstructures.

17. Although martensite is more stable than austenite below M_s, why does a sample of steel only partially transform from austenite to martensite at temperatures between M_s and M_f?

18. In problem 16 determine the relative amounts of proeutectoid and eutectoid ferrite at room temperature after cooling from 700°C (1292°F).

19. Austenite is denser than ferrite. (a) Which phase will be more stable at high pressure? (b) Will high pressure tend to raise or lower the temperature (910°C or 1670°F) of the $\alpha \longrightarrow \gamma$ transformation? (c) Will high pressure tend to increase or decrease the austenite (γ) field in Fig. 10-2?

20. A solid cylinder (25-cm or 10-in. diameter) of steel is quenched from 900°C (1652°F) to room temperature. What is the resulting structure?

21. Compare the effects on mechanical properties of rapidly quenching a eutectoid carbon steel and an age-hardenable alloy. How does the hardness of each respond to further heat treatment?

22. Discuss the differences and similarities between precipitation hardening and the austenite $\longrightarrow$ martensite transformation.

23. In what respect, if any, is the martensitic transformation in iron-carbon alloys unique in comparison with other alloy systems?

24. Discuss differences and similarities between martensite formation and deformation twinning.

25. Describe the mechanism for obtaining bainite by direct quenching.

26. Describe the mechanism by which pearlite can be transformed to martensite without producing austenite as an intermediate step. Can martensite be transformed to pearlite by a similar mechanism?

27. Can pearlite be formed directly from bainite? If so, how? Can pearlite be formed by tempering martensite? If so, how?

28. What happens to pearlite if it is held for a long period of time just below the lower critical temperature?

29. Define tempering. How is it accomplished? What are the general effects of tempering on microstructure and properties?

30. What is interrupted quenching? What practical significance does it have?

31. Discuss the effectiveness of quenching from below the lower critical temperature as a hardening procedure. Explain.

32. Define hardenability. What relation does it have to critical cooling rate? What are the effects of carbon content, alloy content, grain size, and homogeneity on hardenability?

33. Discuss differences and similarities between hardenability and hardening power.

34. If a steel has high hardenability, does this necessarily mean the steel will be hard after rapid quenching? Why?

35. We often hear "Steel is made hard by quenching." List and discuss at least three requirements that must be met to justify the statement.

36. What is meant by "retained austenite"? What practical significance does it have? What procedures exist for eliminating retained austenite? How does each procedure operate?

37. What is meant by "grain size" in steels? What practical significance does it have? How can grain size be controlled? Is it possible in an inherently coarse-grained steel to produce a finer grain size than that found in some inherently fine-grained steels? Explain.

38. If the hardness of martensite is due almost completely to carbon content, then why are various alloying elements found in steel?

39. What is an alloy steel? What are some of the advantages and disadvantages of the use of alloy steels?

40. Alloying elements such as Mn, V, Ni, etc., in steel decrease the rate of transformation of austenite to pearlite. What influence does this have on the possibility of obtaining martensite on quenching austenite?

41. What relationship do the "carbide formers" have to the "alpha stabilizers"? To the "gamma stabilizers"?

42. What are the three types of carbides formed in alloy steels?

43. What is meant by cast iron?

44. What happens during the slow cooling of a pure iron-carbon alloy, with three percent carbon, from 2800°F to room temperature? Describe in terms of microstructure as well as in words (similar to question 6).

45. Cementite (Fe_3C) is metastable. In the reaction $Fe_3C \longrightarrow 3Fe + C$ the carbon may form as flakes or as nodules (spheroids). What is the difference in conditions which will form the two different microstructural arrangements?

11

STRUCTURAL POTPOURRI

11-1 INTRODUCTION

Properties of materials depend upon composition (e.g., Fig. 7-3) and structure (e.g., Fig. 10-7). We are certainly conscious of the differences in properties among a pure metal, an alloy based on that metal, and a ceramic based on the metal. Pure aluminum, for example, is soft and ductile. Addition of Cu (about 4 w/o) to Al produces an alloy with appreciably greater strength and hardness but less ductility. This is also an alloy which can be heat treated (precipitation hardened) to greatly alter the properties from those of the alloy in equilibrium. Aluminum "whiskers" without dislocations are still stronger. Aluminum combines with oxygen to form alumina (Al_2O_3), which is used in grinding wheels for cutting hardened steel. If the same alumina is one large crystal with a little Cr_2O_3, we have a ruby. If TiO_2 replaces Cr_3O_3, we have a blue sapphire.

The information covered in Chapters 7 and 8 applies in general to all materials, at least in kind if not in degree. Chapters 9 and 10, however, which deal with heat treatments, apply specifically to metals. It is the purpose of this chapter to discuss analogous procedures which apply to other materials.

11-2 CERAMICS

As with metals, the properties of ceramics can be altered by changes in chemical composition. Unlike metals, however, there are really no heat treatments, as such, which can be used to modify the properties of the oxides, carbides, nitrides, borides, etc. Basic knowledge that has been gained about microstructure and physicochemical behavior of ceramics has provided principles for their construction and has led to techniques for making them. These techniques represent refinements of the basic procedure outlined in Chapter 12 under the powder process. By paying careful attention to control of chemical composition, purity, particle size and uniformity, and arrangement and packing of atoms, it is possible to synthesize high-quality ceramics in a wide variety. New methods include use of very high pressure to consolidate fine particles prior to firing; decomposition of chemicals at high temperature to deposit a ceramic coating on a substrate; use of a plasma gun that liquefies fine particles at high temperature and sprays a paintlike ceramic coating; precipitation of ceramics from aqueous solution at high pressure and temperature (hydrosynthesis); and slow solidification of liquids to form individual ceramic crystals with "ideal" characteristics.

11-2-1 ORDER-DISORDER TRANSFORMATIONS

Order-disorder transformations in metals were discussed in Secs. 7-2-4 and 9-5-3. Such transformations are common in metals where the nearest neighbors in an AB or AB_3 alloy can be ordered or disordered without a large difference in energy. In ionic materials, however, an interchange of a cation with any one of its nearest neighbor anions requires so much energy that such a change is unknown. Those order-disorder transformations possible in ionic materials are related to anion positions in the anion substructure or to cation positions in the cation substructure. Even then, they involve the second nearest coordination ring, not the nearest.

The most important cases in ceramic systems are found in materials having two different types of cation sites such as in the spinel structure having some cations on octahedral sites and some on tetrahedral sites. Essentially all ferrites, for example, having the spinel structure are disordered at high temperature and ordered at low temperature. A somewhat different form of transformation occurs when unoccupied sites are available in the ordered structure. For example, with Ag_2HgI_4, three-fourths of the cation sites are filled in an ordered manner. Above 500°C, there is disorder with one Hg and two Ag cations randomly arranged on the four available sites.

11-2-2 POLYMORPHIC TRANSFORMATIONS

Polymorphic transformations can be viewed as two general types which differ in the kind of changes which occur in the crystal. They can also be viewed as two general types with respect to the rate of transformation.

Displacive transformations are the less drastic of the two types. If we start with a regular structure, Fig. 11-1(a), it can be changed by displacive transformations, Fig. 11-1(b) or (c), which simply alter or distort the structure without breaking interatomic bonds or changing the basic structure. The

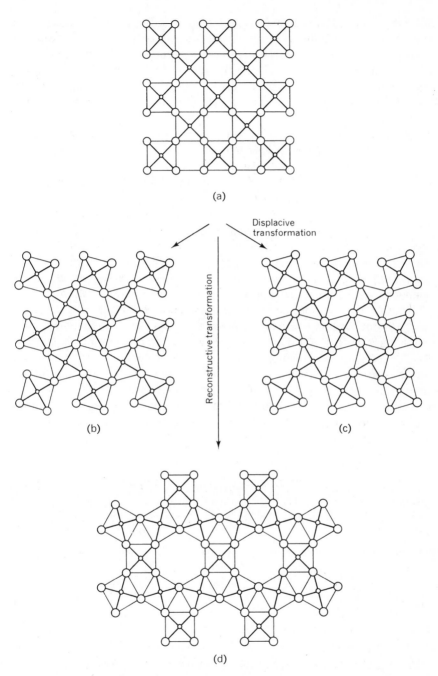

(a)

Displacive
transformation

Reconstructive transformation

(b)

(c)

(d)

Fig. 11-1. Starting structure (a) can transform into (b) or (c) by a displacive transformation or into (d) by a reconstructive transformation.

resulting structure is thus a derivative of the starting structure. Displacive transformations commonly occur rapidly without diffusion. Transformation at any temperature is nearly instantaneous through a shearing motion of atoms. (This is essentially the case for austenite → martensite transformation.) Displacive transformations cannot be controlled by annealing or quenching conditions. The amount transformed is a function of temperature only.

Reconstructive transformations involve not only displacement of atoms but also breaking of interatomic bonds and formation of new ones. If we start with a regular structure, Fig. 11-1(a), it can be changed to a different form, Fig. 11-1(d), by a reconstructive transformation. These transformations tend to be sluggish. They are susceptible to control by quenching since high-temperature forms can often be cooled below the transformation temperature without developing the thermodynamically stable structure. The kinetics of reconstructive transformations show the characteristics of nucleation and growth. Reconstructive transformations require a high activation energy and often will not occur unless there is some way of circumventing the activation energy barrier such as mechanical work or the presence of a solvent.

Silica (SiO_2) provides an excellent example of both types of transformation. The equilibrium phase diagram was shown in Fig. 7-2 and a metastable phase diagram in Fig. 8-1. A better picture of the relationships is given in Fig. 11-2. The seven polymorphs shown involve three basic structures. The transformations among these three structures are reconstructive transformations which, if they occur at all, are very sluggish and need addition of solvents to occur in "reasonable" times. The displacive transformations, on the other hand, are very rapid and cannot be avoided.

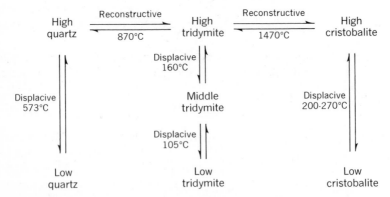

Fig. 11-2. The polymorphic forms of silica and the transformations among them.

11-2–3   GLASS

Glass has been defined as "an inorganic substance in a condition which is continuous with, and analogous to, the liquid state of that substance, but which, as the result of a reversible change in viscosity during cooling, has attained so high a degree of viscosity as to be for all practical purposes rigid."*

Glass is, therefore, something more than a supercooled liquid. The silica (SiO_2) tetrahedron (SiO^{4-}) (Fig. 4-20) in which 4 oxygen atoms are symmetrically arranged about each silicon atom is the major building block in glass. There is definite first order coordination but no long-range periodicity. Comparable structures of a crystal and a noncrystalline network are shown in Fig. 4-21. A random network structure, Fig. 4-21(b), is found in the most important noncrystalline solids such as oxide glasses, hydrogen-bonded inorganic solids, and organic compounds containing hydroxyl groups. The silicate glasses, such as sodium silicate (Na_2O plus SiO_2 in a wide range of proportions), have major importance as commercial glasses. These are modified in a broad spectrum of complexity by adding cations such as Ba^{2+}, Pb^{2+}, Sr^{2+}, Ca^{2+}, Mg^{2+}, Zn^{2+}, Al^{3+}, Ti^{4+}, Zr^{4+}, and so on.

Addition of such fluxing ions often causes the melt, on cooling, to separate into two or more distinct but intermixed phases of clearly different compositions. The distribution of these different liquids in the rigid glass can have a profound effect on mechanical strength, electrical resistance, chemical resistance, and optical clarity.

The major commercial glasses are soda-lime-silica combinations with compositions in the range of 15 Na_2O, 13 CaO + MgO, 72 SiO_2. These glasses are manufactured from inexpensive raw materials, can be easily worked at moderate temperatures, and are chemically stable. We also find a number of special purpose glasses in which silica is either a minor constitutent or missing, although the structure is similar to that found in silicate glasses. These glasses involve compounds such as BeF_2, GeO_2, B_2O_3, Na_2O, phosphates, sulfides, and selenides.

Since glass structure is independent of temperature, it is clearly distinguishable from liquid structure. This is indicated in Fig. 4-1. If a liquid crystallizes on cooling, there is a discontinuous change in volume at the melting point. If there is no crystallization, the specific volume continues to decrease at about the same rate as above the melting point until the temperature reaches a "transformation range" in which the coefficient of thermal expansion decreases. The glass structure does not relax below this temperature range at the cooling rate used and the coefficient of thermal expansion is about the same as that of the crystalline structure. If a slower cooling rate is

*G. W. Morey, *The Properties of Glass*, 2nd ed., New York, Reinhold Publishing Corporation, 1954.

used, the structure has more time to relax, the supercooled liquid persists to a lower temperature, resulting in a higher density glass.

The "point" at which the change in slope of the volume-temperature curve occurs, i.e., the temperature at which the glass becomes rigid and assumes some characteristics of the solid state, is known as the *fictive temperature*. The fictive temperature depends on the composition and cooling rate, the faster rates yielding higher fictive temperatures. The transformation range is thus the range of fictive temperatures determined by the fastest and slowest feasible cooling rates. The fictive temperature is the equilibrium temperature related to the particular noncrystalline structure of a given sample which is obviously not a phase equilibrium. Fictive temperature cannot be used as a quantitative indication of the state of a glass but does provide a useful tool for qualitative understanding of changes occurring.

Annealing processes for metals were discussed in Secs. 9-3 and 10-7. *Annealing* is also used for a somewhat similar process in glasses. If cooled from the working temperature without further heat treatment, glass is usually in an unevenly strained condition resulting in substantial residual stresses. These may be of such magnitude in critical locations that breakage will result during further cooling or in subsequent use. These residual stresses (or strains) are removed by annealing which requires heating to some appropriate temperature so that most or all of the strain will be relieved by viscous flow followed by cooling at a rate which leaves no more than a small predetermined strain in the glass when cold.

A different type of heat treatment is used to produce *tempered* (*thermally toughened*, *case hardened*, or *chilled*) glass. In this case, after forming, glass is reheated to a temperature well above the annealing range (often to the softening range) and then rapidly cooled on the surface by air jets or by immersion in oil or molten salt. The surface layers contract while the glass is sufficiently fluid for the stresses to remain low. The temperature gradient is maintained while the glass cools. When completely cooled, there is a residual stress pattern (approximately parabolic) of compressive stresses on the surface and tensile stresses in the interior. The actual stress distribution is complex and is a function of glass composition, geometry, starting temperature, and cooling rate. Great care is required in this process since the glass will fracture into many small pieces as soon as a crack is started although a rather large force may be needed to start the crack. Tempered glass is used in rear windows of automobiles because the glass is very strong. In addition, if a crack is started, even at a single isolated point, tempered glass will break into a large number of small, interlocking fragments which are far less likely to cause serious wounds than the large, jagged fragments of annealed glass.

Another useful heat treatment converts materials formed as glasses into crystalline materials. The resulting products have useful property combinations which cannot be developed by other processes. These products are known by names such as Pyroceram (Corning Glass Works) and CER-VIT (Owens-Illinois). These products are fabricated by glass forming processes which are rapid, economical, automatic, and can be controlled to small dimensional tolerances with no porosity. Nucleating agents, such as TiO_2 or metal constituents, are added to the melt in which they dissolve. When the glass cools, these agents form nucleation sites for crystals which grow on subsequent heating. By controlling temperature and time, it is possible to produce a ceramic body with the same size and shape as the original glass but approaching 100% crystallinity. This material is much stronger and harder with greater abrasion and impact resistance than the parent glass. By proper choice of starting composition (e.g., $Li_2O \cdot Al_2O_3 \cdot 4SiO_2$) it is possible to have a very low thermal coefficient of expansion and thus high resistance to thermal shock (Sec 13-5-3). It is also possible to vary this coefficient from -20 to $+120 \times 10^{-7}/°C$.

Many chemical reactions, e.g., decomposition, precipitation, dissolution, ion-exchange, and oxidation and reduction can be carried out directly within the glass. These can usually be stopped, or frozen in, at any point desired. Control of these processes has made possible glasses that lase, lighten, or darken in response to light, exhibit semiconduction and photoconduction, fluoresce, and selectively transport ions.

11-2-4 CLAY BASE PRODUCTS

Structural clay products include such well-known items as building brick, tile, sewer pipe, and similar products. These are normally made from local clays and distributed locally since they must be inexpensive to compete with other construction materials. Clays are fine-grained hydrated aluminum silicates which are found in plate or flake form. Basic clay structures have combinations of a $(Si_2O_5)_n$ layer of SiO_4 tetrehedra joined at the corners with an $AlO(OH)_2$ layer of alumina octahedra. The most common clay is kaolinite, $Al_2(Si_2O_5)(OH)_4$. The other major clay type is montmorillonite, typified by pyrophyllite, $Al_2(Si_2O_5)_2(OH)_2$. The structure after firing (Sec. 12-4) is large grains of secondary constituents in a matrix of fine-grained mullite ($3Al_2O_3 \cdot 2SiO_2$) and glass.

Whiteware products include tableware, sanitary ware, electrical porcelain, artware, dental porcelain, hotel china, etc. These are mixtures of clay, feldspar, and flint. Clay provides fine particles, good plasticity for forming, and forms a more-or-less viscous liquid during firing (Sec. 12-4). Feldspar acts as a flux to form a viscous liquid at the firing temperature and aids in vitrification. Flint is an inexpensive filler which is inactive at low firing temperatures but forms a viscous liquid at high temperatures. The final structure is very heterogeneous and depends heavily on the composition and firing conditions. Whiteware is normally finished for the market by applying a glaze. This is commonly prepared by milling clay, frit (premelted silicate glass) and water. The mixture is applied to the ceramic body by brushing, spraying, or dipping. The whole body is then fired to produce a thin silicate glass coating over the surface. Enamels are developed on metallic substrates in a similar manner. Various colors can be obtained by adding appropriate colorants to the glaze or enamel before firing.

High-strength clay base products can be produced with very little porosity. Unfortunately, low porosity products fail readily under application of mechanical or thermal stresses. Thus, some porosity is desirable. Pores can occur in widely differing shapes, sizes and distributions as functions of initial particle size and firing temperature. Duration of firing is not a major factor after some minimum time at temperature. Control of porosity is important since elastic moduli and flexural strength decrease linearly with increased porosity between 5 and 33%.

11-2-5   CONCRETE

Concrete, a widely used construction material, is made by mixing aggregates (sand, rock, and/or various other materials) with water and (Portland) cement. The mixture is poured into an appropriate form, following which setting and bonding result from reaction between water and cement to form a body in which aggregates are "cemented" together by a continuous cement matrix.

Portland cement has major constituents of tricalcium silicate ($3CaO \cdot SiO_2$) and dicalcium silicate ($2CaO \cdot SiO_2$) with smaller amounts of $3CaO \cdot Al_2O_3$, $4CaO \cdot Al_2O_3 \cdot Fe_2O_3$, CaO, MgO, $CaSO_4$, Na_2O, and K_2O in varying amounts. The raw materials are placed in a rotary kiln and sintered to produce a liquid phase in a fraction of the charge. During rapid cooling, a transformation in the dicalcium silicate produces a sufficient volume change to break up the particles into a powder.

Setting of Portland cement results from partial dissolution and hydration in the cement when mixed with water. An initial "flash set" occurs by hydration of tricalcium aluminate to form hydrate crystals which coat the silicate particles and inhibit further hydrations. This process has a high heat of hydration which can lead to thermal cracking in some situations. Cement hardens by hydration of dicalcium silicate and tricalcium silicate. This requires several days to attain reasonable strength (30 days for 70-80% of maximum attainable) but the hardening continues for years. Hydration is accompanied by heat release, creating a major problem in massive concrete structures. In Hoover Dam, for example, many miles of pipe were installed before pouring concrete. Water was circulated through these to remove the heat of hydration. When most of the heat had been removed, the pipes were then filled with concrete.

The properties of concrete depend on the properties of solidified cement paste, porosity (both capillary pores and large ones from excess water or trapped air), and aggregates.

11-3   POLYMERS

As with metals and ceramics, polymers can exhibit substantial property differences as a function of chemical composition, molecule size, and amounts of cross bonding. Heat treatments, in the sense of those used for metals (and to a lesser degree for ceramics), cannot be used to modify structure and properties of polymers. Temperatures and cooling rates obtained in various fabrication techniques (Sec. 12-6) may have some effect.

In general, efforts to produce polymers which are more rigid, more resistant to temperature, more resistant to corrosive chemicals, etc., focus on crystallization, crosslinking, and stiff chains. The first two, crystallization and crosslinking, have long been applied as indicated in Chapter 4. Crystallization is a phenomenon which depends on orientation of the chains, is not temperature dependent, and can be reversed with decomposition of the polymer. Crosslinking depends on a chemical reaction (rather than on chain orientations), is strongly accelerated by increased temperature, and is not reversible. Inherently stiff chains can be produced by a number of methods. One is to "hang" bulky groups of atoms on the chain to restrict bending. Certain polymers are intrinsically stiff in that the chain can twist on its bonds but cannot bend. Some monomers have ring-shaped groups and thus have inherent rigidity. Still other polymers are based on chains made of benzene-type rings which polymerize into "ladders."

It has been clearly established that each of the three phenomena above improves the properties (at least strength) of polymers. It has also been clearly demonstrated that substantial gains in strength, etc., are made by all three pairs of these phenomena. Effort has been directed to producing even more superior results by combining all three phenomena. Even though this effort has been mainly exploratory, some success has been achieved.

11-3-1   GLASS TRANSITION

As a noncrystallizable polymer is cooled from an elevated temperature, there is a significant and rapid change in properties over a narrow temperature range. This is quite similar to the behavior of a glass (Sec. 11-2-3). The change in specific volume, for example, as a function of temperature follows the upper curves in Fig. 4-1. For the change in slope in Fig. 4-1, known as fictive temperature for glasses, there is a corresponding *glass transition temperature*, T_g, for polymers. In addition to a change in specific volume, coefficient of thermal expansion, specific heat and other physical properties, mechanical properties also have changes over the same narrow temperature range. Of major importance is the fact that the polymer undergoes a ductile to brittle transition at T_g on cooling. The term glass transition temperature derives from this ductile-brittle transition.

Table 11-1 lists glass transition temperatures for a number of polymers. In general, materials with bulky side groups or highly polar groups have relatively high transition temperatures. Polymers with small or no side groups and good backbone mobility or with long and flexible side chains tend to have relatively low transition temperatures. Glass transition has been found in semicrystalline, as well as noncrystalline, polymers. T_g increases with crystallinity in semicrystalline polymers.

The glass transition appears to be a dynamic phenomenon as with glass, since T_g can be raised by increasing the rate of heating. In addition, ductility can be partially restored by holding the polymer for long periods at temperatures below, but near, T_g.

Table 11-1. Glass transition temperatures of some polymers

Polymer	$T_g(°C)$
Polybutadiene	−85
Polyisobutene	−77
Polyisoprene (natural rubber)	−75
Neoprene	−50
Silicone rubber	−123
GR-S (78% butadiene, 22% styrene)	−67
Copolymers of butadiene with acrylonitrile (T_g varies linearly with percentage of acrylonitrile from 0 to 52% acrylonitrile)	(−85)
Poly(vinylidene fluoride)	−39
Atactic polystyrene	75–100
Poly(methyl methacrylate)	72
Poly(ethyl methacrylate)	47
Poly(vinyl chloride)	82
Poly(n-butyl methacrylate)	22
Poly(vinyl acetate)	29
Poly(n-propyl methacrylate)	33
Poly(methyl acrylate)	0
Poly(ethyl acrylate)	−23
Poly(n-propyl acrylate)	−51
Poly(n-butyl acrylate)	−63
Poly(n-butyl methacrylate)	17
Polyethylene	−110 (nominal)
Polypropylene	−18 (nominal)
Nylon 6	47 (nominal)

11-3-2   COPOLYMERS

It is often possible to combine two monomers into a single material known as a copolymer. In fact, nearly all commerically important synthetic polymers are copolymers. The molar ratio of monomers can vary from 1:1 to 98:2. Important variations in properties are produced by different types and quantities of monomers. Copolymerization obviously increases the number of possible polymer products and offers outstanding potentialities for production of polymers well suited to a wide variety of specific applications. It can also greatly extend the range of usefullness of some monomers of otherwise limited use.

Copolymerization can result in four different structural arrangements as shown in Fig. 11-3. Random and alternating copolymers usually are homogeneous single phase systems with properties intermediate to those of the parent homopolymers. Block and graft copolymers, in contrast, form heterogeneous multiphase systems. They have properties which not only differ from the parent polymers but are directly due to the multiphase character. Multiphase copolymers have glass transition behavior which

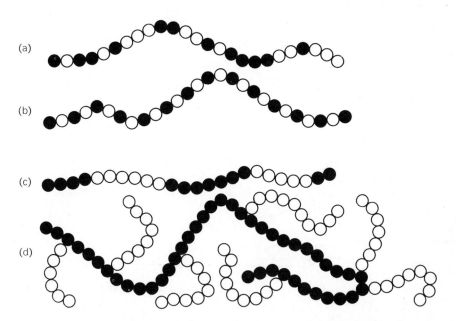

(a)

(b)

(c)

(d)

Fig. 11-3. Copolymer arrangements: (a) two different units distributed randomly along the chain; (b) two different units alternating regularly along the chain; (c) block copolymer with alternating segments of the homopolymers; (d) graft copolymer with a backbone of one homopolymer and side chains of the other homopolymer.

clearly differs from that of single phase homopolymers or copolymers. Multiphase copolymers normally are superior in impact resistance to single phase systems. Copolymers can have three, four, or more monomers involved. For example, an important type of terpolymer has two monomers present in major amounts while the third is a minor constituent added to develop some modification of the basic copolymer.

Individual molecules in a polymer contain the order of 10^4 mers. Individual molecules are not identical but vary in molecular weight (perhaps a tenfold range), in configuration, and in structural features such as chain branching. Despite this, a homopolymer has a well-defined melting point (T_f) for a given average molecular weight. If two types of monomers, such as two nylon monomers (Fig. 11-4) are copolymerized into a random structure such as Fig. 11-3(a), the melting point of the copolymer will have a minimum at some ratio of the monomers. Even though this diagram looks much like a binary equilibrium diagram, it is not one because each copolymer composition ratio is a distinct substance with its own characteristic melting point just like either homopolymer. Construction and interpretation of an equilibrium phase diagram for two monomers are difficult since the two components tend

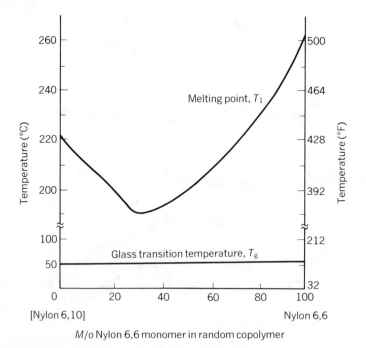

Fig. 11-4. A "component" diagram for copolymers of two nylon monomers, showing the variation of melting point and glass transition temperature as a function of composition.

to react chemically, especially when molten. In the solid state, however, homopolymers are almost always immiscible. As a result, "alloying" of polymers is done by mechanical mixing, i.e., *blending* the homopolymers.

11-3-3   ORGANOMETALLICS

This is an additional group of materials which are sometimes classified as polymers. While these are more like polymers than like metals or ceramics, they differ greatly from polymers. The organometallics are hybrids formed by combining metal atoms and organic groups. Thus they join the two large areas of chemistry, organic and inorganic. Many of these organometallics are highly reactive and are especially useful as polymerization catalysts. They are used as fuel additives (tetraethyl lead), antiseptics (mercurochrome, merthiolate, metaphen), bacteriocides (Salvarson, the first effective anti-syphilitic), and fungicides. Other organometallics are especially valued for their lack of reactivity, which is a necessary condition for stability under extreme conditions. Specific properties of organometallics are largely deter-mined by the metal and can range from spontaneous combustion in air by organoaluminums to thermal stability in oxygen at 500°C by organosilicons.

The organosilicons, better known as silicones, are the results of efforts to develop unreactive polymers which remain flexible, resilient, and elastic at very high or very low temperatures and under other extreme conditions. Silicone oils remain fluid at very low temperatures. Silicone rubbers which are elastic at −80°C are resistant to flame at 2800°C.

The silicones, which are polymers, differ from the organic polymers in that the chain backbone is composed of recurring silicon and oxygen atoms and contains no carbon atoms. The chain is modified to produce special pro-perties by attaching various organic radicals to the silicon atoms. A typical silicone is polydimethyl siloxane, which has the chemical formula illustrated in Fig. 11-5 in which n, indicating the chain length, can be as large as 10,000 in a "rubber."

Liquids of varying viscosity are made from linear silicone polymers such

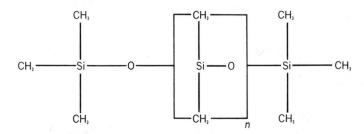

Fig. 11-5. Chemical formula of polydimethyl siloxane.

that viscosity depends on chain length. Unlike hydrocarbon oils, they retain fluidity at low temperatures and do not decompose at high temperatures.

Some crosslinking between long linear chains gives rubberlike properties. Silicone elastomers are flexible and elastic from at least $-110°C$ to over $500°C$, a range unmatched by natural or other synthetic rubbers. Increased crosslinking and alteration of side chains produce a moldable hard material.

Silicones are "unnatural" since they are different from any material found in nature. They are incompatible with most organic materials because of their radically different character. They are biologically inert and do not dissolve in or stick to most organics. Silicone surfaces are highly water-repellent. Their distinctive surface properties make them useful as surfactants, i.e., substances which suppress foam by reduction of surface tension of the liquid surrounding the bubbles.

11-4 COMPOSITES

Composite materials are among the oldest and the newest structural materials. Men discovered long ago that using two or more materials in combination can result in a material which has superior performance in comparison with each individual material. For example, combination of straw with clay to make bricks was done in biblical times. With a few notable exceptions, the potential of composites was virtually untapped for centuries while monolithic materials such as copper and iron served mankind's needs. Even with rather recent plywood, plaster board, linoleum, and reinforced concrete, composites appeared to be *ad hoc* solutions to specific problems and thus out of the main current of materials technology development. Development of materials such as lightweight honeycomb structures and polymers reinforced with glass fibers starting in the 1930's marks the start of the current era of composite engineering materials.

A composite can be regarded as a combination of two or more materials which are used in combination to rectify a weakness in one material by a strength in another. In this sense, structural steel and the paint normally applied to it for oxidation protection can be regarded as a composite. Early cast cannon were often wound with steel wire under tension to withstand the increase in internal pressure from a larger powder charge. Tungsten carbide particles, which are very hard and stable under high temperatures, can be used in tools for cutting strong alloys or for resisting abrasion when cemented together with a small amount of cobalt. This is a composite and is also considered a cermet.

From a somewhat different viewpoint, pearlite formed upon cooling austenite can be regarded as a composite. In this case, the alternate platelets of ferrite and cementite, each having its own properties (quite different from each other) combine to give a composite structure appreciably different from each. If we have eutectoid steel and allow isothermal transformation in one specimen at 1300°F and in a second at 1100°F, we will have pearlite as a final structure in both cases. We know, however, that pearlite resulting from the lower transformation temperature is stronger than the other. This obviously is not due to differences in composition but must be due to differences in structure, not of ferrite and cementite, per se, but the manner in which the two are arranged. We know that the platelets in the stronger pearlite are thinner than in the other.We also know that strength is related to resistance to movement of dislocations. The thinner plates give an effective mean free path for movement of dislocations which is less than for the thicker plates, and thus the strength is greater in the second specimen.

This example of pearlite illustrates the recently developed concept of composites; i.e., not only is a composite a combination of two materials, but the combination has its own distinctive properties. In terms of strength, heat resistance, or some one (or more) desired characteristic, the composite is better than either component alone—it is radically different from either.

A number of composites are found in nature. Wood, for example, is a composite of cellulose and lignin. Cellulose fibers are strong in tension and are flexible. Lignin cements these fibers together to give them stiffness. We are also aware of an appreciable spectrum of woods, for example, with densities ranging from balsa to ebony. Even though woods are in one generic class, the differences in composition and structure have substantial effects on the properties. Bone is a composite of strong but soft collagen (a protein) and hard but brittle apatite (a mineral). Man-made composites not only imitate but seek to improve upon nature. For example, fiberglass, which has a wide variety of applications, consists of glass fibers in a polymer.

The bimetallic strip used in devices such as thermostats is a simple example of the current concept of composites. Such a strip might be made of a thin strip of brass and a similar piece of iron. If the two pieces are separate and heated simultaneously, the brass would expand more than the iron. If the two are welded together and the composite heated, the brass with its greater expansion forces the iron to bend. At the same time the iron forces the brass to bend with it. The deflection can be used to indicate temperature or to operate an on-off switch.

The example of the bimetallic strip emphasizes two points about composites: (1) either material alone would be useless in the specific use, and the combination has a new and different property; and (2) the behavior of the composite (the combined action) is vital in the utilization of the composite.

The major constituents of composites are fibers, particles, laminae, flakes, fillers, and matrices. The matrix can be viewed as the "body" constituent which gives the composite bulk form. The other five are structural constituents which establish the characteristics of the internal structure and resulting properties. Behavior of composites obviously depends on properties of constituents and interactions among these constituents. Some collective properties, such as weight, thermal conductivity, and electrical conductivity can be determined from the law of mixtures. A second category of property occurs when each of the constituents contributes its specific property to total composite performance such as a surface layer of low-strength high-corrosion-resistance material bonded to a substrate of high-strength material. The third category of property, perhaps the most important, occurs when the whole is something different from the sum of the parts. In this case, the properties of the constituents are interdependent and function together synergistically.

The constituents are always intermixed in some manner. Thus there are contiguous regions which are analogous to grain boundaries. These regions can be interfaces formed by surfaces in contact or they can be interphases involving a separate added phase such as adhesive bonding plywood laminations together. Much remains to be learned about these interface and interphase regions.

Composites can be classified in three groups: particle composites, layered composites, and fiber-reinforced composites.

Particle composites consist of particles of one material dispersed in a matrix of a second material. Particles may have any shape or size. They may be added to a liquid matrix material which later solidifies; they may be grown in place by a reaction such as age-hardening; or they may be pressed together and then interdiffused via the powder process. Particles may be treated to be compatible with the matrix, or they may be incorporated without such treatment. Probably the most important engineering aspect is to extend the range of strength (especially) or other properties of inexpensive materials by subtle additions of other materials.

The largest improvement in mechanical properties is found in composites having finely divided structural constituents dispersed in a matrix. The matrix, usually more ductile than the particles, carries some of the applied load. It also can flow somewhat to more evenly distribute applied stresses to the particles and to redistribute stresses from incipient cracks. The particles frequently carry the major portion of the load. They also serve as barriers to dislocation movement through the matrix.

Particle composites of one type, known as *cermets*, are composed of ceramic particles in a metallic matrix. The particles are generally larger than one micron with a volume fraction generally between 25 and 70%. Cermets are produced by the powder process (Sec. 12-4). A second type of particle composites is known as dispersion-strengthened with particles smaller than one micron in a volume fraction less than 15%. Particle composites are not particularly new. Tungsten thoria, for example, has been used for incandescent lamp filaments for over thirty years. SAP (sintered aluminum powder), a composite of metallic aluminum and aluminum oxide, is finding rather wide use because of good resistance to corrosion and oxidation while retaining mechanical strength at temperatures which severely reduce the strength of aluminum alloys.

Layered composites are found in many familiar applications. We are all aware of plywood, which is a laminated composite of thin layers of wood in which successive layers have different orientations of the grain or fiber. The result is a more-or-less isotropic composite sheet which is stronger in any direction than it would be if all the fibers were aligned in one direction. The stainless steel in a cooking vessel with a copper clad bottom provides corrosion resistance while the copper provides better heat distribution over the base of the vessel. In a small heat exchanger where weight, corrosion resistance, and thermal conductivity were critical considerations, an answer was found by permanently bonding stainless steel to both sides of a copper core (a three-layer composite) and forming it into a helical-coil geometry.

Safety glass, universally used in automobiles, is a layered composite of two sheets of flat, tempered glass with an intermediate layer of a polymer. The "sandwich" is subjected to heat and pressure to produce the final product having excellent flexibility and resistance to abrasion, impact and other mechanical shock. If it does fracture, it crumbles into small, rounded pieces which are relatively harmless. Safety glass also has a pronounced damping effect on sound as it will reduce noise level in a room 7–15 decibels below that obtainable with ordinary glass under the same conditions.

It might be noted in passing that layered composites, in contrast with the other two groups, have no matrix.

Fiber-reinforced materials are typified by fiberglass in which three components are involved: glass filaments (providing mechanical strength); a polymer matrix (encapsulating the filaments); and a bonding agent (tying glass to polymer). Fibers need not be limited to glass, however. Metal, ceramic, and polymer fibers are used to provide much wider ranges of properties than glass can. The fibers can be used as continuous lengths, in staple-fiber form, or as whiskers. Fibers of circular cross-section have been most popular because of ease of production, but a wide variety of cross-sections is possible. In one application, i.e., large vessels to contain gas at high pressure, continuous glass filaments are wound on a mandrel after passing through a liquid polymer bath. The winding pattern can be controlled to place more fibers along the directions of greatest stress. At the same time, the fibers can be stretched, if necessary. Obviously, such a technique is completely different from normal metallurgical techniques for making strong alloys and has the great advantage of not requiring high temperatures or high pressures.

It must be emphasized that fiber-reinforcing does not produce a material as such but rather gives a product in the shape of a specified part. It is a sophisticated combination of materials whose characteristics are determined as much by the fabrication procedure as by the properties of the component materials. Fiber-reinforced materials can be made quite anisotropic through directional control of the strong fibers in the relatively weak matrix. This makes it possible to produce parts where strength control is developed in different directions. This implies techniques for producing the most efficient structural sections.

Concrete (Sec. 11-2–5) certainly qualifies as a particle composite with the particles of aggregate in a matrix of hardened cement. At the same time, reinforced concrete with steel reinforcing rods or wire mesh qualifies as a fiber-reinforced composite.

Composites, then, provide properties unattainable in the constituents themselves. The combined action is often of a different order of magnitude, or different in character from the behavior of the constituents. Certain questions must still be answered as use of composites advances. These include: What takes place at the interface between constituents? What are the internal mechanics (the micromechanics) of these materials? How can they be tested in situ to determine combined action? How can properties of a complex composite be tailored to meet desired conditions?

QUESTIONS AND PROBLEMS

1. Compare the treatment possibilities of metals, ceramics and polymers. This should include cold working and various heat treatments.

2. Abrasive products usually use aluminum oxide or silicon carbide grains for the cutting phase. Individual grains are bonded to a wheel, paper, or cloth. Bond materials include fired ceramic binders and a variety of organic resins and rubbers. Describe the general characteristics of this composite material. Should a grinding wheel have a large void fraction? Why?

3. A quartz rod is quenched from 800°C in water at room temperature. What is the maximum tensile stress developed? (Assume Poisson's ratio $= \frac{1}{4}$.)

4. Ordinary window glass is fragile. How can glass which is quite shock-resistant be produced?

5. Under its own weight, glass flows like glycerine at 1200°C. At about 700°C, it deforms slowly (viscosity $= 10^{-1}$ Newton-seconds/m²). Estimate the force required to slowly deform glass with a viscosity of 10^4 N-sec/m², assuming a density of 3 g/cm³. If the viscosity is as great as 10^{11} N-sec/m², is the concept of viscosity still valid? Why?

6. Polyisoprene (natural rubber) caps are stretched over steel containers. At −30°C, the caps do not remain good seals and even fall off. Why?

7. Why do the silicones have such drastically different properties from other polymers?

8. Can a glass state be produced in synthetic polymers and natural organic compounds such as alcohols, glycerol, and glucose? If so, how?

9. Can a glass state be produced in elements such as sulfur, selenium, and tellurium? If so, how?

10. Why are metals and polymers so much more resistant to cracks (and resulting brittle fracture) than ceramics? How is this combined with the fact that most ceramics are very hard and strong to produce various composites?

11. A refractory brick for furnace lining weighs 4.0 lb dry, 5.0 lb when saturated with water, and 2.5 lb when suspended in water. Determine the open porosity, bulk density, and apparent density.

12. You are a manufacturer of polymeric ball bearings. You wish to make a product with 50 v/o of very fine glass fibers (density = 2.5 g/cm^3) in nylon (density = 1.15 g/cm^3). What relative weight of glass fibers should be mixed with nylon to make the desired composite?

13. A paint weighs 15 lb/gal and has 30 v/o thinner (8 lb/gal) with the 70 v/o being pigment and a reactive polymer. One gallon is rated to cover 450 ft^2. What is the average thickness of the paint film after drying? (1 ft^3 = 7.5 U.S. gallons.) (b) If one pint of thinner is added to the original formulation, the paint will cover 900 ft^2. How thick is the film of wet paint? How thick is the dry film?

14. Polystyrene "glasses" are used for serving cold drinks on commercial airplane flights. How suitable are they for hot coffee?

15. Stretch a 2-in. rubber band over the length of an unsharpened pencil. (a) Push it off the pencil at room temperature. (b) Place in a freezing compartment of a refrigerator and leave until it is cold. Remove from the refrigerator and push the band off. (c) Compare the two results and explain.

16. A porcelain coating and its steel substrate (as on a refrigerator or stove) are stress free at 425°C. Upon cooling to room temperature (25°C), stresses are set up because of a differential in coefficients of thermal expansion. Assuming both porcelain and steel behave elastically, what stress is developed in the porcelain? Assume porcelain has a coefficient of thermal expansion of 4×10^{-6}/°C, an elastic modulus of 750 kbar (11×10^6 psi) and a Poisson's ratio of 0.25. What does the result of your calculation tell you about expected behavior of the composite structure?

17. An "ideal" load-bearing member in supersonic aircraft should be very light, very strong, very stiff, corrosion resistant, have a low coefficient of thermal expansion, a high resistance to abrasion, and a high melting point (or softening point). Most light, stiff, and strong materials are ceramics (like glass, graphite, sapphire, alumina, beryllia, and boron). Compare metals and polymers with these on strength/weight, stiffness/weight, and temperature stability bases. Can you suggest a compromise to obtain the best of "two worlds"?

18. Precipitation-hardening Al-4%Cu alloy can have a variety of microstructures depending on the heat treatment. Which, if any, of these could be considered composite structures? Why?

19. Slowly cooled hypereutectoid steel is very difficult to machine. The same steel, after spheroidizing, is relatively easy to machine. Discuss the reasons for this including necessary details of heat treatment.

20. The fuel element in the Army Package Power Reactor (nuclear reactor) has fissionable UO_2 particles dispersed in a stainless steel matrix. This dispersed particle composite is then clad on all sides with stainless steel (sort of like a sandwich) to make a layered composite in plate form. Since the resulting fuel element is rather lightly loaded (mechanically) why all the effort and expense to make such a relatively complex composite?

12

FABRICATION OF MATERIALS

12-1 INTRODUCTION

In the preceding chapters we have discussed crystal structure, microstructure, phase relationships, reactions, transformations, and so on, that affect properties and behavior of materials. All this knowledge may be of little practical value, however, unless we are able to form the specific material into the desired internal structure, and external shape, i.e., fabricate the actual hardware. It is the purpose of this chapter to consider general fabrication processes and their characteristics.

12-2 CASTING PROCESSES

Casting is the oldest process known for forming materials. In essence, casting consists of introducing molten material into a cavity or mold of the desired form and allowing it to solidify. There is practically no limit to size, shape, or alloy from which metal castings can be made. Castings range from tiny dental inlays of rare metals to complicated steel castings exceeding 200 tons. Using proper technique and design, almost any article may be cast. Casting does not represent the best solution to all production problems, but it has possibilities of such significance that it should not be overlooked when considering a new or revised design.

Most metallic objects begin their history by being cast. If used as-cast, or with only machining required for finishing, they are called castings. If subjected to subsequent deformation, i.e., mechanical working, the cast shapes are called ingots or blanks. Both castings and ingots must meet the same requirements, i.e., good density, good strength, and least possible waste in casting and subsequent operations.

Molding material has an important influence on ease and cost of making the mold, permanency of the mold, rate of production, rate of cooling of molten metal, surface roughness, dimensional tolerances, and mechanical strength. Choice of mold material and casting process for a component is influenced by three interrelated factors: (1) material to be cast—strength, corrosion resistance, and other properties will be influenced by the casting process and mold material selected, as well as ease and cost of producing the casting; (2) design—whether the part should be cast in one piece or should be produced in small parts and then assembled; (3) number of parts required —unit cost of parts produced by different methods varies widely with the number of parts.

12-2-1 SAND CASTING

Casting in sand molds is the oldest process known and is still used for the largest tonnage of castings produced. Complex parts weighing up to many tons may be produced using a wide variety of alloys. Very intricate forms can be produced. To assure soundness, however, all steps from pattern to final cleaning must be closely controlled.

Sand casting requires a pattern from which molds can be made. The pattern is similar in shape to the desired part, but is larger to compensate for shrinkage which occurs during solidification and cooling. If machining is to be performed, further allowance must be added to the pattern. The pattern is placed in a flask and molding sand is rammed around it. When the mold is finished, the pattern is removed, and the mold prepared for pouring. The pattern may be made in one or many pieces, but each piece must be so placed in the flask that it can be removed without disturbing the impression in the sand. A typical example of a sand mold is shown in Fig. 12-1.

Many types of sand are used for molds. All contain bonding agents such as fire clay, bentonite, cereal or liquid binders, and moisture to promote cohesion. After molding, dry sand molds are dried thoroughly by baking before closing and pouring. Green sand molds are closed and poured without drying. The types and relative quantities of sand grain, binder, and moisture depend on the procedure used for the particular metal. Binders are chosen for their ability to burn out at molten metal temperatures, so that molds and cores will collapse under compression and allow metal shrinkage without damage to the casting.

Sand casting has the great advantage of being extremely flexible and suitable for job lots. It has the disadvantage that each mold can be used only once.

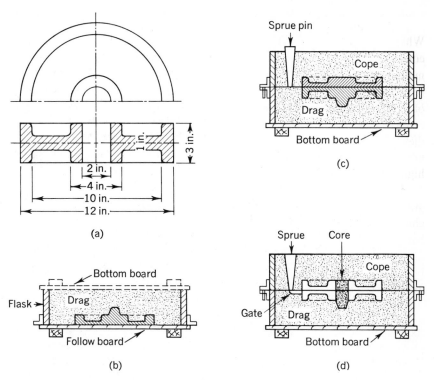

Fig. 12-1. Mold for a gear blank: (a) The machined blank; (b) The drag; (c) The cope; (d) The finished mold.

12-2-2 PLASTER OF PARIS CASTING

Castings made from plaster molds have superior surface finish, better dimensional accuracy, finer detail, and a more solid structure than sand castings, although at a higher price before machining. The process allows thinner sections than sand castings. Tolerances on plaster castings compare well with tolerances on metal mold castings. As with sand molds, plaster molds can be used but once. The advantage of the plaster mold is found in product quality.

The mold is made by mixing plaster of Paris with excess water. This mix is poured around the pattern and allowed to partially set before the pattern is removed. After the mold impressions are completed, the two halves of the mold assembly are baked separately. This baking operation completes setting and drives off excess moisture, leaving many small passages in the mold to serve as vents for air and gases trapped in the mold.

12-2-3   PERMANENT AND SEMIPERMANENT
MOLD CASTING

When a large number of parts is needed or when better surface or dimensional control is required than obtainable in sand molds, castings are made in permanent metal molds. If only a part of the mold is metal, and the remainder is sand, the castings are called semipermanent mold castings. Such molds, of course, can be used a great number of times. Obviously, they are not economical when only a few castings are to be made, since the cost of metal dies greatly exceeds the cost of individual sand or plaster molds. In addition, any metal mold process is unsuitable for large castings or for alloys having high melting temperatures.

The permanent mold method provides castings with: (1) dense, fine-grain structure free of shrink holes or blow holes; (2) relatively low tool charges; (3) lower mold cost than for other metal mold methods; and (4) better surface finish and closer tolerances than sand castings. Its limitations are: (1) an inability to maintain as close tolerances and as thin sections as pressure or plaster of Paris casting; (2) the fouling of molds by some alloys; and (3) a lower production rate than other metal-mold methods.

The same metals are suitable for both permanent mold and semipermanent mold casting, but production rate in the latter is somewhat lower. An additional disadvantage of semipermanent molds is that grain structure of metal cast in sand portions is substantially the same as in normal sand casting, and thus differs from grain structure in the metal portions.

12-2-4   DIE CASTING AND COLD CHAMBER
PRESSURE CASTING

In these two processes, highly fluid metal is forced, under pressure, into metal molds, and pressure is maintained until solidification is complete. The molds or dies consist of a minimum of two steel blocks, each containing a part of the mold cavity, which are locked together while the casting is being made. One half of the die is stationary, and the other half is movable.

Die castings are produced in machines which consist of a basin holding a considerable quantity of molten metal, a metal mold or die, and a metal transferring device which will automatically withdraw a certain amount of molten metal from the basin and force it into the die (Fig. 12-2). The cold chamber process differs in that the molten metal reservoir is separated from the casting machine and just enough metal for one shot is loaded into a small chamber or pouring well from which it is forced into the die.

Pressure castings of this type share certain properties, but in varying degrees. They have close tolerances, sharp outlines and contours, a fine smooth surface, and a high production rate with low labor cost. They also

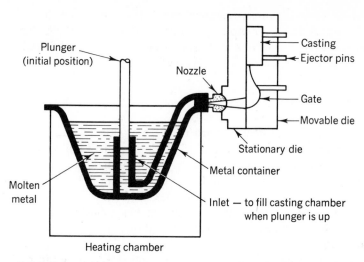

Fig. 12-2. Typical piston-type die-casting machine. Plunger, heating chamber, metal chamber, and die carriage are parts of the same machine.

have a hard "skin" with a softer core which results from rapid chilling of the metal in contact with the cooled metal mold. Machining the surface reduces strength and causes warping by release of internal stresses.

12-2-5 CENTRIFUGAL CASTING

A true centrifugal casting process consists of pouring molten metal into a revolving mold. Centrifugal action forces the metal tightly against the mold. The metal solidifies having an outer surface which conforms to the mold's shape and a surface of revolution on the inside.

Centrifugal casting has many advantages, such as finer grain size, slightly increased density, uniform distribution of alloying elements, and elimination of weakness at the boundary of columnar grain growths which occur when the casting is cooled simultaneously from two or more sides. A major advantage is the elimination of dross from the body of the casting since impurities, e.g., slag, oxide, pieces of refractory, etc., float to the inside surface. Other advantages are low-cost molds and applicability to large and long hollow shapes, especially in metals difficult to work by forging. A major disadvantage is the basic shape of a thick-walled hollow cylinder. Some departures are possible in an outer surface which is square, hexagonal, octagonal, or round, with or without flanges or grooves. Occasional protuberances which break the symmetry of the perimeter are permitted provided they are not too high or too thin.

Centrifuged castings are a form of pressure casting. Centrifugal force is used to fill molds of nonsymmetrical pieces that cannot be spun around a central axis. A mold (or molds) is placed off center in a rotating fixture. During rotation the metal is channeled from the center and forced into the mold. Sand, plaster, or metal molds may be used.

12-2-6 INVESTMENT CASTING

Investment casting is often referred to as "precision casting" or "lost wax casting" and is a versatile method of producing small and/or complicated parts. This method has been quite widely used for making plates, inlays, and other dental fittings. Large quantities of castings can be produced in a relatively short time. Millions of turbine blades and buckets have been produced from "super alloys." Investment casting makes small and intricate parts that cannot be machined economically even though they are made of machinable alloys.

A model is made of the desired product and is used to produce a permanent plaster or glue mold. Wax parts are then made from this mold. The casting mold is made by pouring investment material (a specially bonded sand) around the wax pattern and allowing it to harden. The mold is inverted, placed in an oven, and baked. The baking hardens the mold and melts the wax, which runs out.

12-2-7 SHELL MOLD CASTING

Molds are thin shells of sand bonded with a thermosetting phenolic resin, made by sprinkling the sand and resin mixture over a heated pattern. The heat causes the resin to partially set so that a thin shell is formed. The excess sand and resin can be reused for additional molds.

The shell is removed from the pattern and baked at 300–400°F to completely set the resins. A baked shell can then be assembled with other shells to produce a complete model. These shells, with the necessary risers, gates, etc., are placed in position and backed with steel shot to hold the mold in place. It is also relatively easy to reprocess sand for reuse.

Shell molding can produce molds rapidly while eliminating the problem of handling the vast quantities of sand required in sand casting. Better dimensional control and surface finish can be obtained than with sand casting and there are no limitations on metal to be cast. Cost of resin is a limiting factor.

12-2-8 MELTING

Metal for casting may come directly from reduction of ore (e.g., blast furnace reduction of iron oxide to iron), from an open-hearth or other remelting furnace (e.g., steel production), from an electroreduction process (e.g., aluminum), or from remelting and alloying. For remelting, a heat source and refractory container are needed. Evaporation losses, oxidation, and contamination of the melt are minimized by using a protective coating or "flux" on the surface of the melt. Fluxes can be charcoal, molten salts (e.g., zinc or ammonium chloride), or an inactive or inert gas (e.g., nitrogen or argon). A typical example is the melting of aluminum by direct or indirect flame in a refractory-lined furnace using chlorine or nitrogen as a flux.

12-2–9 GAS ABSORPTION AND EVOLUTION

Solution of gases in metal generally increases with increasing temperature and with time at temperature. Thus, casting is usually done at the lowest possible temperature and with the least possible delay. During cooling and solidification, dissolved gases are expelled from solution and try to leave the metal. If gases are trapped, the resulting porosity produces an unsound casting.

12-2–10 FILLING A MOLD

To obtain a perfect casting, liquid metal must completely fill every part of the mold before solidifying. Liquid metal should, therefore, have high fluidity and low surface tension at the melting point. Experience indicates that vacuum melting permits higher casting temperatures, better fluidity, and lower surface tension conditions without high gas concentration. The vacuum melting process is relatively expensive, however, and is normally used only when no other method proves suitable.

12-2–11 DIRECTIONAL SOLIDIFICATION

During solidification, a casting commonly freezes on the exterior while the center is still liquid. Since metal contracts on solidification, this is conducive to formation of a shrinkage cavity. If, on the other hand, the casting progressively solidifies, starting at the point most removed from the source of molten metal, without leaving any liquid regions behind, then all sections of the casting have continuous access to liquid and no cavity can form.

This desired directional solidification can be obtained in many ways. A thick section, for example, can be made to solidify rapidly by "chilling" it with a metal mold or with pieces of metal close to this section. The mold for a thin section can be preheated or made from material having poor heat conductivity. A thin section can be "padded" to prevent early solidification, but extra material must be removed later by machining.

12-2-12   SEGREGATION

Many impurities are soluble in liquid metal but are relatively insoluble in solid metal. Thus, as metal freezes, these impurities remain in the liquid as well as light insoluble impurities such as slag, sand, refractory, and so on. This difference in impurity content between the first and last portions of a casting to solidify is known as segregation. It is sometimes called ingot segregation to differentiate it from crystal segregation (Sec. 9-4-1). This segregation gives the region which freezes last properties which can be greatly different from the region which freezes first.

Regional segregation can be minimized by providing a "hot top" on an ingot mold or a riser on a casting. Both are essentially reservoirs above the useful part of the casting. This reservoir freezes last and contains most of the impurities. It can be cut off and remelted.

A related example is grain boundary segregation. Impurities are concentrated in the final film of solidifying liquid which constitutes the grain boundaries. Certain impurities, even in very small quantities, can thus form a brittle film around the grains and make the entire structure brittle. Sulfur in excess of a few hundredths of one percent in iron, for example, concentrates at grain boundaries as iron sulfide and makes the iron brittle at forging temperatures. This can be cured by eliminating the impurity.

12-2-13   GRAIN STRUCTURE IN CASTING

During solidification under quasi-equilibrium conditions, we obtain a structure of equiaxed grains having random orientations. In practice, however, we have different structures.

If molten metal is poured into a relatively cold mold (e.g., an ingot mold), the metal nearest the wall is the first to cool below the freezing point. Therefore, the first crystal nuclei form near the wall and begin to grow in all directions. Circumferential growth about each crystal nucleus soon results in obstruction by neighboring crystals. These crystal nuclei continue to grow inward until solidification is complete. The surface has long (columnar) crystals which are essentially perpendicular to the mold surface while the interior has grains having more random orientations.

If a mold has a sharp corner, a plane of weakness develops at the corner since both gaseous and solid impurities are concentrated in this region due to rejection by the growing crystals. To avoid this, we use molds with rounded corners.

If a sand mold, a preheated metal mold, or a relatively large mold is used, columnar grains are found in a relatively thin "skin" whereas the bulk of the metal has equiaxed grains. These castings will have an essentially random orientation. In a chill casting, however, with an extensive columnar structure, there is preferred orientation.

A crystal withdrawn from the melt normally has dendritic structure (Fig. 4-34) and a "uniform" orientation within the dendrite. As these crystals grow into grains, they do not interlock but must accommodate themselves to their neighbors. This makes the exterior of each grain irregular while preserving the regular arrangement of atoms in the interior. Grain size cannot be predicted very accurately. It is generally true, however, that smaller grains are formed as the heat extraction rate is increased. In contrast to segregation (Sec. 12-2–12), some impurities may separate out before solidification occurs. These often serve as nuclei for crystal formation and are located within the grains rather than at boundaries. This tends to promote a fine grain size.

12-3 MECHANICAL WORKING PROCESSES

Mechanical working of a metal is simply plastic deformation performed to change dimensions (shape), properties, and/or surface condition. The phenomena of plastic deformation and the cold-work-anneal cycle are discussed in Chapters 6 and 9, respectively. Plastic deformation below the recrystallization temperature is cold work. Plastic deformation above the recrystallization temperature, but below the melting point, is hot work. Some of the differences are indicated in Fig. 12-3.

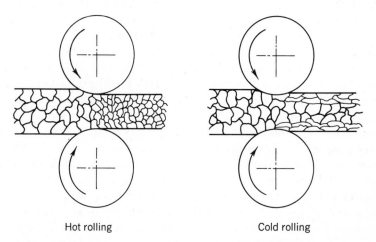

Hot rolling   Cold rolling

Fig. 12-3. Schematic representation of hot and cold rolling. Hot rolling refines grain structure, and cold rolling distorts grain structure.

12-3-1   COLD WORKING

Cold working gives good surface finish and close dimensional tolerance and no weight loss during working. It causes a considerable increase in strength and hardness and a reduction in ductility as indicated in Fig. 9-1. These changes in properties are often desired. When the metal has been hardened as much as possible without cracking, the piece can be annealed and cleaned, and working can continue. In many cases, several cold-work-anneal cycles are required before a piece of metal is reduced the desired amount or formed to the desired shape. For relatively ductile metals, cold working is often more economical than hot working. In addition, if an alloy cannot be strengthened by one of several different treatments (Chapters 9 and 10), cold working may be the only method by which strength can be increased.

12-3-2   HOT WORKING

Hot working combines the working and annealing processes by deforming metal above the recrystallization temperature. Ductility is constantly restored by recrystallization and grain growth during, or immediately after, recrystallization. Since most commercial metals have relatively high recrystallization temperatures, they must be worked at high temperatures. Hot working can be performed until the metal cools to the recrystallization temperature at which point reheating is required before further hot working. As in cold working, several cycles of operation may be necessary before a piece of metal is reduced to the desired size or shape.

Hot working has many effects on metal characteristics or properties in addition to a mere change of shape. Among these effects are: (1) General densification of the metal if blow-holes or other cavities are present, providing cavity surfaces are not oxidized. (2) Refined grain structure. Since grain size is reduced by recrystallization during hot working, working can be continued until the metal has cooled to the lowest possible recrystallization temperature. Little grain growth is then possible and the recrystallized grains remain small. (3) The metal is made more homogeneous since inclusions tend to be broken up, and if plastic, become elongated in the direction of the metal flow. Secondary phases, e.g., carbides in steel, behave likewise. Inhomogeneities in composition tend to be reduced through increased diffusion. (4) Hot working may develop preferred orientation. For best service performance the flow direction in hot-worked metals should coincide with direction of maximum stress. (5) Hot-working processes require less power than cold-working operations for the same job. (6) Exposure of heated metal to air usually causes scaling. This may not be a serious disadvantage, but it is difficult to obtain close tolerances and good surface finish without further operations.

Each metal has a characteristic hot-working temperature range over which hot working may be accomplished. There are economic and metallurgical reasons why hot working should be done at correct temperatures. We desire to work metal as hot as possible since power requirements vary inversely with metal temperature. The upper limit of working temperature is determined by metal composition, and impurities play an important part. For example, iron sulfide segregates at grain boundaries in steel and causes brittleness (hot shortness) at too high a temperature. Obviously, the upper limit for working must be somewhat below the melting point.

The manner of heating large pieces for hot working is also important. If heated too rapidly, large pieces tend to crack because of large thermal gradients and/or possible phase transformations. Also, pieces which have two sections of considerably different thickness may present the problem of appreciably different grain size in the two sections.

12-3-3 DEFINITIONS OF WORKING PROCESSES

Bending: A plastic deformation of a sheet, strip, or bar performed in a device which clamps the material along one side of a bending line and lifts it along the other side forcing it around the edge of the clamp at a specified radius of curvature.

Blanking: A press operation of cutting the outside contour of a piece which is usually processed further in some later operation, e.g., deep drawing.

Broaching: A consecutive shearing of a hole or contour by a series of stepped cutting edges similar to those on a saw. While this is generally considered a machining operation, it is a cold-working process when used in a slow-acting press for accurate sizing of holes or contours such as gear teeth and keyways.

Burnishing: A smoothing or polishing operation accomplished by compression and/or pressure; for example, blanks may be polished and sized by being forced through an opening which has sides tapered slightly inward. Another example is the plastic smearing of surface metal during buffing.

Coining: The compressive cold sizing of a metal used to obtain a smooth finish, close tolerance, or fine detail at less cost than machining.

Crimping: A process which produces flutes or corrugations. For example, this process is used to gather metal as in a stovepipe joint.

Cutoff: An operation which shears a piece from a longer piece, such as cutting a strip or bar into desired lengths.

Dinking: An operation similar to blanking and piercing which is done by pressing a sharp, thin, steel edge through sheet material lying on a flat hardwood platen. Rubber, fiber, paper, felt, and thin metal foils may be blanked in this manner.

Drawing: A stretching process. In tube, rod, or wire drawing, the metal is pulled through a die; in cup, shallow, or deep drawing, the bottom of a blank is pressed into a die, and the walls are pulled in.

Embossing: A shallow drawing operation which is rarely deeper than three thicknesses of metal. An example of this is an embossing of raised letters in thin strip.

Flanging: Any process which produces a flange.

Forming: A process similar to bending except that one or more bends are accomplished about linear axes (usually but not necessarily parallel) by making the metal conform to a die shape. Forming is often combined with other stamping operations such as piercing, notching, cutoff, etc.

Heading: A process which gathers or upsets the metal. The name derives from its original use for the production of screw and rivet heads. The process is used to produce a wide variety of shapes.

Knurling: A cold-working process in which a series of sharp serrations on a hardened steel roller are pressed in the material being knurled. Its principal use is to roughen surfaces for thumb screws, handwheels, etc.

Lancing: A special form of piercing in which the entire contour is not cut out, and the blanked material remains as a tab.

Piercing: (1)—An operation similar to blanking with the difference that piercing applies to inside contours such as holes and slots.
(2)—The process of spinning and rolling a billet over a mandrel in such a manner that a hole is opened in the center. This also includes opening a hole in a billet by hot punching.

Punching: A general term applied to all cutting operations performed on the punch press, including blanking, lancing, and piercing.

Roll Threading: A method of producing a thread on a bolt or screw by pressing and rolling it between two grooved die plates.

Seaming: A process of bending and flattening an interlocking fold such as a stovepipe seam.

Shaving: A refinishing operation which cuts a small amount of material off the edge of a punching or stamping to gain accuracy and/or a square edge. This may also be considered a machining operation.

Shell Drawing: *See* Drawing.

Spinning: A process or technique of kneading sheet material into a desired hollow shape about a rotating form.

Stamping: A general term to cover all punch press operations, which includes the squeezing operations such as embossing, coining, swedging, forming, and shallow drawing as well as the cutting operations usually considered as punching processes.

Swaging: A hammering process in which the material is not fully confined. For example, a rod may be reduced in diameter or tapered by repeated blows between two halves of a die.

Upsetting: A process of gathering metal in a rod or bar to achieve a larger cross section. (*See* Heading.)

The working processes defined above are primarily cold-working processes although several of these are used as hot-working processes. The difference lies in detail of operation rather than in principle.

Rolling: Rolling (Fig. 12-3) is a process which is performed in much the same manner whether hot or cold. Cold rolling is usually applied to sheet or plate to make it thinner, to give it a better surface finish, to maintain a closer tolerance, and to increase strength and hardness. Hot rolling is often used to form ingots into smaller sections known as billets. Hot rolling is used for producing plate, sheet, and strip, but can also be used to produce rods, bars, and various structural shapes by making successive passes through rollers which are grooved to produce the desired shape.

Pipe and Tube Processes: Round tube or pipe may be made by drawing a strip, heated to the proper temperature, through a bell-shaped die. The die curls the strip or skelp into tubular form, forces the edges of the seam together, and welds them. Most seamless pipe or tubing is made by piercing a heated rod. Two large rollers run slightly askew and rotate the rod rapidly while advancing it slowly along its own axis. This rolling action ruptures the rod at its center, and the tube advances over a mandrel which enlarges and smoothes the break. Final dimensions are obtained by further rolling or drawing operations.

Extrusion: Pressure is applied to metal causing it to flow through a restricted orifice thereby forming an elongated part of uniform cross-section. A classic analogy is squeezing tooth paste from a tube. A large variety of shapes, both solid and hollow, which are practically impossible to produce in any other fashion, can be produced in this manner. In the Hooker extrusion process a cup-shaped blank is extruded ahead of a punch through an annular orifice between the punch and die. Impact extrusion is quite similar to the Hooker process, but the flow of metal is in the opposite direction. In this case a flat blank is placed on a solid bottom die, and the impact of the punch causes the metal to flow back over the punch. Typical examples of products formed by the latter process are tooth-paste, shaving-cream, and paint-pigment tubes.

Forging: Forging is a general term for working by a localized pressure or force applied by hammers, presses, or forging machines. Blacksmithing was undoubtedly the first use of forging. Practically all hand forging operations have been replaced by various kinds of power forging. Drop forging forms a metal part by repeated hammer blows on a bar or billet placed between a pair of dies containing an impression of the desired shape. Press forging is a variation of the drop forging process, in which the material is deformed with a single sustained, slowly applied loading instead of several hammer strokes. Press forgings generally have greater accuracy and fewer design limitations than drop forgings.

12-4 THE POWDER PROCESS
(METALS AND CERAMICS)

The powder process is used, in general, to fabricate parts which can be made no other way, or products which can be made more economically than by other forming operations. The process can produce parts from powders of a single material or from mixtures of several materials, either metals or nonmetals.

Nearly all metal products, whether cast or wrought, originate from a casting, yet the powder process sometimes is the only method by which a desired metal structure can be produced. *It is the only fabricating process possible for most ceramics and cermets.*

Processing of metal powders dates back to at least 3000 B.C. when the Egyptians utilized part of the technique. It is only in the last few decades, however, that the process has been of much commercial significance for metal processing. Use of the process for fabricating ceramics goes back much further than for metals since pieces of pottery have been found which are several thousand years old. The process has served for many years for making whiteware, grinding wheels, etc.

12-4-1 THE BASIC PROCESS

There are four fundamental steps involved in fabricating parts by the powder process:

1. *Production of powder.*
2. *Weighing and mixing* of the necessary powders (and lubricants) to arrive at a composition which processes satisfactorily and which produces desired properties in the fabricated part.
3. *Pressing or forming* of the powder mixture into the approximate size and shape required. This expels air from the powder, compacts it,

and gives sufficient "green strength" to the "briquette" or "compact" to permit handling.

3a. *Presintering* (sometimes used with metals) or *drying* (commonly used with ceramics) by heating and holding at a moderate temperature to develop additional green strength and drive off mixing lubricants and/or moisture.

4. *Sintering or Firing* by heating to an elevated temperature (below the melting point of the major constituent) to promote a permanent bond between powder particles and to develop desired final properties.

These steps are all that may be required to fabricate the part. In some cases, especially in metals, one or more of the following treatments may also be required:

1. Repressing for greater density or closer dimensional control.
2. Annealing.
3. Rolling, forging, drawing, or other working processes for special products, e.g., tungsten wire, beryllium products.
4. Machining.
5. Hardening by heat treatment.
6. Oil impregnation for lubrication applications.
7. Buffing or polishing.
8. Various surface treatments for corrosion resistance.
9. Grinding to final dimensions (particularly in ceramics).

12-4-2 POWDER PRODUCTION

Properties of the final product are closely related to characteristics of the powders used. We must control powder production to regulate particle shape, size, size distribution, and purity. Particle shape is basically dependent on the production process. A typical mixture of ceramic powder would be 60% coarse, 20% medium, and 20% fine.

Several methods exist for producing metal powders. Among them are:

1. Reduction of powdered oxides by carbon or hydrogen at temperatures below the melting point of the metal.
2. Decomposition of compounds (e.g., hydrides) under vacuum at elevated temperatures.
3. Electrolytic deposition at high-current densities followed by ball mill grinding.
4. Condensation of metallic vapors to a powder, e.g., carbonyl process for iron or nickel.

5. Atomizing by entraining liquid metal in an air jet.
6. Ball mill grinding of cast pellets or shot.
7. Granulation (stirring during solidification).
8. Stamping.

Ceramic powders can be obtained by grinding larger bodies of the desired ceramic followed by screening and sizing. Other methods are precipitation, calcination, and decomposition of compounds.

The powders are carefully graded (sized) by screening, air separation, or sedimentation methods. Fine powders are commonly stored under controlled atmospheres to avoid contamination or reactions with the atmosphere due to relatively large surface area.

12-4-3 POWDER MIXING

Powders are weighed in correct proportions and mixed, wet or dry, in a ball mill. Mixing with water or an organic liquid is sometimes used to obtain better mixing, prevent explosions, prevent surface oxidation, and minimize dust. Various lubricants are often added to the mixture to facilitate pressing. If prealloyed metal powders are used, it is still customary to use mixing or blending since this results in a closer approach to uniformity of particle size and shape throughout a large amount of powder with better control of pressing and sintering operations. Mixing is similar for both metal and ceramic powders.

12-4-4 FORMING METHODS

Forming of metals and cermets is substantially limited to pressing although slip casting is sometimes used. Ceramics are formed by pressing, extrusion, jiggering, and slip casting.

For *metal pressing*, the proper amount of powder is mixed, placed in a die, and compressed under pressures ranging from 5 to 100 tons per square inch (Fig. 12-4). It is possible, with ductile powders such as tin, copper, and lead, to obtain briquette densities approaching the theoretical, but this is seldom advantageous. Usually only enough pressure is used to obtain a briquette which will sustain normal handling with density about 50 to 80% theoretical. The use of lower pressures generally gives greater die life.

Powder movement in the dies tends to be unidirectional for both metals and ceramics. In other words, there is little tendency for lateral motion of powder (as we find in fluids), and compacts pressed from one side only tend to be denser on that side than on the other. Such compacts are not usually self-sustaining if the ratio of height to diameter is greater than about three. Better compaction can be obtained by using highly polished carbide die inserts, die wall lubricants, and simultaneous pressing from both sides.

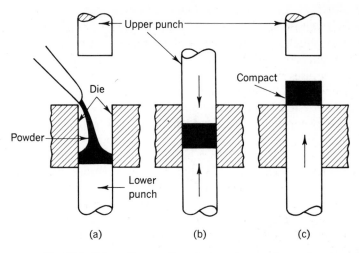

Fig. 12-4. Schematic operations for pressing powder parts.

"Green strength" (i.e., as pressed) is due partly to mechanical interlocking of deformed powder particles and partly to "cold welding" between particles. Green strength is relatively small, but is noticeably greater for briquettes of ductile powders than for briquettes of brittle powders. For instance, compacting tungsten powder requires use of wax or a similar substance as an adhesive since the metal has no appreciable cold ductility, whereas metals like aluminum and zirconium compact readily in a dry condition. The adhesive used is one which volatilizes and disappears during sintering.

Although a great variety of parts can be formed by powder metallurgy there are definite limits as to shape, size, and thickness. Die design is extremely important. Parts which have cavities can be readily formed only if the opening lies in the pressing direction. Extensive elastic deformation of the die during pressing can lead to cracking of the briquette when it is ejected. Extensive differences in thickness make it difficult to obtain adequate uniformity throughout the compact.

Ceramic pressing is considered in three categories. In *"dry" pressing*, the powder mixture can contain about 5–15% water. Pressure is normally between 5000 and 10,000 psi although as much as 100,000 psi may be used in special circumstances. Small amounts of various materials, e.g., wax, are sometimes added as a binder. Dry pressing is similar to pressing metal powders. *"Isostatic"* pressing can be used for pieces up to about 15 inches in diameter requiring uniform density. Length is limited by die or press size. A rubber or plastic bag surrounds the piece, and pressure is applied to the outside of the bag. This is more complicated and expensive than dry pressing. A binder may be necessary at low pressure but not at relatively high pressure.

"Hot" pressing with induction heating can be used for pieces up to 12 inches in diameter and 18 inches long. Dies and plungers for hot pressing are usually made of graphite (also a ceramic). In hot pressing, as in dry pressing, there may be nonuniform density with lowest density in the center.

Extrusion uses powder mixed with plasticizer to give a plastic, but very stiff, paste which is forced through a die to form a continuous section which can be cut to desired length. Small pieces can be formed by injection molding, a process developed for high speed molding of organic plastics. A finely ground ceramic powder mixed with about 15% thermosetting and thermoplastic resin can be forced from a heated chamber through a cooled mold. It is necessary to "heat treat" the chilled piece to remove organic matter.

Jiggering is the oldest method of forming ceramics. This can be done by hand alone (still done by primitive potters) or by the potter's wheel. Jiggering is used largely in the whiteware industry to form plates and some types of hollow ware, either by hand or automatically, using modern versions of the potter's wheel.

Slip casting is used to form pieces having large dimensions, thin walls, or complicated shapes. A mold is made first. Powder is then mixed with liquid to the consistency of cream and poured into the mold. Porous molds allow liquid to penetrate and be absorbed while powder in the slip dries. If a hollow part is desired, the mixture is allowed to partially set and the excess poured out. If a solid part is desired, it is necessary to supply additional slip until the piece is solid since the volume shrinks considerably as liquid is lost.

12-4-5   PRESINTERING OR DRYING

Presintering of metal or cermet compacts is short-time heating at relatively low temperature to volatilize the pressing lubricant and to develop additional green strength yet leave the compact relatively easy to machine. This operation is not usually performed unless considerable machining is required.

Drying is required on most ceramic bodies since they often contain appreciable moisture. Drying should be as fast as possible. If it is too rapid, however, there may be enough differential shrinkage to crack the piece. Drying is generally carried out by passing hot air over the pieces or by radiant heat directed on the pieces. Moving air supplies heat to compensate for evaporative cooling and carries away water vapor. Drying develops additional "dry" or "green" strength which allows pieces to be handled readily.

12-4-6   SINTERING OR FIRING

Sintering consists of heating pressed metal or cermet compacts in batch or continuous furnaces to a temperature below the melting point of the major constituent in an inert or reducing atmosphere, where time, temperature, heating rate, and cooling rate are automatically controlled.

Sintering produces finished strength in the compact. Even though it is possible to obtain strengthening without measurable densification, shrinkage (or densification) may occur during sintering and is always accompanied by strengthening. The amount of shrinkage varies with the metal but increases with higher sintering temperature and lower briquetting pressure. Although some metals can be prepared with clean surfaces which allow sintering at room temperatures, the surface of most metal powder is contaminated by absorbed gases and oxides which prevent autogenous bonding. The use of reducing atmospheres and elevated temperatures in practical sintering operations removes these films and improves interparticle contact.

Bonding of powder particles during sintering can take place in any of three ways:

1. Melting of a minor constituent.
2. Diffusion.
3. Mechanical bonding or entrapment of nonmetallic particles in a metal matrix.

Movement of material during sintering is caused by "volume diffusion" or "body diffusion," although surface diffusion, evaporation, and condensation also contribute to bonding. It is probable that all of these mechanisms can take place individually, successively, or simultaneously during sintering. The exact mechanism depends on the type of metal, method of powder fabrication, sintering temperature, time, and atmosphere.

Firing of ceramics follows drying and is similar to sintering of metals. Most ceramics are fired at temperatures about $\frac{2}{3}$ of the melting point to produce the final, dense, strong ceramic body.

The forces responsible for bonding during sintering or firing are the same ones which hold the atoms together in normal structure. In the interior of a powder particle there is a balance or equilibrium between the forces of attraction and repulsion. On the surface, however, there is an imbalance. Consequently, if atoms of the same kind are brought into sufficiently close proximity, they will be mutually attracted and form a bond. If two clean powder particles are brought together in close contact at several points, the probability of obtaining a bond is good. To strengthen the bond or to obtain a bond over a wider area, it is necessary to improve mating of surfaces. This can be done by heat and/or pressure. Some bonding takes place during compacting due to "cold welding" and becomes complete at elevated temperatures. The simultaneous application of heat and pressure (hot pressing) in a properly controlled atmosphere accelerates the whole process and is particularly effective with hard powders.

12-4-7 ADVANTAGES

Principal advantages of the powder process are:

1. Minimum scrap. Only the required amount of powder involved is processed.
2. Minimum of machining operations. After sintering, the part may be within allowable tolerances or may require only a sizing or coining operation.
3. Suitability for mass production. Only one stroke of the press is required for compacting with speeds up to about 60 strokes per minute. Several bodies can be sintered in one batch.
4. Composition, structure, and properties can be controlled more closely than in other fabricating processes.
5. "Impossible" parts can be produced. This includes:
   a. Production of a strong part without fusion.
   b. Molding to a definite form or shape which can be retained during subsequent operations.
   c. Cementing together of particles of high-melting-point metals or ceramics by fusion of lower-melting-point metals. No distortion of the finished product occurs since the fusing metal is present in minor quantities.
   d. Production of a composite structure with uniform distribution of two (or more) metals or nonmetals which do not otherwise alloy or dissolve to any appreciable degree.
   e. Obtaining mixtures of metals and nonmetals without limitations imposed by solubility.
   f. Stopping the process and obtaining a partially rather than completely sintered product.
   g. Fabrication of structures of controlled porosity ranging from "none" to about sixty percent porosity. Proper control gives void sizes ranging from very fine to coarse with a uniform distribution of voids.
   h. Production of structures of controlled porosity with interconnecting pores forming a capillary network.

12-4-8 DISADVANTAGES

Principal disadvantages of the powder process (particularly with metals) are:

1. Relatively high tool and die cost. With average compacting pressures of 30 tons per square inch, wear on dies is quite high. Relatively large expensive presses are required.
2. Generally lower physical properties due to inherent porosity.

3. Relatively high cost of raw materials. The cost of metal powders is several times that of metal for wrought products even though scrap metal is a satisfactory source of metal powders.
4. Design limitations. The size is usually limited by the available press size. Since powders do not flow or transmit pressures like fluids, this constitutes a severe design limitation.
5. Extreme care must be taken in handling pyrophoric powders (e.g., Mg, Th, Zr) to prevent fires or explosions and with toxic powders (e.g., U, Be, Th) to minimize health hazards.

EXAMPLE 12-1

A powder compact has a uniform linear shrinkage of 15%. What is the volume shrinkage?

Solution:

$$1 + \frac{\Delta V}{V} = \left(1 + \frac{\Delta L}{L}\right)^3$$

$$\frac{\Delta V}{V} = (1 - 0.15)^3 - 1$$

$$\simeq -39\%$$

12-5   WELDING PROCESSES

We think of welding as a joining process, but it is also a forming process and, in fact, the earliest known applications of welding were forming rather than joining processes. As a forming process, welding competes with (and complements) casting, working, and machining. As a joining process, welding competes with riveting, bolting, and other mechanical methods which depend on rigidity of parts and connectors. Welding differs from mechanical joining methods in that welded parts are held together by interatomic forces.

Welding has been used for many years, and its use is continually expanding. There are many fields, e.g., ship building, and building construction, which formerly relied entirely on mechanical joining but now make extensive use of welding. Welding involves more conventional scientific disciplines and more variables than nearly any other industrial process.

One broad definition of welding is: *Welding is the localized union of metals by fusion, diffusion, or surface alloying accomplished by applying heat and/or pressure with or without a filler metal.* This includes three subdivisions known as welding, brazing, and soldering. The process is welding if it is performed above 800°F (425°C—this temperature is arbitrary but widely used) and if the filler metal, when used, melts at about the same temperature as the base metal. Brazing and soldering always use filler metals although some welding processes do not. The filler metals used in brazing and soldering always melt at lower temperatures than the base metals and usually differ greatly in composition; for example, various nonferrous alloys are used to braze steels. The major distinction between brazing and soldering is that brazing is performed above 800°F (425°C) and soldering below that temperature.

12-5-1 WELDING PROCESSES

We find a variety of methods used to classify welding processes. This variety is due, at least partially, to differences in usage of terms and in names or descriptions of processes. It is also partly due to overlapping and combinations of terms and conditions of welding. This variation, especially when one is being introduced to welding, causes confusion. One method of classification is given in Fig. 12-5. A somewhat different classification is presented in Table 12-1.

Table 12-1. Classification of welding processes

A   Heat source
    1) Electrical
        a) Arc
        b) Resistance of pieces being welded
    2) Chemical
        a) Combustion of fuels
        b) Exothermic reactions
    3) None
B   Mechanics of making the joint
    1) Pressure welding
        a) With fusion
        b) Without fusion
    2) Fusion welding
        (implies no pressure)
C   Similarity of base and filler metals
    1) Autogenous—same base metals, no filler metal
    2) Homogeneous—same base metals, same filler metal
    3) Heterogeneous—dissimilar base metals and/or dissimilar filler metal

---

Arc welding applies to processes in which local union is accomplished by *heating with an electric arc, without applying pressure,* and *with* or *without filler metal.* The heat source is usually a continuous electric arc discharge between the end of a conducting "electrode" and the surface of the work. The name of a specific process is derived from identifying features, such as the electrode material, means of transferring heat, or the protection used. For instance, we talk about carbon-arc welding, atomic-hydrogen welding,

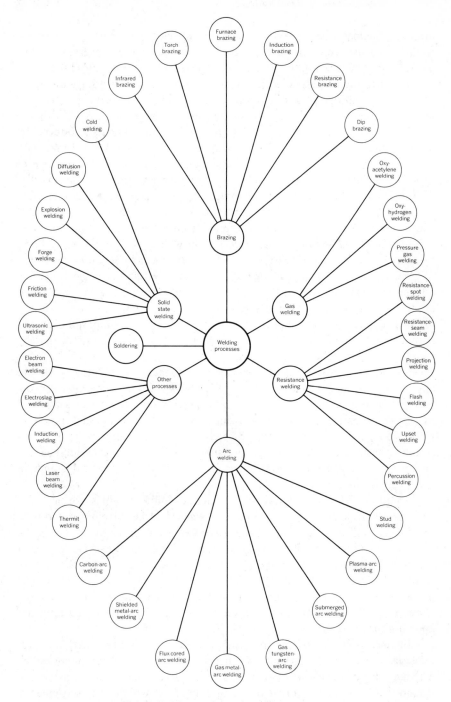

Fig. 12-5. Master chart of welding processes.

inert-gas-shielded arc welding, consumable-electrode-inert-gas shielded welding, plasma arc welding, etc. A continuous laser can be programmed like gas tungsten-arc welding to accommodate heat buildup and crater filling. For welding pressure vessels, electroslag welding, electrogas welding, and high heat input variations of gas metal-arc and submerged arc processes are used as well as shielded metal arc and gas tungsten-arc processes.

Gas welding applies to processes in which local union is produced by *heating* with a *gas flame* or flames, *with* or *without applying pressure*, and *with* or *without filler metal*. The gas used generally determines the name of process, for instance, oxy-acetylene or oxy-hydrogen welding. The quantity of heat is controlled by varying the torch tip size, gas pressure, and position of the torch. Filler metal, when required, is added separately in rod form. Fluxes in powder or paste form are applied to either the work or the filler rod.

Resistance welding applies to processes in which local union is produced by *heating* due to *resistance* to *passage* of an *electric current*, *with* application of *pressure*, and *without filler metal*. Resistance welding has the following advantages: (1) it is faster; (2) it permits more accurate regulation of heat application; (3) mechanical pressure is used to forge the weld with a resultant grain structure and mechanical properties somewhat comparable to the parent or base metal; and (4) no filler metal or fluxes are required, thereby simplifying the metallurgy of the weld.

In most applications, electrodes carry electrical current to the work and apply pressure to the work. Heat is generated at the interface between the two pieces because of the high electrical resistance, and the pieces are joined by their respective structures growing together under applied pressure. Fusion occurs where temperature exceeds the melting point. In some materials, however, a sufficiently high pressure (with proper prewelding surface cleaning and treatment) makes acceptable welds without melting, and, in some cases, without heating.

In addition to the material to be welded, major variables are pressure, current, electrode material, and timing. In general, large currents are required to generate sufficient heat in a relatively short time. Whereas each type of resistance welding has its unique features, timing of the cycle generally requires application of pressure before application of current with pressure maintained briefly after release of current. At the same time, pre-pressure is not as necessary nor as general as post-pressure. The electrodes may be the work pieces, or they may be separate pieces used for carrying current and applying pressure.

In *electron beam welding*, the kinetic energy of a high-velocity beam of electrons is converted into sufficient thermal energy to melt the metals which are then joined in fusion. A high vacuum is necessary to provide a focused stream of electrons and prevent arcing between the high-voltage elements. A principal use is obtaining ultrahigh purity welds in reactive and refractory metals, nickel and stainless steel.

Friction welding can produce high-quality welds between dissimilar metals as well as similar ones. At least one component to be joined must be approximately circular and one must be rotatable at high speeds. A rotating mechanism is disengaged and braked to a stop during the upset portion of the weld cycle. Inertia welding brings a part up to rotational speed while storing energy in a flywheel which is used to rotate one part (while the other is braked) when the two pieces are brought together. The stored energy is dissipated in friction to form a weld when rotation ceases. A major use is joining wear-resisting stems to heat-resisting exhaust valve heads.

Thermit welding applies to processes in which local union is produced by *heating* with *superheated liquid metal* and slag resulting from an exothermic reaction between a metal oxide and aluminum, *with* or *without applying pressure.* Filler metal, when used, is obtained from the liquid metal. In one nonpressure process a mechanical mixture of finely divided aluminum and iron oxide is ignited to yield highly superheated molten iron and aluminum oxide slag. Suitable alloying elements are added to the powder mixture to produce steel of a desired composition. Welds made by this process are metallurgically sound and strong with a minimum of distortion and residual stress. Application of this process is limited to relatively large joints and is used principally for repair work, e.g., railroad rails.

Brazing applies to processes in which local union is produced by *heating* to a suitable temperature *above* 800°F using a suitable *nonferrous filler metal* or alloy having a melting point below that of the pieces to be joined. The molten filler metal must "wet" the surfaces to be joined and must be distributed by capillary action to fill the entire space between the pieces. Some alloying between base metals and filler metal usually occurs resulting in a strength in a properly brazed joint intermediate between cast filler metal and the metals being joined. We normally think of brazing as being used to join metal components, but alloys are used to braze ceramic components.

Strength of a brazed joint depends on several factors:

1. Nature and strength of metals being joined.
2. Nature and strength of filler metal.
3. Amount of alloying occurring in the joint.
4. Clearance of pieces being joined. (If the pieces are held together too tightly, filler metal will not enter the joint. If the opening is too great, strength will be low. There is an optimum spacing for each combination of base and filler metals.)
5. Cleanliness of materials involved.
6. Use of flux.
7. Heating and cooling of the assembly. The joint should not be disturbed in any manner while the joint is cooling through the plastic range of the filler metal, and no load should be placed on the joint until it has cooled appreciably below the softening point of the filler.

Unfortunately, considerable confusion has resulted from careless use of terms, which is principally a matter of historical accident. For instance, "silver soldering" and "hard soldering" are actually brazing with silver base alloys, and "silver brazing" is proper terminology.

Soldering (or "soft soldering") applies to processes in which local union is produced by *heating* to a suitable temperature *below* 800°F using a suitable *nonferrous filler metal* or alloy with a melting point below that of the metals to be joined. There is less alloying between base metals and filler metal than in brazing. Joint strength depends mainly on adhesion and mechanical bonding. Most solders are based on lead and tin. Since these are weak with low melting points, a soldered joint should carry little or no load. Solders are generally used as fillers to stop leaks in mechanically locked seams, to seal against corrosion, and to carry electricity. Lead pipe and cable sheathing may be joined by a wiping technique in which a thick layer of solder is applied and, while in a mushy state, smoothed out or "wiped" into a smooth and "streamlined" form. (See question 7-37.)

12-5-2 METALLURGICAL ASPECTS OF WELDING

Obviously, there are a large number of specific welding processes, each having a number of interesting technical features and practical applications, but we limited ourselves to a brief discussion of the general classes of welding processes. The actual joint, regardless of process, will differ in structure and properties from the base metal. This implies a number of metallurgical considerations which we consider in two parts: (1) weld metal, i.e., the weld deposit or the metal in the actual joint, and (2) the heat-affected zone, i.e., metal immediately adjacent to the weld deposit.

12-5-3 THE WELD DEPOSIT

In *pressure welding*, metal in the joint is heated to the plastic condition, or to higher temperatures, and compressed while hot. Consequently these processes are more-or-less analogous to a hot-working operation, and the welded joint is essentially a "forging." This involves the principles of finishing temperature, recrystallization (with the possibility of grain refinement), grain growth, physical property effects, and so on.

In *fusion welding*, metal connecting the two parent pieces (whether melted from the edges to be joined or supplied from filler rod) is liquid, and cools to a lower temperature. The structure obtained under these conditions is essentially a "casting." In addition, since metals are good heat conductors, a nonpressure fusion weld is essentially a chilled casting having the associated columnar crystals. Further working operations or heat treatments may be employed in limited circumstances to modify weld structure. As in all casting operations, action of the atmosphere (including weakening and embrittlement due to trapped gas), fluidity of the molten metal and its surface tension, solidification process (including crystal nucleation and growth, segregation and shrinkage), and cooling rate all affect final structure and properties. The interplay of these variables may give different effects in the various welding processes and even between different applications of the same process.

12-5-4 HEAT AFFECTED ZONE

The effects of the welding process upon the metal immediately adjacent to the weld (the "heat-affected zone") are equally as important as the weld metal proper and depend on composition and previous history of the metal. For instance, in fusion welding, temperature changes from the melting point at the joined edges to room temperature over a relatively short distance. Since metals are good heat conductors, the heated zone undergoes rapid cooling. This implies a gradation of heating and cooling rates (i.e., quenching treatments) for the successive regions. Consequently, grain growth may occur near the weld, or recrystallization may occur if the metal were previously cold-worked. New grains will be formed if the material is allotropic. In steels, austenite is produced in the high temperature regions and may only partially transform during cooling, or may transform into several of its transformation products. The structure obtained in any given heat-affected zone will depend upon the alloy and its response to various heat treatments as discussed in Chapters 9 and 10. A typical example in a welded steel plate is shown in Fig. 12-6.

The major factors affecting temperature distribution in a weld are: (1) the welding process (e.g., arc welding gives steeper temperature gradients and smaller affected zones than gas welding); (2) thermal constants; (3) preheating; and (4) dimensions and shape of pieces. Velocity of travel of the heat source and heat input are also factors; however, heat-source velocity is more important than heat input.

In some cases, it is possible to heat treat a weld to substantially improve the properties. This is generally limited to relatively small pieces because of physical size limitations of furnaces. One notable exception was the stress relief anneal given the container vessel of the Calder Hall reactors. Each vessel, about 60 feet in diameter, was welded, blanketed with electrical heating elements and insulating material, and stress-relieved as one piece.

12-5-5 DISTORTION AND CRACKING

The possibility of distortion and cracking is a major problem. Expansion occurs during heating. This volume change is reversible upon heating or cooling regardless of rate. A metal is generally softer, weaker, and more

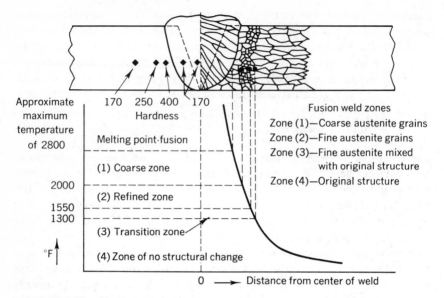

Fig. 12-6. Survey of hardness and microstructure across a welded steel plate.

ductile when hot. This softness and ductility can be advantageous in distributing stresses over the heated area and thus reduce peak values. At the same time, strength must be sufficient to sustain stresses which are developed. Some metals and alloys are relatively brittle within certain temperature ranges and thus are susceptible to "hot-cracking."

Heating during welding has a strong tendency to be localized and uneven. This means uneven expansion and contraction which tend to warp the piece and often produce residual stresses. Volume changes accompanying allotropic transformations provide an additional complication.

Distortion and cracking are ever-present problems. Both arise from volume changes, caused by heating and cooling, which cannot be accommodated (or dissipated) by surrounding material. Residual stresses (and distortion) can often be reduced or relieved by pre- and/or post-welding heating.

12-5-6 WELDABILITY

Weldability is the ability of a material to be welded. The American Welding Society defines *weldability* as: "The capacity of a metal to be welded, under the fabrication conditions imposed, into a specific, suitably designed structure and to perform satisfactorily in the intended service." Probably all metals can be welded by some process, provided adequate precautions and sufficient control are exercised. It is seldom practical, however, to employ materials which require very special treatment and complicated procedures except in unusual circumstances. In welded construction, the designer must consider the weldability and decide, not so much if a joint can be made, but if it is practical and economical to do so. The question of weldability is thus primarily a question of how much precaution is required for successful welding and how drastic the aftereffects may be.

12-6 FABRICATION OF POLYMERS

Polymers are used for an almost infinite number and variety of purposes, and new ones are added continuously. They have become construction materials just as have the ceramics (including glass). In small formed objects, they rival metals in variety and number of applications. When the number of polymers and their possible variations are also considered, it seems almost hopeless to attempt a brief description of their fabrication. Even so, some basic methods can be outlined.

Fabrication of thermoplasts generally follows a basic pattern of heating (to soften), mechanical deformation (to obtain desired form), and cooling (to harden).

Fabrication of thermosets is generally accomplished by heating the uncured polymer, forming it, curing in the mold to form the 3-dimensional network, and cooling.

Fabrication of elastomers is similar except that vulcanization by formation of cross-linked bonds (by chemical means) takes place after achieving the desired shape.

Various machine operations include extrusion, injection molding, compression molding, transfer molding, blow molding, sheet formation, casting, etc. Various cutting and joining operations—machining, grinding, slitting, heat welding, adhesive bonding, etc.—can be performed on appropriate polymers. These may be primary fabrication techniques but are also often secondary in other operations such as trimming flash from a molded object.

Fluid polymer solution and film-forming latexes can be coated on substrates such as paper and fabrics or can be converted to continuous films by evaporation of solvent (either organic or water). Such film-forming operations are the bases for a broad class of manufactured products.

Polymers can also be formed into fibers by forcing molten or dissolved material through tiny spinnerettes (melt spinning, solution spinning). The fibers, in turn, can be twisted into yarns and then woven into fabrics.

Many polymers can be "expanded" to form low-density foams, i.e., cellular structural materials, in which the cells may either be open or closed.

The degree of rigidity depends on the polymer and any modifiers or plasticizers that may be incorporated, and varies from high rigidity to high flexibility.

12-6-1 EXTRUSION

Extrusion of thermoplasts is similar to extrusion of metals with some differences. Metal extrusion is effectively a batch process. Polymer extrusion is a continuous process. Granules of polymer are hopper-fed into the rear of a cylinder in which there is a rotating screw. As the screw turns, the granules are compacted, mixed, heated, forwarded, and eventually forced through a die at the front end of the cylinder to emerge as ribbon, tube, sheet, or in some other shape.

Two extruders can be used in tandem to produce special products. In producing a special gasoline hose, for example, one extruder handles a soft polyvinyl chloride while the other extrudes a rigid polyvinyl chloride. The two machines feed to a special crosshead where two concentric tubes are extruded as a single unit. A rotating mechanism makes the rigid material spiral around the soft lining to produce a flexible hose which has a soft, gasoline-resistant inner tube protected by a mechanically strong outer tube.

12-6-2 COMPRESSION MOLDING

Compression molding is the process in which polymers flow under pressure into cavities between two parts of a hot mold. The charge is predetermined by weight or volume and placed in the lower half of the mold cavity. The mold is slowly closed until the polymer starts to flow and then the closing rate is reduced. When the excess material has flashed out and the mold is completely closed, the mold is kept closed for a definite time period. During this period, the mold must be cooled if molding a thermoplast. If molding a thermoset, curing takes place during this period. At the proper time, the mold is opened and the object removed for deflashing and any other finishing operations which may be required.

12-6-3 INJECTION AND TRANSFER MOLDING

Injection molding is essentially a process of transforming a granular thermoplast into a fluid, plasticized mass by applying heat and then forcing it under high pressure into a cooled mold where it hardens. Although the process is almost entirely automatic, it is an intermittent rather than continuous process. Rapid cooling and solidification occur wherever the melt touches the cold surface of the mold. Filling is thus accomplished by pushing hot material through the middle of a channel. Excess melt is often pushed in under high pressure to ensure complete filling.

Transfer molding is the analogous operation for thermosets. Granules are subjected to heat and pressure in an outside chamber and forced into a closed mold cavity. Heat is maintained on the closed mold until curing is complete, after which the object is ejected, and the mold chamber is cleaned before another cycle.

Both of these processes for molding polymers have a clear similarity to die casting and cold chamber pressure casting for metals (Sec. 12-2–4).

QUESTIONS AND PROBLEMS

1. Give a definition of casting. What four basic steps are involved?

2. What are the general characteristics of castings? What is the distinction between castings and ingots? What are the general requirements of castings? What factors influence the choice of mold material for a casting?

3. List and discuss four economic or metallurgical reasons for fabricating metal parts by casting.

4. Discuss the similarities and differences of the various casting processes: sand mold, plaster mold, permanent mold, die casting, centrifugal casting, investment casting, and shell mold. Find one unique characteristic that differentiates each process from the others. What are the advantages and disadvantages of each?

5. Can steel castings be made by die casting? If not, why not? What metals can be die-cast?

6. Discuss the general procedure in melting metals, with special emphasis on the particular requirements of some of the metals peculiar to reactor use.

7. What are the effects of solubility of gas, low fluidity, and high surface tension in molten metal on the quality of a casting?

8. What is "directional solidification," and what is its influence on casting quality?

9. What is ingot segregation? Is this related in any manner to quality and if so, how?

10. Discuss grain structure in castings. What effect does mold material have on grain structure? What effect do impurities have on grain structure? What are dendrites?

11. What is mechanical working, what use is made of it, and why?

12. Discuss the differences and similarities of elastic and plastic deformation.

13. Discuss the differences and similarities of hot and cold working.

14. What are the two general areas of industrial use of the powder process? What justification exists for the use of powder metallurgy in these applications?

15. Is low porosity desired in some metal-powder parts and not in others? Why?

16. Is it possible to obtain high density without high shrinkage? If so, how might this be done? Why is it not always done when high density is required?

17. Would a cylinder pressed from one end show different degrees of shrinkage in different directions during sintering? Why?

18. What necessitates the use of multiple punches in some compacting operations? What effect does addition of a lubricant have?

19. Is it possible to press parts with undercuts or re-entrant angles from powder? Why?

20. In what manner does the powder metallurgy process differ from casting? What relative advantages and disadvantages are found between the two processes?

21. What are the four basic steps in the powder process? What part does each play? What mechanisms are involved in each?

22. What is the effect of atmosphere on powder behavior and the sintering or firing process?

23. Name some applications where the powder process is commercially employed.

24. Type 304 stainless steel is 18–20% Cr, 8–10% Ni, balance Fe. Stainless steel filters are desired. There are at least two ways this can be accomplished using the powder process. What are the two ways? How would you do this if very high purity stainless steel were desired?

25. Metallic filters are used in many applications to remove foreign particles from a fluid. How would you produce these filters and control the porosity?

26. Define welding. What different meanings does the term have? What is the significance of each meaning?

27. Discuss the differences and similarities among soldering, brazing, and welding.

28. What are the dangers of allowing a weld to solidify very rapidly?

29. Of what importance or significance is microstructure in a weld? Why?

30. In a two-pass weld, if the first weld deposit is rapidly cooled, what effect does the deposition of the second layer have on the structure and properties of the first layer? Why?

31. Is there any connection between "hardenability" in a steel and the "maximum safe cooling rate" in welding? If so, what is the connection, what is its significance, and how does it operate?

32. What is the distinction between stress and distortion in a weld? Is there a relationship between them? If so, what is the relationship?

33. The electric arc is much hotter than the oxyacetylene flame. Is it possible that a steel which cannot be welded with the electric arc can be welded with an oxyacetylene torch? Why or how?

34. In gas welding, it is possible to adjust the flame to obtain an oxidizing, a neutral, or a reducing flame. What would be the differences among these flames and what practical differences would result from using them for welding? Why?

35. What is the significance of the use of pressure in some welding processes? What does this mean in terms of properties developed in the weld area?

36. Is it necessary that a brazing alloy "wet" the surfaces to be joined? Why?

37. What are the factors upon which the strength of brazed joint depends? Of a soldered joint? How do these factors operate?

38. What factors are involved in the "cold-welding" of a metal such as aluminum? What preliminary precautions would be required? How is the joint made? How do the interatomic forces act to form a bond?

39. What is weldability? Of what significance is it? Compare the weldability of the metals in each of the following pairs: (a) low-alloy steel and cast iron, (b) low-alloy steel and copper, (c) aluminum and copper, (d) aluminum and low-alloy steel, (e) low-alloy steel and high-alloy steel. What are some of the specific factors in each pair that determine the relative weldability?

40. Discuss the similarities and differences of extrusion of metals and extrusion of polymers.

41. Discuss the similarities and differences of compression molding of polymers and casting (e.g., sand casting) of metals.

42. Discuss the similarities and differences of injection and transfer molding of polymers and die casting of metals.

13

EFFECTS OF ENVIRONMENT

13-1 INTRODUCTION

In the preceding chapters we have discussed structure and properties of materials and how these characteristics may be changed by changes in composition, by various heat treatments, and by fabrication techniques. Properties and behavior may be further modified by the immediate environment; i.e., the response of a material can be influenced (drastically in some cases) by its ambient atmosphere. This requires consideration of items such as corrosion, oxidation, temperature, and radiation.

Corrosion is a major problem in the everyday world of our personal lives as well as in technological areas. It is estimated that, in the United States alone, the annual economic loss from corrosion is about 15×10^9. It must be recognized that every material undergoes chemical changes of one kind or another when exposed in any environment even though the degree may be small rather than great.

13-2 CORROSION OF METALS

Corrosion is often defined as: "Destruction of metal by chemical or electro-chemical action." This implies transfer of electrons as, for example, in the rusting of iron and low-alloy steels. Corrosion by liquid metals and fused salts does not necessarily fall within the above definition. Destruction of solid metal by liquid metal, for example, does not involve transfer of electrons but involves factors such as solubility of solid in liquid metal. Thus a broader definition might be: "Corrosion is the destruction of material by: (1) chemical, electrochemical, or metallurgical interaction between the environment and material, or (2) dissolution of material in the environment."

We will consider corrosion independently of mechanical factors such as abrasion, erosion, or cavitation. At the same time, corrosion and mechanical action often operate together with the effect of promoting each other. Corrosion may merely disfigure appearance under service conditions, or it may reduce strength to a level at which failure will occur. Corrosion may be reasonably uniform, or it may be concentrated along grain boundaries and other lines of weakness. It may result in pits on the surface. Corrosion of material depends not only on the material and its environment but also on the internal structure of the material and the nature of the service. Corrosion rate is influenced by: (1) structural factors, e.g., composition, distribution of secondary phases, residual stresses, voids, inclusions, and dissolved gases, and (2) environmental factors, e.g., concentration, temperature, movement of corrodant, and the presence of electrochemical couples, surface films, and applied stresses.

Corrosion of metals by chemical or electrochemical reaction with the environment occurs because metals are inherently unstable and tend to revert to a more stable form. Naturally occurring metallic ores are common examples of these more stable forms, and in a sense, corrosion of metal reverses the process by which it was obtained from ore. Energy must be supplied to obtain metal by such processes as smelting, electrowinning, etc. During corrosion, energy is released. Most common metals and alloys are attacked by ordinary environments such as the atmosphere, water, aqueous solutions, and other fluid media to form compounds of the metal. These corrosion products are usually oxides (more or less hydrated), carbonates, and sulfides.

Since most metals are attacked by common environments, we might consider that any contact between metal and fluid results in corrosive action. In some cases, corrosive attack proceeds so slowly it can be neglected. In other cases, corrosive attack proceeds at a rapid rate and then stops before serious damage occurs. In still other cases, attack may proceed rapidly and continue unabated until the piece fails. The rate and degree of attack in any given case depend on the metal and its environment. *There are no metals which will withstand corrosive attack in all environments.* Any metal corrodes under certain conditions.

All chemical or electrochemical corrosion phenomena are electrochemical in nature for they require transfer of electrons whether the environment is a hot gas, an aqueous solution, or an organic medium. Electrochemical corrosion, however, usually refers to reactions in which electric currents flow for perceptible (although possibly very small) distances in metals, i.e., when *relatively anodic and cathodic areas are coupled to form a galvanic cell,* e.g., a brass fitting with a steel pipe.

A second type is known as direct attack, for example, acid pickling of steel during fabrication. Reactions of dry chlorine, hydrogen sulfide, oxygen, or other dry gases with dry metals are also examples of direct chemical attack, although they are also sometimes considered as oxidation-reduction reactions.

The phenomena involved are essentially the same for all metals with differences in degree rather than in kind. Our discussion will be specifically in terms of iron (in a general sense, not in the sense of commercial iron as opposed to steels).

In the electrochemical theory of corrosion, which implies a transfer of electrons, three conditions must be fulfilled: (1) at least two spatially separated areas of the metal surface must act as electrodes for flow of electrons (electric current) from the metal into the solution and back again; (2) both the solution and metal must be capable of conducting current; and (3) a driving force or electrical potential must exist to make the electrons flow. These conditions are illustrated in Fig. 13-1.

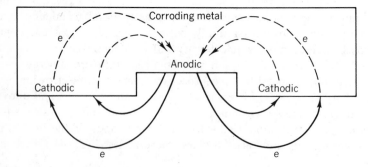

Fig. 13-1. Schematic galvanic cell for electrochemical corrosion.

Most metallic elements in solution enter the solution as ions. In other words, a metallic atom gives up one or more valence electrons to become a positively charged ion which enters the solution, making the solution an electrolyte and leaving the remaining metal negatively charged with an equivalent potential. The difference in potential thus established between metal and solution is termed solution potential or solution pressure. It is the driving force which causes electrons to flow. It also determines the tendency of a metal to corrode.

It is desirable to consider cathodic and anodic reactions separately although both reactions proceed simultaneously. Even so, the slower one controls the overall rate of corrosion. In the iron-pure water system the possible anodic reactions are:

Ferrous hydroxide formation

$$Fe + 2(OH)^- \longrightarrow Fe(OH)_2 + 2e^- \tag{13-1}$$

Ferrous oxide formation

$$Fe + 2(OH)^- \longrightarrow FeO + H_2O + 2e^- \qquad (13\text{-}2)$$

Ferrous ion formation

$$Fe \longrightarrow Fe^{++} + 2e^- \qquad (13\text{-}3)$$

The only cathodic reaction is discharge of hydrogen ions:

Hydrogen gas formation

$$2H^+ + 2e^- \longrightarrow H_2 \qquad (13\text{-}4)$$

Knowing the anodic and cathodic reactions, we can fit them together to see how the overall corrosion reaction proceeds.

Metallic ions enter into solution only if an equivalent number of positive ions of some other element are displaced, since the solution must remain electrically neutral. For iron in water, hydrogen ions are displaced and gather on the iron surface as a thin, invisible film which tends to obstruct the reaction by: (1) insulating or separating iron from the solution, and (2) the increasing tendency of hydrogen in the film, as it becomes thicker, to reenter the solution and oppose the reaction.

For corrosion to continue, the film must be removed. This can be done by combination of hydrogen with dissolved oxygen to form water, escape of hydrogen as gas bubbles, or dissolving hydrogen in the metal. In any case, the hydrogen film is removed, and the reaction continues. In other words, more iron goes into solution, more hydrogen "plates out," and the process continues at a rate dependent on the rate of hydrogen-film removal. For iron and water, the rate of film removal generally depends on the effective concentration of dissolved oxygen in water adjacent to the metal. This effective concentration, in turn, depends upon the degree of aeration, amount of motion, temperature, presence of dissolved salts, and other factors.

Corrosion is generally more rapid in acid than in alkaline solutions, because the tendency of hydrogen to plate out increases with acidity of the solution. This tendency can increase to such an extent that film removal is predominantly by bubble rather than by water formation. The reaction is, however, identical to that discussed above.

In solutions containing ions other than iron or hydrogen, the principle is the same, but the actual end result depends on the integrated effect of several different reactions which may occur between the various kinds of ions present.

For iron in pure aerated water, the typical primary reaction is:

$$\underset{\text{(metal)}}{Fe} + \underset{\text{(ions)}}{2H^+} \rightleftharpoons \underset{\text{(ion)}}{Fe^{++}} + \underset{\text{(atoms)}}{2H} \qquad (13\text{-}5)$$

This reaction is followed by (1) destruction of the film by combination with oxygen:

$$\underset{\text{(atoms)}}{2H} + \underset{\substack{\text{(dissolved} \\ \text{gas)}}}{\tfrac{1}{2}O_2} \longrightarrow \underset{\text{(liquid)}}{H_2O} \qquad (13\text{-}6)$$

(2) removal of film as bubbles of gas:

$$2H \longrightarrow H_2 \qquad\qquad (13\text{-}7)$$

$$\text{(atoms)} \quad \text{(gas)}$$

or (3) solution of hydrogen in the metal. All these reactions permit continuance of the primary reaction resulting in accumulation of Fe^{++} in solution. These ions, in turn, combine with dissolved oxygen and water to form ferric hydroxide which is insoluble and which precipitates.

$$4Fe^{++} + O_2 + 8OH^- + 2H_2O \longrightarrow 4Fe(OH)_3 \qquad (13\text{-}8)$$

The mechanism of electrochemical corrosion can be summarized as follows: The metal being corroded is oxidized (i.e., its valence is increased by loss of electrons) while some substance in the environment is reduced (i.e., its valence is decreased by gain of electrons). This implies that the two parts of the process take place at different, though not normally widely separated, areas. *The area of metal attacked or oxidized is the anode, and the area of metal reduced is the cathode.* For a continuous process, an electrical current must flow through the metal and the environment (usually an aqueous solution) between these two areas. This combination of anode, cathode, and electrolyte constitutes a galvanic cell. The corrosion reaction proceeds much like the generation of electrical current by chemical action in a primary cell or discharge of a storage battery.

Obviously, we must be able to determine which areas are anodic and which are cathodic, since anodic areas suffer corrosion. A decrease in anodic area must be accompanied by at least a proportionate decrease in the total amount of corrosion, or there will be a strong tendency toward localized corrosion. The principal factors (alone or in combination) determining anodic and cathodic areas are summarized in Table 13-1.

The fourth item in Table 13-1 deals with metals and impurities which are anodic or cathodic relative to the metal under consideration. Whether a metal is anodic or cathodic to another can be determined from the electromotive force series given in Table 13-2 for the more common elements. This

Table 13-1. **Factors determining relative anodic and cathodic areas in corrosion couples**

Anodic areas	*Cathodic areas*
1. Strained metal	1. Unstrained metal
2. Areas with a low oxygen concentration	2. Areas with a high oxygen concentration
3. Areas at which a protective film has been broken	3. Impurities (in the metal) having a low oxygen or low hydrogen overvoltage
4. Impurities in metal (or in contact with metals) which are anodic to the metal	4. Impurities in metal (or in contact with metals) which are cathodic to the metal

Table 13-2. Electromotive force series

	Element	Symbol	Standard electrode* potential, in volts at 25°C
	Potassium	K	−2.922
	Calcium	Ca	−2.77
	Magnesium	Mg	−2.34
	Beryllium	Be	−1.70
	Aluminum	Al	−1.67
	Manganese	Mn	−1.05
Anodic; electrode negative to Hydrogen	Zinc	Zn	−0.762
	Chromium	Cr	−0.71
	Iron	Fe	−0.440
	Cadmium	Cd	−0.402
	Cobalt	Co	−0.277
	Nickel	Ni	−0.250
	Tin	Sn	−0.136
	Lead	Pb	−0.126
	Hydrogen	H_2	0
	Antimony	Sb	+0.1
	Bismuth	Bi	+0.226
	Copper	Cu^{++}	+0.345
	Copper	Cu^+	+0.522
	Mercury	Hg	+0.799
Cathodic; electrode positive to Hydrogen	Silver	Ag	+0.800
	Palladium	Pd	+0.83
	Mercury	Hg^{++}	+0.854
	Platinum	Pt	+1.2
	Gold	Au^{+++}	+1.42
	Gold	Au^+	+1.68

*For unit activity in water. A change in environment may mean a change in relative order.

electromotive force series indicates relative tendency to go into solution. Every metal has a tendency to put ions into solution, accompanied by establishment of a potential difference between metal and solution. The potential difference is known as solution pressure and has a characteristic value at a specified concentration of ions in solution. A change in concentration will give a change in potential. Table 13-2 lists potentials between metals and hydrogen for unit activity in water. A lower concentration of hydrogen ions tends to move hydrogen upward relative to metals in this series, or a higher concentration of metal ions tends to move metal downward relative to hydrogen. A given metal is anodic to one lying below it in the electromotive force series. In other words, the greater (more negative) the solution potential of a metal, the more readily it corrodes.

It should be emphasized that the EMF series given in Table 13-2 indicates the relative tendency of metallic elements to go into solution but under controlled laboratory conditions. Such data, while sometimes useful, can be

Table 13-3. Corrosion galvanic series

in Sea Water

Magnesium

Magnesium alloys

Zinc

Aluminum

Cadmium

Aluminum 2024

Mild Steel

Cast Iron

Ni-Resist

18–8 Stainless (Type 304) (active)
18–8–3 Stainless (Type 316) (active)

Lead-tin Solders
Lead
Tin

Nickel (active)
Inconel (active)

Brasses
Copper
Bronzes
Monel

Silver solder

Nickel (passive)
Inconel (passive)

18–8 Stainless (Type 304) (passive)
18–8–3 Stainless (Type 316) (passive)

Silver

Graphite
Gold
Platinum

Increasingly Anodic ⟶

misleading. It is desirable to obtain potentials of metals and alloys in a common environment such as shown in Table 13-3. If two metals from this list are electrically connected in sea water, the metal higher on the list will be relatively anodic and will corrode. In addition, the farther the two metals are from each other, the greater the tendency to corrode. It must be emphasized that a galvanic series of the same metals in another environment may have a radically different order of ranking.

EXAMPLE 13-1

Assume copper and zinc are connected in an electrically conducting medium.
(a) Will corrosion occur? If so, what is the chemical reaction?
(b) What is the maximum emf possible?
(c) If there is corrosion, which metal is the anode? Which is the cathode?

Solution:

(a) There is corrosion.
$$Zn + Cu^{2+} \longrightarrow Zn^{2+} + Cu$$
(b) From Table 13-2,
$$0.762 + 0.345 = 1.1 \text{ volts}$$
(c) Zinc is the anode, copper is the cathode.

EXAMPLE 13-2

Three galvanic cells, Mg—Fe, Zn—Fe, and Cu—Fe, are compared for electrochemical corrosion. Show which is the anode and which is the cathode in each cell. Indicate the relative tendency to corrode.

Solution:

From Table 13-2,

	Anode	*Cathode*	*Maximum emf*	*Corrosion rate*
Mg–Fe	Mg	Fe	1.90	Greatest
Cu–Fe	Fe	Cu	0.78	Intermediate
Zn–Fe	Zn	Fe	0.32	Least

13-2-1 SUPPRESSION OF ELECTROCHEMICAL CORROSION

There are various methods by which electrochemical reactions can be reduced or eliminated, as indicated in Table 13-4. All efforts depend on suppression or retardation of anodic reactions.

Table 13-4.   Methods of suppressing
electrochemical corrosion

---

1. Cathodic protection
2. Isolation by films
   a. development of protective corrosion product films
   b. applied films, e.g., paints, metal plate
3. Inhibitors
4. Alteration of environment
   a. oxidizing capacity
   b. temperature
   c. velocity or agitation
   d. concentration of corrodant
5. Purification of metal
6. Alloying

13-2-2   CATHODIC PROTECTION

Cathodic protection (Fig. 13-2) adequately suppresses the anodic reaction by shifting the reaction between two areas of the same metal (Fig. 13-1) to a reaction between exposed metal and an auxiliary anode. Protection of this type can be achieved by using: (1) a sacrificial anode, or (2) an impressed current (a battery or a rectifier). If two dissimilar metals are in electrical contact in a corrosive environment in which the more active metal (the one higher in Table 13-2 or Table 13-3) does not become passive (Sec. 13-2-3),

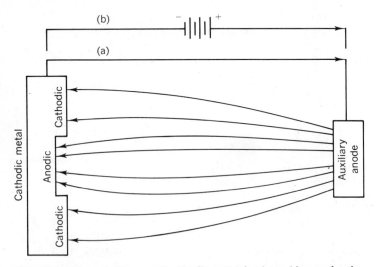

Fig. 13-2. Schematic diagram of cathodic protection by making a galvanic cell (Fig. 13-1) relatively cathodic by (a) using an auxiliary anode or (b) an impressed current.

the more active metal will be anodic; i.e., it will corrode and cathodically protect the less active metal. Cathodic protection is thus provided by using a sacrificial anode which is consumed by the corrosive environment. Its corrosion provides the desired protection, however, sacrificial anodes must be replaced periodically. The amount and region of protection depends on conductivity of the environment, solution potential of the anodic metal, shape and dimensions of the system, and other factors.

Galvanized iron is a common example of cathodic protection. In most environments the zinc coating provides protection for iron and is consumed (or sacrificed) under service conditions. It is not necessary for the zinc coating to be perfect since it provides protection as long as the exposed areas of iron are relatively small. Once the exposed areas of iron become too large, corrosion of zinc is accelerated.

Use of an active metal for cathodic protection may be unsuccessful if the environment tends to make the active metal become passive without affecting the activity of the less active metal. For example, zinc protects iron in most environments, but iron becomes anodic relative to zinc in certain types of hot water and corrodes quite rapidly.

13-2–3 ISOLATION BY FILMS

The electrochemical reaction often forms a film of corrosion product on the surface of corroding metal. Behavior of this film is most important in determining the ultimate result of action of the environment. The corrosion product film tends to stifle the corrosion process by increasing anodic polarization, cathodic polarization, or resistance between the two areas. The protective power of a given film is relative rather than absolute. A film which provides excellent protection in one environment may be completely useless under different circumstances. Films range all the way from a spongy mass (films at cathodic areas are usually of this type) to an adherent, practically insoluble, and practically nonporous film. A spongy film merely hinders diffusion of reactants while an adherent, insoluble, and nonporous film may completely passivate the metal or alloy.

All metals and alloys, except the "noble" metals (e.g., Au, Pt), are attacked by oxygen upon exposure to air at room temperature or higher. If the metal is completely covered by a film which is only a few molecular layers thick, it may be relatively protective, such as on Cu, brass, Sn, Pb, Fe, Ni, monel, stainless steel, Zn, Cr, Al, and many other metals and alloys exposed to relatively dry air. The relative permanence of these metals and alloys at room temperature is due to this oxide film formation. If temperature is increased sufficiently, however, the film grows by diffusion of oxygen or metal ions through the film, leading to continued oxidation. A common example is scaling of iron and steel at hot-working temperatures.

Films which give complete protection are invariably thin, in fact, so thin that they are ordinarily invisible. Passivating films are most commonly formed in air, but similar films can be formed in some aqueous media, e.g., aluminum and stainless steels in many aqueous solutions.

An item of considerable interest in nuclear reactors is corrosion of iron in high-temperature water. There is an initial period where there is an intact and protective oxide film (at 600°F, this period is about 150 to 200 hours for an electrolytically polished surface; the film becomes about 0.5 microns thick). The postulated oxidation-reduction reaction proceeds in the manner illustrated in Fig. 13-3.

Ferrous ions formed at the metal-oxide interface diffuse through the oxide (Fe_3O_4) by interchange with vacant lattice sites. Upon reaching the oxide-water interface, these ferrous ions combine with hydroxyl ions (or absorbed water molecules) to form ferrous hydroxide [$Fe(OH)_2$] which, in turn,

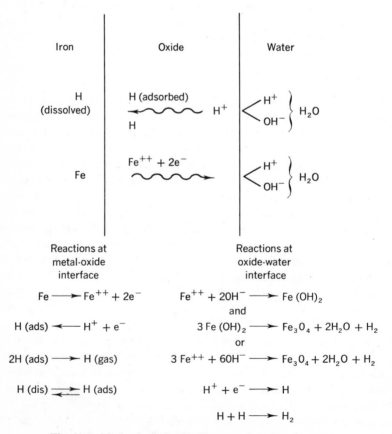

Fig. 13-3. Mechanism of oxide film growth on iron in water.

gradually changes into Fe_3O_4, water, and hydrogen. To maintain electrical neutrality, electrons must also migrate. These electrons neutralize protons at the oxide-water interface to form hydrogen molecules. The fact that hydrogen enters the metal, sometimes in large amounts, means that either hydrogen atoms or protons migrate from the water through the oxide to the metal.

This mechanism forms new oxide at the oxide-water interface. In some other metals, e.g., zirconium, a similar mechanism is postulated, although new oxide is formed at the metal-oxide interface. This requires diffusion of oxygen rather than metal ions through the oxide.

In many metals, oxide film is formed in the same manner as for iron. This means the reaction proceeds at a decreasing rate until the reaction has practically stopped, and the metal is passivated. In iron, however, the corrosion rate becomes constant at a low value, and a linear rate law is followed. This implies that corrosion rate comes under some form of "mechanical control" where the oxide is no longer completely protective. At least two possibilities exist: (1) the scale breaks down at a fairly constant rate, and overall corrosion is controlled by an interface reaction, or (2) only the outer portion of the scale breaks down, leaving a more or less constant thickness of protective scale. Of these, the second appears more likely.

Corrosion of metal exposed to water (or aqueous solution) depends greatly on the effect of the medium on the film. An air-formed film undoubtedly has weak spots due to impurities, mechanical handling damage, or other chance variations. If the solution tends to repair such defects, the film provides protection. On the other hand, if the environment does not repair defects, or if it causes deterioration, the film is not protective, and corrosion increases with time. In general, oxidizing anions and anions which form insoluble compounds with the metal tend to aid film formation and film repair. The use of soluble chromates, phosphates, silicates, and borates in the proper amounts gives protection to iron and steel in aqueous media. Other anions, such as the halogens, cause film breakdown and accelerate corrosion. The relatively rapid corrosion of aluminum and magnesium in marine atmospheres are examples.

From one viewpoint, we can regard the film as a means of isolating metal from the environment. Within this interpretation, we can consider plating, cladding, painting, and other types of corrosion resistant coatings. These coatings can be considered in two classes: (1) those providing mechanical protection by resisting the environment and (2) those providing cathodic protection. Materials in the first class either are not attacked or form a protective film. In providing mechanical protection, a completely unbroken surface must be maintained, or corrosion may be drastically accelerated in that location if the protective surface is broken.

At least one costly instance occurred at Los Alamos Scientific Laboratory in the LAPRE-1. This reactor was fueled with enriched UO_2 dissolved in aqueous phosphoric acid. This solution is highly corrosive, which required all equipment to be gold-plated or made of platinum. This reactor was shut down after five hours of operation due to a failure. Examination revealed the gold plating on the cooling coils had been damaged, presumably during assembly, and corrosion had occurred, resulting in failure. LAPRE-2, using the same fuel, was much more successful.

13-2-4 INHIBITORS

Any chemical substance or mixture which, when added to an environment (usually in low concentration), effectively decreases corrosion rate is called an *inhibitor*. Anodic inhibitors decrease corrosion rate by impeding formation of metallic ions, while cathodic inhibitors increase total cathodic polarization. Use of chromates, phosphates, silicates, etc., to decrease corrosion rate of iron and other metals is anodic inhibition. These anodic inhibitors apparently function by adhering to the metal surface to form a tightly bound mono-molecular film or monolayer of ions. Effectiveness as an inhibitor is apparently due to more effective formation and repair of passive oxide films.

Successful use of inhibitors requires considerable knowledge of their action and a thorough understanding of the specific corrosion process in the system under consideration. For example, a substance which appreciably decreases rate of attack, or even completely stops attack by one medium may stimulate rate of attack in another. Under other circumstances, a substance may decrease overall attack while increasing intensity of attack at localized, restricted, anodic areas.

Oxygen may act as an inhibitor or as a stimulator, depending on conditions. Oxygen normally tends to make a surface cathodic since it is a reactant in the cathodic reaction and because it helps maintain passivating oxide films. The higher the oxygen concentration on an iron or steel surface, the lower the probability of attack. Once attack starts at a weak point, however, rate of attack is stimulated by increased oxygen concentrations.

In practice, prevention of localized corrosion is of more concern than a decrease in overall corrosion rate. Failures are seldom due to "uniform" corrosion over large areas. They generally result from relatively intense attack over a small or localized area. Consequently, the influence of an inhibitor on the area attacked is at least as important as its influence on the rate of corrosion. In general, corrosive attack is intensified when anodic areas are relatively small and the corrosion process is under cathodic control, a condition which usually results when too little anodic inhibitor is added. For example, addition of too little chromate to completely stifle corrosion of iron, steel, zinc, and aluminum causes intensification of attack. Larger amounts of chromate will generally bring the process under anodic control and provide good protection. The actual amount of chromate required depends on concentration of other ions, e.g., sulfate and chloride, in the aqueous medium since these interfere with formation of passivating films.

13-2-5 ALTERATION OF ENVIRONMENT

Oxidizing Capacity

Formation, by an oxidation-reduction reaction, of a passivating film is a common corrosion protection measure. Thus we expect that the oxidizing capacity of an environment will have appreciable influence on corrosion rate. Oxidizing capacity may increase, decrease, or have no effect on corrosion rate as indicated in Fig. 13-4. We cannot be sure of the effect until we know the specific system and conditions.

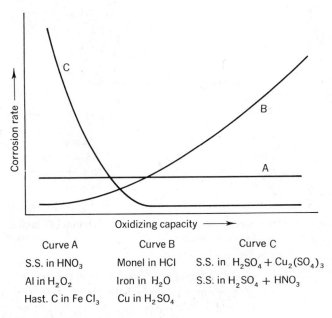

Curve A	Curve B	Curve C
S.S. in HNO_3	Monel in HCl	S.S. in $H_2SO_4 + Cu_2(SO_4)_3$
Al in H_2O_2	Iron in H_2O	S.S. in $H_2SO_4 + HNO_3$
Hast. C in Fe Cl_3	Cu in H_2SO_4	

Fig. 13-4. Schematic curves showing effect of oxidizing capacity of solution on corrosion rates of various metals.

Temperature

Considering the great effect of temperature on most rate processes (Chapter 8), we might anticipate a similar effect on corrosion rate. We know the principal factors in determining corrosion rate are: anodic polarization, cathodic polarization, electrical conductivity of the environment, and diffusion rates. In general, temperature increases diffusion rate and electrical conductivity but decreases polarization. The resultant effect, in general, is an increase in corrosion rate. Some metals and alloys are passive in certain environments at room temperature but lose this passivity at higher temperatures and become quite active. These general effects are indicated in Fig. 13-5. There are a few cases, however, in which corrosion rate decreases with increase in temperature.

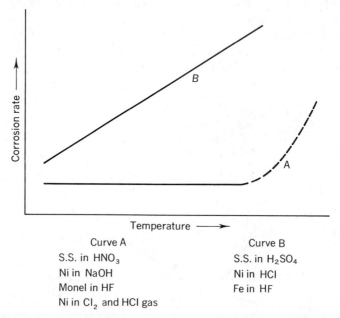

Curve A	Curve B
S.S. in HNO_3	S.S. in H_2SO_4
Ni in NaOH	Ni in HCl
Monel in HF	Fe in HF
Ni in Cl_2 and HCl gas	

Fig. 13-5. Schematic curves showing effect of temperature of solution on corrosion rates of various metals.

If we hold some metals at constant temperature in a given environment, we find that corrosion rate may increase with the rate at which heat is transferred through the metal.

Velocity

Velocity affects corrosion through its influence on factors which control corrosion rates. In general, corrosion increases with velocity (Fig. 13-6) although the reverse is true in a few cases. Type 347 stainless steel, exposed to aqueous, oxygenated, uranyl sulfate solution at 250°C and higher, forms a stable protective coating up to a certain flow rate. Above this flow rate (critical flow rate) the protective coating does not form, and corrosion proceeds rapidly.

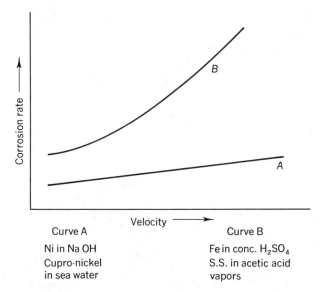

Curve A   Velocity   Curve B

Ni in Na OH   Fe in conc. H_2SO_4
Cupro-nickel   S.S. in acetic acid
in sea water   vapors

Fig. 13-6. Schematic curve showing effect of velocity of solution on corrosion rates of various metals.

Motion or agitation generally tends to make conditions more uniform, thereby rendering corrosion more uniform and reducing local attack. Corrosion results are more reproducible when there is some motion in the solution. On the other hand, motion can set up turbulence which in turn can produce nonuniform conditions and localized corrosion.

Motion generally increases total weight loss by supplying corrosive agents at a faster rate. Motion also thins the quiescent protective layer making it easier for corrodants to reach the metal. On the other hand, the character of the corrosion products may change with high relative motion and become more protective.

Under extremely high flow rates, mechanical effects may add to corrosion. For example, erosion may mechanically remove protective films and keep corrosion going at a relatively high rate, while turbulence and vibration can cause mechanical damage through cavitation.

Concentration

We expect a change in corrosion rate with a change in concentration of the corrodant. We can postulate arguments for either an increase or a decrease with increasing concentration. In practice, we find both effects as well as situations in which concentration has little effect as shown in Fig. 13-7.

Concentration is directly related to acidity or alkalinity of the solution,

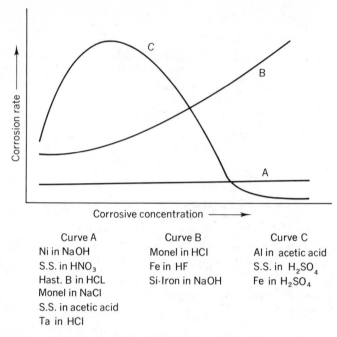

Curve A | Curve B | Curve C
Ni in NaOH | Monel in HCl | Al in acetic acid
S.S. in HNO$_3$ | Fe in HF | S.S. in H$_2$SO$_4$
Hast. B in HCL | Si-Iron in NaOH | Fe in H$_2$SO$_4$
Monel in NaCl
S.S. in acetic acid
Ta in HCl

Fig. 13-7. Schematic curves showing effect of concentration of corrosive solution on corrosion rates of various metals.

normally stated in terms of pH, i.e., an indication of concentration of hydrogen ions in a solution. The pH of any solution is defined as

$$pH = -\log_{10} a_{H^+} \qquad (13\text{-}9)$$

where a_{H^+} is the hydrogen ion activity. A neutral solution has a pH of 7. Lower pH indicates acidity and higher pH indicates alkalinity. The more the pH differs from 7, the greater the acidity or alkalinity, respectively.

With iron, a sufficiently low pH prevents formation of hydroxide or oxide films, and the anodic reaction is simple dissolution of iron.

We expect pure water to be noncorrosive, and this is essentially the case at relatively low temperatures. As temperature is increased to 500–600°F, pure water becomes corrosive with a behavior suggesting that it acts like a strong acid and a strong base at the same time. This behavior is particularly troublesome with aluminum and zirconium.

13-2–6 PURIFICATION OF METAL

Control of the anodic reaction can often be achieved by using highly purified metal. The effect comes from reduction or elimination of elements which serve as cathodes in the electrochemical reaction. Purification is particularly effective with aluminum and magnesium. This increase in corrosion resistance through use of purer metals is obtained at the expense of lower strength since alloys are stronger than pure metals.

13-2-7 ALLOYING

Alloying is often more effective than purification in controlling the anodic reaction. Adding certain elements in specified amounts to base metal often provides an adequate solution to corrosion control where the alloying elements achieve their effects through mechanisms discussed above.

Addition of chromium to iron forms alloys which are particularly resistant to oxidizing acidic environments. These alloys are "stainless steels" whose corrosion resistance is developed by formation of a passive surface film rich in Cr_2O_3. The oxide film is very stable in oxidizing media but much less so in reducing media, where, once breakdown occurs, corrosion can be rapid. Addition of nickel and/or molybdenum increases the ability of these alloys to maintain corrosion resistance in reducing media.

13-2-8 OVERALL (GENERAL) CORROSION

The foregoing discussion has tacitly assumed uniform, overall, corrosion. By way of summary, Table 13-5 indicates a general comparison of corrosion

Table 13-5. General comparison of the corrosion resistance of several materials

General comparison of the corrosion resistance of several materials

	Nickel	Monel	Inconel	Ni resist	Copper base alloys	Iron and steel	Chrome iron	Austenitic stainless
Caustic alkalies	●	●	▲	▲	◬	◬	△	⊘
Sulfuric acid	▲	●	◬	◬	◬	△	△	◬
Halides	▲	▲	◬	◬	◬	△	△	⊘
Alkaline salts	●	●	▲	◬	◬	△	△	◬
Neutral salts	▲	●	◬	◬	▲	△	△	▲
Organic acids	◬	◬	▲	△	◬	△	◬	●
Acid salts	◬	◬	⊘	△	◬	△	△	▲
Ammonia solutions	△	△	●	◬	⊘	◬	▲	●
Oxidizing acids	△	△	▲	△	△	△	◬	●

●—Excellent resistance
▲—Good resistance
◬—Fair resistance
△—Poor resistance
○—Susceptible to pitting or stress corrosion

resistance of several materials in different media. This summary is not sufficiently detailed to serve as a selection guide but does indicate relative resistance to corrosion.

Example 13-3

Table 13-4 lists a number of ways in which electrochemical reactions can be reduced or eliminated. All of these depend upon suppression or retardation of anodic reactions. From a somewhat different viewpoint, there are two principal (major) methods (ways) in which corrosion resistance is established for a given metal or alloy. What are these?

Solution:

(a) Development of a protective film, for example, as in aluminum, austenitic stainless steel, or electroplated surface films.
(b) Establishment of thermodynamic quasi-equilibrium, for example, uranyl phosphate in a gold or platinum container.

13-2-9 Pourbaix Diagram

The Pourbaix diagram is analogous to the equilibrium diagram in that it represents pertinent equilibrium relations in corrosion. During corrosion of iron (Eq. 13-3), liberation of electrons tends to build up an electrical potential. Taking an Fe^{++} concentration of 10^{-6} mole/liter (-0.62 volt potential) as one tolerable in practice without harm to iron, we see shaded areas in Fig. 13-8 representing regions in which corrosion occurs (Fe^{++} concentration greater than 10^{-6}) and unshaded areas representing regions in which there is "immunity." Any combination of electrical potential and acidity which places the point thus determined in the shaded region will cause corrosion unless some kinetic feature of the process interferes.

Since rate of corrosion is a vital factor in practical corrosion control, an important use of the Pourbaix diagram is as a base for kinetic studies.

Use of the Pourbaix diagram for corrosion control can be shown by considering the problem of iron in tap water. In this case, a potential of about -0.45 volt is developed (Point 0 in Fig. 13-8). Three types of corrosion control are illustrated. The most effective is cathodic protection, which increases the negative potential sufficiently to move the point into the lower immunity region. This can be accomplished by electrically connecting a piece of zinc to iron in the same solution.

An alternative is anodic protection, achieved by applying sufficient positive potential to move the point into the upper region of passivity. Two possible difficulties may be found: (1) iron may undergo severe corrosion if this procedure falls short since the point is being moved through regions of strong corrosion; (2) passivity depends on formation of a protective film

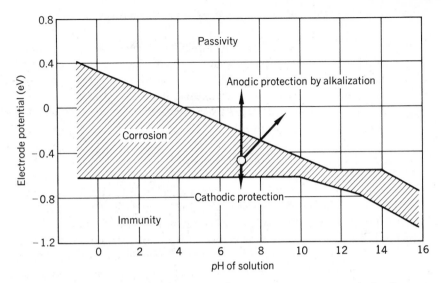

Fig. 13-8. Pourbaix diagram for iron in water showing three methods of corrosion control.

which, if not formed, will allow continued corrosion, possibly with formation of deep pits at unprotected areas.

Passivity can also be reached by increased alkalinity of the solution. This is often the most practical approach, but it is subject to the difficulties noted above.

Pourbaix diagrams give information about equilibrium among a metal, its ions, the ions in water, and the reaction products. This is information which can be very useful. We must remember, however, that practical corrosion problems often depart significantly from equilibrium. The Pourbaix diagram also does not give direct information on corrosion rates.

EXAMPLE 13-4

(a) How much overvoltage is required to provide cathodic protection to iron in water? What are some possible metals which might be used for this purpose?

(b) How does one provide corrosion protection to iron by passivation?

(c) How does one provide corrosion protection to iron by alkalization?

Solution:

(a) From Fig. 13-8, iron has a potential of about −0.45 volt. Fig. 13-8 indicates that the overpotential must be at least −0.15 volt (−0.25 volt or more is better). From Table 13-2, zinc or any element above it in Table 13-2 should be effective. Zinc and magnesium are commonly used.

(b) From Fig. 13-8, it is seen that applying an overpotential of about $+0.25$ volt (minimum) will place iron in water in the passivation region. An overpotential of $+0.40$ (or more) would be much better.

(c) If the potential of iron is not changed from -0.45 volt, changing the environment from a pH of 7 to a pH of about 11 (or greater) will provide effective protection.

13-2–10 LOCALIZED CORROSION

Corrosion may be "uniform" over a given surface and thus be relatively superficial. If not superficial but still uniform, it may be relatively easy to predict expected service life under a given set of conditions. Unfortunately, corrosion is seldom limited to reasonably uniform attrition but all too often localizes and results in premature (often catastrophic) failure. Consequently, while we are concerned with overall corrosion, we are usually much more concerned with localized corrosion which operates by the same mechanisms but more intensively. In other words, *major difficulties arising from corrosion are generally due to localized accelerated corrosion* rather than overall, relatively uniform, corrosion.

Corrosion data in handbooks are commonly given in units of mpy (mils per year), mdd (milligrams per square decimeter per day), ipy (inches per year), and others. Note well that these data are generally calculated from weight changes and similar measurements on the assumption of uniform removal of material from the surface. Such data are completely misleading when there is localized corrosion.

The most common forms of localized corrosion are pitting, crevice corrosion, stress corrosion, grain boundary corrosion, and corrosion fatigue.

13-2–11 PITTING

Pitting is a common example of localized corrosion through the electro-chemical reaction. Protective films are generally continuous, but in some environments they break down, and localized corrosion (pitting) takes place. The point of film breakage becomes an anode while the surrounding unbroken film acts as a cathode, effectively creating a small galvanic cell (Fig. 13-1). The supply of oxygen and similar factors determine the amount, while localizing factors determine the distribution of corrosion. It has been conclusively demonstrated that areas which are shielded from oxygen in any way become anodic relative to areas which are richer in oxygen (Table 13-1 and Sec. 13-2–12).

Pitting is caused by the concentration of electric currents resulting from potential differences on the metal surface. Factors which promote pitting are: (1) those associated with the metal proper—heterogeneous structure (e.g., coring) and impurity concentrations, grain boundary effects, differential metallurgical or thermal treatments, and possible grain orientation and carbide structure in steel; (2) those associated with metal surface—surface roughness or nonuniform finish, scratches or cut edges, differential strain, mill scale or irregular corrosion products, or openings in applied coatings (see LAPRE-1, Sec. 13-2-3); (3) those associated with environment— differential concentration or composition of solution, differential aeration, differential heating, differential illumination, differential agitation of solutions, contact with other metals, or stray electrical currents from other sources.

13-2-12   CONCENTRATION CELLS AND CREVICE CORROSION

There are two principal forms of concentration cells: metal-ion and oxygen. The oxygen concentration cell is a case of differential aeration as demonstrated in Fig. 13-9. Oxygen has relatively easy access to the outside, and we find that iron at the center of the drop is relatively anodic. Referring to the primary mechanism (Eq. 13-8), we see that ferrous ions are produced at the anodic center and hydroxyl ions at the cathodic perimeter. These ions diffuse toward each other, meet at some intermediate position, and deposit a ring of rust. This deposit of insoluble corrosion product around the anodic center

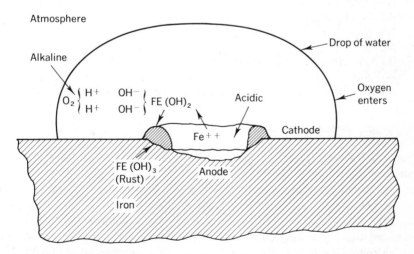

Fig. 13-9. Oxygen concentration cell set up by a drop of water on a piece of iron.

tends to more completely exclude oxygen, resulting in an increased electrical potential. If the cell operates long enough, a pit forms at the center. As action continues, metal at the bottom of the pit becomes still more anodic and the rate of penetration accelerates.

This mechanism for pitting applies equally well to crevice corrosion, which is a general term including accelerated attack at the junction of two metals exposed to a corrosive environment. Crevice corrosion can be a problem in journal and sleeve bearing combinations, or, in general, in any region with small clearances whether the parts are static or in intermittent motion. Complex mechanisms with small clearances and limited operating forces are particularly susceptible to crevice corrosion.

In the oxygen concentration cell, accelerated attack occurs because of a differential in oxygen concentration. Corrosion occurs at the region of oxygen deficiency. Metal-ion concentration cells operate in a similar fashion since corrosion occurs at the region of metal-ion deficiency. Both types depend on concentration differences.

Another form of corrosion is related to crevice corrosion and is worthy of mention. This is stagnant area corrosion which is accelerated attack at any stagnant area. Acceleration comes from increased corrosivity of the environment as a result of accumulation of both soluble and insoluble corrosion products formed in the area.

EXAMPLE 13-5

Concentration cells accentuate or accelerate corrosion. Indicate where this accentuation occurs in
(a) metal-ion cells.
(b) oxygen cells.

Solution:

(a) Corrosion is accentuated where the concentration of electrolyte is lower.
(b) Corrosion is accentuated where the concentration of oxygen is lower.

13-2–13   STRESS CORROSION

Stress corrosion occurs through the combined action of tensile stress and corrosive environment. Stress may be residual or applied, and it may be low enough to cause no concern in the absence of the environment. Aggressiveness of the environment, in turn, may vary greatly. It may be relatively mild, causing little or no corrosion in the absence of stress, but it may also cause localized attack (cracking) in the material when stress is applied. In a relatively strong environment, cracking may occur in the same material at much lower stress levels.

Stress corrosion attack results in cracks which grow slowly or rapidly, depending on the environment. The cracks may be intergranular, transgranular, or a combination of both. Occurrence and rate of stress corrosion are influenced by stress level, corrosive agent, time and temperature of exposure, structure of the metal, amount of plastic strain, behavior of protective films, and possibly other factors.

A major stress corrosion problem occurs with austenitic stainless steels in chlorine-bearing environments. Both chlorine and oxygen must be present. Keeping chlorine concentration low is helpful, but it is not effective if localized chlorine concentration can occur. Keeping oxygen concentration as low as possible, even by using oxygen scavengers, if necessary, is effective.

Electrochemical theory applied to stress corrosion postulates that intergranular cracking is related to the presence of anodic constituents at grain boundaries. Tensile stresses pull the metal apart at corrosion crevices and thus expose fresh anodic material. Continuous paths of anodic material may also be found on slip planes and on planes of precipitated material. Thus the theory applies equally well to transgranular and intergranular cracking.

Susceptibility of an alloy to stress corrosion can be minimized by one or more of the following: (1) treatment to relieve internal residual tensile stresses; (2) adjustment of composition (including elimination of certain impurities) to minimize development of marked composition differences between grains and grain boundaries; and (3) heat treatments to produce homogeneity.

13-2–14   GRAIN BOUNDARY CORROSION

Many metals, under certain combinations of heat treatment and corrosive environment, are susceptible to intergranular or grain boundary corrosion. This looks like intergranular stress corrosion cracking, but it is distinctly different since it depends on the metallurgical condition of the alloy and not on the presence of stress. A well-known example occurs in austenitic stainless steels (e.g., Type 304) in which corrosion resistance depends on the free (i.e., uncombined) Cr content which should be uniform throughout the alloy. If temperature becomes high enough, due to various heat treatments, welding, etc., complex chromium carbides have a pronounced tendency to precipitate at grain boundaries. This causes local depletion of chromium along grain boundaries and local susceptibility to corrosive attack by various media (e.g., Cl^- ions are especially bad). Attack is accelerated by tensile stresses but is found in their absence. The problem can be overcome by: (1) reducing carbon content to the point where the amount of chromium carbide is negligible; or (2) adding alloying elements (e.g., Nb, Ti, Ta) with which carbon forms stable carbides in preference to chromium.

13-2-15   CORROSION FATIGUE

Fatigue is a problem in the design of any mechanical component. A large number of factors influence fatigue characteristics, including environment. Most fatigue data available in handbooks, etc., are from tests conducted in air where corrosion has little effect. Unfortunately, these data are of little help in designing for fatigue in a corrosive medium.

We expect that exposure to a corrosive environment would reduce fatigue strength, and this is the case. For most corrosive media, corrosion products and films tend to retard or eliminate attack. Cyclic stresses tend to rupture these films or render them more permeable, thereby accelerating corrosion. Roughening and pitting of the surface by corrosion form effective notches which concentrate stresses and thus further accelerate corrosion.

In dealing with corrosion fatigue, the designer uses a "corrosion fatigue limit" or "corrosion fatigue strength," which is the maximum repeated stress a piece can endure without failure for a specified number of stress cycles under specific conditions of corrosion and stressing. Corrosion fatigue strength may, in many cases, be lower than fatigue strength (in air) by a factor of 10 or more.

13-2-16   FRETTING CORROSION

Fretting corrosion occurs at contact regions between materials subjected to vibration and slip while under load. It is observed as pits or grooves in metal surrounded by corrosion products. It is a special case of erosion-corrosion occurring in the atmosphere rather than in an aqueous medium. It is highly detrimental because of destruction of metallic parts, possible seizing or galling, possible loss of machined tolerances with subsequent loosening of mating parts, and potential initiation of fatigue failure at fretting pits which act as stress raisers.

Fretting corrosion will develop under very small relative motions. Displacements as small as 10^{-8} cm can cause damage. Repeated relative motion is necessary. Fretting corrosion sometimes occurs on automobile axles or ball bearings during long-distance shipping by boat or rail which gives "continual" vibration under load. An automobile in normal operation does not show fretting corrosion since the relative motions are extremely large.

13-3   OXIDATION OF METALS

Whenever a chemical element combines with another element or a molecular group by losing electrons, an oxidation reaction has occurred. Oxidation, therefore, implies the transfer of electrons by an element and an accompanying increase in valence. The reaction of oxygen with other elements, particularly metals, which will be discussed here, is thus only a rather small aspect of oxidation. In addition, our discussion will be a cursory introduction to a very complex situation which is poorly understood.

When an element is oxidized, the element which reacts with it is reduced; i.e., it gains electrons. Some common examples of oxygen reactions are:

Oxidized species	Reduced species	Resulting product
$4 Cu$	O_2	$2 Cu_2O$
$2 Cu_2O$	O_2	$4 CuO$
$2 Fe$	O_2	$2 FeO$
$2 Ni$	O_2	$2 NiO$
$4 Cr$	$3 O_2$	$2 Cr_2O_3$

A univalent metal is expected to form only one oxide, whereas a multivalent metal is expected to form as many different oxides as it has valences. During oxidation, however, some metals exhibit unusual valences, either temporarily or permanently, so that the usual number of valences of a metal does not always correspond to the usual number of oxidized states. In addition, an oxide may exhibit polymorphism (as do metals); e.g., titania (TiO_2) is known to exist in three different structures, two tetragonal and one orthorhombic.

Metal oxides are ionically bonded, and each oxide has a definite crystal structure. Two different oxides may have the same crystal structure, in which case the lattice parameters will normally be different. Oxides can recrystallize, exhibit grain growth, and may deform plastically at high temperatures.

In essence, the mechanism of oxidation appears to be very simple. As soon as the metal oxidizes, the oxide film becomes a barrier to further reaction. It is then obvious that some sort of diffusion mechanism is involved as oxidation continues. The oxide film appears to grow by ionic transport, but there must also be a movement of electrons in the same direction as positively charged metal ions or in the direction opposite to the movement of negatively charged oxygen ions. Electron movement is associated with defects in the oxide lattice. Ionic diffusion is determined by the movement of ions between vacant lattice sites and is much slower than electron movement. In some metals, oxidation takes place at the oxide-oxygen interface when oxygen reacts with metal ions which have diffused through the oxide, e.g., iron. In other cases, oxidation occurs at the metal-oxide interface when the oxygen ions have diffused through the oxide, e.g., zirconium.

Some oxides are protective, i.e., they inhibit further reaction, whereas others are nonprotective. At one time, it was thought that if the Pilling-Bedworth ratio (the ratio of the volume of oxide formed to the equivalent volume of metal) were less than one, the oxide film would be nonprotective whereas if this ratio were greater than one, the film would be protective. While this is true in many cases, there are many situations in which the Pilling-Bedworth ratio is less than one and protective oxides exist.

Structures and properties of oxide films are often different from those of the same oxide in bulk form. Cuprous oxide, for example, has a certain conductivity, lattice structure, etc., in bulk. In cuprous oxide film (on copper), however, some of these properties vary through the film thickness. This should be expected since the amount of oxygen is a maximum at the oxide-oxygen interface and decreases to a minimum at the metal-oxide interface. This variable composition is specifically responsible for some of the important oxide film characteristics.

In addition, many oxides have an intrinsic nonstoichiometric composition. Cuprous oxide is denoted as Cu_2O yet chemical analysis indicates that $Cu_{1.8}O$ is more accurate. This is possible since some cupric ions are contained in the oxide. To accommodate these cupric ions, there must be some vacant cation sites, since two cuprous ions are electrically equivalent to one cupric ion. As a result, there must be one vacant cation (Cu^+) site for each Cu^{++} present. Since ionic conduction is a diffusion process and depends on the existence of vacant sites, ionic conductivity would be expected to increase in cuprous oxide as the number of cupric ions increases. The ionic arrangement in cuprous oxide might be somewhat like that shown in Fig. 13-10. If impurity cations are introduced into the cuprous oxide film, the conductivity will be further altered. If a trivalent cation, e.g., Cu^{+++}, is substituted for a Cu^+ cation, the number of vacant cation sites will increase.

It is well-known that chromium is often added to other metals, e.g., Fe and Ni, to improve corrosion and oxidation resistance, as indicated earlier in this chapter. It is also true, however, that addition of Cr to Ni can increase the

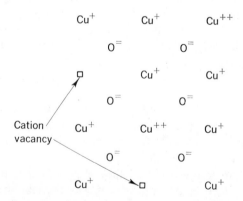

Fig. 13-10. Arrangement of ions in cuprous oxide.

oxidation rate. NiO is a cation-defective oxide like Cu_2O, and has both Ni^{++} and Ni^{+++} cations. When Cr is added to NiO in small amounts, the oxidation rate increases since the vacancy concentration in NiO is increased by Cr^{+++}, as indicated in Fig. 13-11. The oxidation increases to a maximum and then

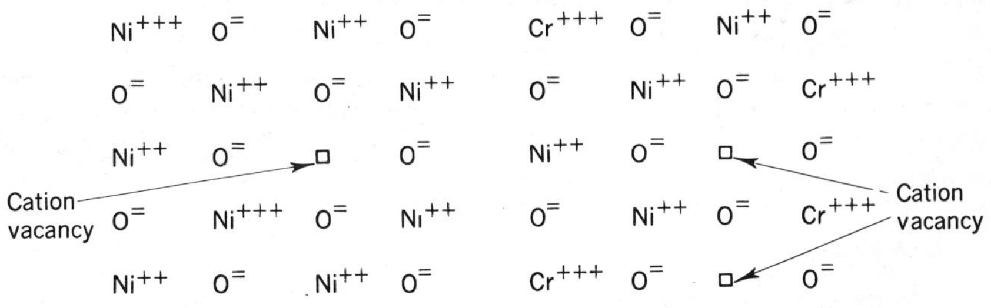

Fig. 13-11. Arrangement of ions in NiO. Effect of chromium creates additional cation vacancies.

decreases as the Cr begins to oxidize to Cr_2O_3 rather than substitute in the NiO lattice. At the same time, substitutional solution of a highly active element such as Li (which is monovalent) in NiO will initially decrease the oxidation rate. This occurs because two Li^+ cations are required for electrical equivalence to one Ni^{++} cation, and thus the vacancy concentration is reduced and there is an accompanying decrease in oxidation rate. Continued addition of Li will lead to a minimum oxidation rate after which oxidation will increase as the Li begins to oxidize rather than substitute in the NiO lattice.

It is obvious that even though a diffusion mechanism is involved in oxidation of metals, it can be greatly modified by the presence of various contaminants in the metal. It is equally true that oxidation rate can be greatly modified by contaminants in the gas or on the surface.

It has been clearly demonstrated (e.g., by J. V. Cathcart at the Oak Ridge National Laboratory) that differences of orders of magnitude in oxidation rates exist on different crystal planes. It has been found in Cu that (100) oxidizes more rapidly than (111) which is more rapid than (110) which in turn is more rapid than (311). There is no relationship to the density of packing. Cathcart has concluded that paths of "easy diffusion" across the oxide are crucial in determining the rate of diffusion through the oxide film. In thin Cu_2O films, for example, there are many paths of easy diffusion, but the extent of these paths varies with the crystal plane. Cathcart has been able to correlate the variation in oxidation rate in crystal planes with the variation in extent of paths of easy diffusion in the oxide. Current theory cannot account for this behavior.

Another factor which current theory cannot explain is the effect of epitaxial relationships on oxidation rate. In general, both metal and the oxide have characteristic crystal structures. Oxide being formed, at least within certain thickness ranges, tends to assume an orientation relative to the substrate such that the mismatch between the two structures is minimized. In most cases, a thin oxide film has a preferred orientation and is often virtually a single crystal. In the case of Cu, the (111) plane in Cu_2O normally forms parallel to (100) and (111) planes of Cu. It is interesting to note that the (100) plane in Cu oxidizes the most rapidly and is the least oriented. It is not known whether the lack of orientation produces the rapid oxidation rate or vice versa.

EXAMPLE 13-6

The oxidation of copper to cuprous oxide is parabolic at temperatures above 500°C. If 7.5 micrometers of oxide are formed in ten minutes at 550°C, what thickness of oxide will be formed in five minutes at 600°C? The activation energy is 37.7 Kcal/mole.

Solution:

For parabolic behavior,

$$x^2 = Ct$$

where C is a "constant" of the form

$$C = C_0 e^{-(Q/RT)}$$

From the data at 550°C,

$$C_0 = \frac{x^2}{te^{-(Q/RT)}} = \frac{(7.5 \times 10^{-4})^2}{(600)(e^{-37,700/(2 \times 823)})} = 9.35 \text{ cm}^2/\text{sec}$$

at 600°C

$$x = (Ct)^{1/2} = [9.35(e^{-(37,700/2 \times 873)})(300)]^{1/2}$$
$$= 10 \text{ micrometers}$$

13-4 DETERIORATION OF CERAMICS

Compared to metals, especially iron, nonmetallic materials are generally resistant to deterioration. While they are immune to electrochemical corrosion, they are not impervious in all situations.

Many ceramics are compounds, or combinations of such compounds, and are generally resistant to attack by aqueous media. Carbides, borides, and nitrides have good oxidation resistance. They, as well as oxides, are often used in high temperature applications. Many ceramics are slowly dissolved at high temperatures by molten salts or molten metals.

Whiteware and porcelain (Sec. 11-2–4), especially with a good integral glaze, are used extensively in many applications because of good corrosion resistance. Whiteware is used in sanitary applications, absorption towers, pipes, valves, pumps, etc. Porcelain is used in similar equipment, e.g., acid nozzles, and in insulators and spark plugs.

Carbon and graphite have good resistance to alkalies and many acids (other than oxidizing acids like nitric, concentrated sulfuric, and chromic acid). Graphite reacts readily with steam to form CO, CO_2, H_2, and CH_4.

Corrosion resistance in glass is more commonly known as *chemical durability*. It is well known that some glasses provide superior containers for hot water, fruit juices, and weak chemicals of many kinds. Glass made entirely of silica and soda is soluble in water while borosilicates are extremely resistant. The necessity of producing a glass sufficiently stable for the intended use places practical limits on composition. For example, a number of glasses which have desirable mechanical or optical properties may not be acceptable in the specific use because of susceptibility to deterioration.

Decomposition of glass by water is not a simple case of solution. It is a highly complex process, involving penetration of glass by water, followed by decomposition of the complex silicate mixture to give substances entirely different from those originally present. Presence of an acid or alkali will often markedly affect the rate of decomposition with the effect depending on glass composition and dissolved material. In testing chemical durability of glass, it is the rate of reaction that is measured and compared. Rates of reaction, however, are highly sensitive to small differences in experimental conditions. Since many of the available data have been obtained under rather poorly defined conditions, such data should be used very, very cautiously.

Concrete is a multiphase ceramic which is exposed to atmospheric weathering and to chemicals in manufacturing or sanitary installations. Concrete can deteriorate from external attack or from internal reactions. Concrete must be impermeable to reactive solutions to resist external attack. Concrete has low resistance to salts used for highway deicing. Internal reactions can occur between aggregate and cement and/or between reinforcements and cement. Flint or volcanic rock aggregates can react with soda in cement to form bulky silicates which crack the concrete. Steel reinforcing rods are initially protected by an alkaline environment (pH about 12). With time, however, moisture containing CO_2 penetrates the overlying concrete. The CO_2 reacts with free lime in the concrete to decrease the pH. As steel corrodes, bulky oxides are formed which fracture the concrete and accelerate the process. Aluminum pipe or tube is not compatible with concrete which may surround it.

Ceramics are resistant to oxidation at moderate temperatures. Many of them are thermodynamically unstable in oxygen atmospheres at higher temperatures, e.g., high-melting sulfides, nitrides, phosphides, carbides, silicides, and borides. Some of these do show moderate to good oxidation resistance by formation of a metal oxide protective coating. Silicon carbide, for example, can be used for many hours in oxygen or air at 1600°C. While the oxides might be expected to be highly stable in oxygen, many of the metals which have more than one valence will absorb oxygen and form a higher order oxide. For example, UO_2 will change to U_3O_7 at 150°C and to U_3O_8 at about 375°C.

Even an uncombined ceramic such as graphite is quite stable in air up to about 200°C and has been used in several nuclear reactors. At 300°C, however, oxidation of graphite by air is serious.

Example 13-7

Most ceramics are corroded much less than metals when exposed to aqueous media. Briefly indicate why.

Solution:

Most ceramics (graphite is an exception) are not electrical conductors and thus an electrochemical cell cannot operate. As an alternative statement, most ceramics (eg., oxides, sulfides, etc) are in a relatively stable (thermodynamically) state and are close to a minimum free energy.

13-5   DETERIORATION OF POLYMERS

Polymers, as well as ceramics, are not subject to electrochemical corrosion but are subject to corrosion or deterioration through different mechanisms. All three types of polymers are used in many applications where resistance to environment is a major factor. Temperature, chemical environment, applied stress, and irradiation are important factors in determining ability of a polymer to resist deterioration. Resistance of polymers to various chemicals varies widely, not only from polymer to polymer, but in some cases within different grades of a single polymer. A few, such as polytetrafluoroethylene (e.g., du Pont's Teflon), are not attacked by even the more corrosive chemicals or powerful solvents. At the other end of the scale are polymers which are dissolved by water or mild solvents.

At the risk of vast oversimplification, deterioration of polymers occurs through reactions that are basically chemical, i.e., bond breaking. These reactions range from *scission* (breaking) of polymer chains to *dissolution* and *swelling*.

Polymers with large molecules are stronger and more resistant to deformation than those with small molecules. Thus there is much interest in reactions (Sec. 4-7-2) which increase the degree of polymerization. If the reaction reverses, *depolymerization* occurs. Substantial energy input is required to break an average C-C bond in depolymerization. The necessary energy can come from heating, light photons, neutron radiation, or gamma radiation. Oxygen and ultraviolet light, acting jointly, can cause deterioration of polymers at room temperature. In polymethyl-methacrylate, for example, depolymerization can occur by successive scissions of monomers until the entire chain is consumed. In polyethylene, on the other hand, scission occurs randomly along the polymer chain to give a rapid and significant reduction in the average chain length or molecule size. Scission breaks *intra*molecular bonds.

Polymers are degraded by a gaseous or liquid environment through solution of polymer in a solvent or by incorporation of gas or liquid into the polymer. Glassy polymers (not crosslinked and nonpolar) are most susceptible to dissolution by organic liquids. Dissolution can be complete as in the case of polystyrene in benzene.

It is more common, however, to find limited solubility (e.g., acetone in polystyrene). In such cases, there is penetration of the environment into the polymer. This results in *swelling* accompanied by changes in dimensions and properties. For example, the glass transition temperature of a polymer decreases with increasing amounts of solvent so that a polymer which is initially strong may become rubberlike as swelling progresses.

The micromolecules of solute enter between the macromolecules of the polymer and make direct polymer-to-polymer contact impossible. This breaks *inter*molecular bonds although these are far weaker than the intramolecular bonds broken by scission. Absorption of water molecules in polyvinyls provides a good example of swelling leading to weakening of the polymer. Gasoline hoses are made of rubber only after processing to give the structural changes necessary to avoid absorption of petroleum molecules. Both crosslinked and crystallized polymers are less subject to swelling than amorphous polymers. Almost all polymers will degrade in the presence of oxygen at very moderate temperatures, e.g., the oxidative aging of rubber at room temperature. Property changes are accompanied by formation of carbonyl, hydroxyl, carboxy, and peroxy groups along the polymer chain. In some cases, there is a decrease in molecular weight accompanied by softening and increased solubility, while in other cases, there is increased molecular weight accompanied by embrittlement and decreased solubility. These changes are common to all polymers in varying degrees. Although the details are not clear, it is generally agreed that molecular oxygen leads to formation of unstable hydroperoxides which decompose to form free radicals. These, in turn, participate in further polymerization, depolymerization, scission, crosslinking, and other free-radical-catalyzed organic reactions.

A rather special, but very important, aspect of deterioration of polymers is the question of flammability. Burning of material creates potential danger to humans through oxygen depletion, flame, heat, fire, gases, and smoke. In addition, there may be a loss of structural integrity in the material leading to collapse of large structures. In combustion of a polymer, decomposition occurs first as a result of heat supplied from some source. Burning then depends on combustible gases formed during polymer degradation being ignited in the presence of an adequate oxidizing agent. Burning provides additional heat to propagate further decomposition and ignition. Once initiated, spreading of flame to engulf the surroundings is a self-sustaining continuous decomposition and ignition process.

In addition to combustible gases (e.g., hydrocarbons, hydrogen, alcohols, aldehydes, etc.), noncombustible gases (e.g., H_2O, CO_2, HCl, HBr, etc.) are generated as well as some liquids (e.g., partially degraded polymer), solids (e.g., carbonaceous char or residue) and entrained solid particles (smoke). Elimination of all combustible gases would preclude burning but it is seldom possible to completely eliminate release of volatile gases.

Approaches to reducing flammability of polymers can be grouped in six categories: (1) dilution of polymer with nonflammable materials, e.g., inorganic fillers; (2) incorporation of materials which decompose on heating to give nonflammable gases such as CO_2; (3) addition of flame retardants which catalyze formation of char rather than flammable product; (4) design of polymer structures which favor char formation; (5) incorporation of materials which terminate free radical chain reactions occurring during combustion; and (6) formulation of polymers which decompose thermally to give a net endothermic (rather than exothermic) reaction.

13-6   HIGH-TEMPERATURE BEHAVIOR

Elevated temperatures can produce great effects on properties as indicated in Chapters 7 through 10. In addition, strain rate, which (within broad limits) is relatively unimportant at room temperature, has a marked influence on mechanical properties at high temperature. Elevated-temperature tests are conducted with special provisions for heating to obtain strain readings during loading.

Increasing temperature generally decreases short-time tensile and yield strengths, and there is an accompanying increase in ductility unless structural damage occurs. For many applications, however, we desire usefulness for a long period. Unfortunately, a long time at temperature still further reduces strength. For example, reduction in strength in annealed steels after extended time at temperature under no load is probably due principally to spheroidization of pearlite and to some reduction of section due to scaling. Similar effects are found in other alloys, e.g., those which are precipitation-hardenable where softening is due primarily to overaging and precipitate agglomeration. In alloys strengthened by strain-hardening, a similar effect is found where softening is due to recrystallization and grain growth (if continued long enough). It should also be kept in mind that rate of structural change, e.g., agglomeration of precipitates, may be appreciably altered under loading.

13-6-1 CREEP AND STRESS RUPTURE

Many materials, both metallic and nonmetallic, exhibit flow (or a gradual change in dimensions) over an extended period under applied load. This flow is called *creep*, i.e., time-dependent strain occurring under stress. Creep can occur at stresses well below the proportional or elastic limit at elevated temperatures. Whether a given temperature is "elevated" or not depends on the material in question, since one material may creep more at room temperature than another at 1500°F.

Temperature at which creep occurs at appreciable rates is related to minimum recrystallization temperature which also has a direct bearing on tensile testing. Strain rate (i.e., rate of deformation) is relatively unimportant at room temperature, provided minimum recrystallization temperature is at least somewhat above room temperature. Below minimum recrystallization temperature we expect to find little effect of strain rate on stress-strain characteristics for stresses less than yield.

At higher temperatures, however, the situation is different. We find a significant change in tensile properties depending on the time over which the test is conducted. To illustrate the effect of strain rate and time upon tensile properties, consider Table 13-6. Because of these effects, the tensile test (as such) loses significance at elevated temperatures whereas creep and stress-rupture tests become meaningful. Tests commonly used are the long-time creep test under constant load and the stress-to-rupture (stress-rupture) test. Both tests are conventionally conducted by applying constant load to a tensile specimen maintained at constant temperature. Stress rupture tests normally

Table 13-6. **Effect of time and strain rate on tensile
strength of steel at 1100°F**

Rate (in./in./hr)	Tensile strength (psi)	Elongation (%)	Red. of area (%)	Time for failure (hr)	Type of failure
7.5	38,100	37.5	86.0	0.058	Transcrystalline
1.0	34,000	56.5	80.5	0.58	Transcrystalline
0.1	28,100	51.0	62.5	5.1	Transcrystalline and Intercrystalline
0.01	22,500	35.0	42.5	36.0	Intercrystalline
0.001	14,300	40.0	55.0	393.0	Intercrystalline

take from 10 to 1000 hours although they may run anywhere from 1 hour to 2000 or 3000 hours. Creep tests usually take about 1000 hours minimum but are normally run 10,000 hours or longer. In other words, creep tests are run at relatively low stress levels which will not cause failure during the test, whereas stress-rupture tests are run to failure under higher stresses. The two tests are related, and results are frequently shown together on common graphs which provide design data.

Elongation is always measured as a function of time in a creep test. Although desirable, it is not always done in a stress-rupture test because of higher loads and shorter duration. A typical schematic stress-rupture curve is shown in Fig. 13-12. This curve is for a constant nominal stress and constant temperature. A creep curve has a similar appearance but seldom extends beyond initiation of tertiary creep.

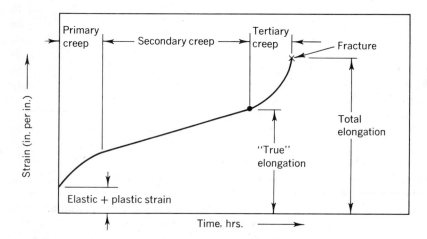

Fig. 13-12. Schematic typical stress-rupture curve (nominal stress and temperature are specified and constant).

This typical curve (Fig. 13-12) is usually considered in four parts: (1) elastic and plastic deformation occurring on load application; (2) primary (transient) creep in which creep rate diminishes with time; (3) secondary creep in which creep rate is virtually constant (steady state or minimum creep rate); and (4) tertiary creep in which creep (and creep rate) increases with time and usually terminates with fracture.

Creep rate, $\dot{\epsilon}$, is a function of stress level, temperature, and the period over which both are applied. At relatively low temperatures and stresses, primary creep may gradually decrease and stop. At higher temperatures and stresses, creep generally continues until fracture occurs. Thus, the design of a structural element for use at elevated temperature must be based on anticipated useful life. This, in turn, depends on the particular application. For equipment such as turbines and other power generation equipment, it is normal to design for a very long life whereas items such as torpedoes, rockets, and missiles may be designed for lives measured in hours or minutes.

For the designer, the essential information obtained from creep testing is creep strength, i.e., (1) constant nominal stress which produces a specified elongation in a specified time at a specified temperature, e.g., the stress which produces 1% creep in 10,000 hours at 1000°F or (2) constant nominal stress which produces a specified creep rate at a specified temperature. Data may be reported in tabular form or in graphical form as in Fig. 13-13. A similar figure

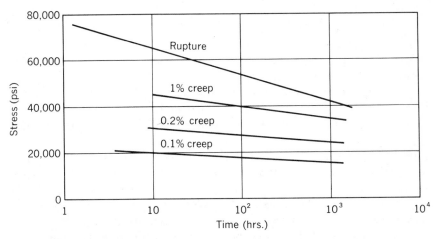

Fig. 13-13. Creep and rupture data for forged low-carbon N-155 alloy at 1200°F.

is required for each temperature. By cross-plotting, we can make an equivalent figure for a series of temperatures with only one value of creep strain as indicated for stress-rupture in Fig. 13-14. These figures may be log-log or semi-log plots.

Two less frequently used creep tests are constant-strain-rate and relaxation tests. In the constant-strain-rate test, we measure the load, i.e., the reaction, which is produced by applying a strain at a constant rate to the specimen. The relaxation test is usually performed by maintaining total strain (elastic plus plastic) at a constant level and measuring the decrease in load (or stress) as a function of time. This latter test has particular application to bolted joints and assemblies with shrink or press fits. In a bolted flange connection used at elevated temperatures, creep deformations produced with time in both bolts and flanges produce a relaxation of stress in the bolts. If

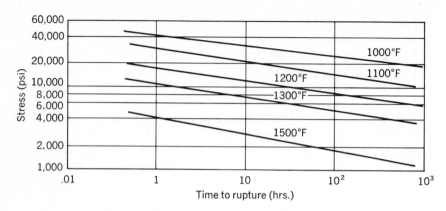

Fig. 13-14. Stress-rupture curves for a Cr-Mo-Si steel.

initial tension is incorrect, the bolt may elongate and loosen the joint, with the result that leakage eventually occurs.

Data for selecting creep-resistant material are often limited because of the long periods required to obtain meaningful data. In addition, little correlation is possible among data obtained from different types of creep testing. Stress-rupture tests of 10 to 1000 hours are often used to select or discard alloys for further creep testing. This practice is based on the fact that stress-rupture tests often give two alloys the same relative rating as creep tests at usable loads. Unfortunately, it also often happens that the reverse is true. *We must be extremely wary of extrapolating existing data to longer times or higher temperatures* for it is all too easy to get into serious difficulties from this practice. A further design difficulty is that nearly all creep data have been determined for simple tension. Many applications have stresses in shear, bending, compression, and combinations of these. Use of available data may require extensive ingenuity on the part of the designer.

13-6-2   CREEP MECHANISM IN METALS

Rapid and extensive deformation of a metal occurs primarily by slip and is accompanied by strain hardening and appearance of slip bands. During primary and tertiary stages of creep, deformation occurs by normal slip with some minor contribution by twinning. Accelerated diffusion, as a result of temperature increase, recrystallization, etc., may noticeably increase creep by normal slip.

In most applications we can tolerate creep only if: (1) the primary stage is short; (2) the secondary stage has a very low creep rate; and (3) the tertiary stage does not occur during service life. In other words, once the material has "settled down," it should be in the secondary stage of creep, and should have very slow deformation during the rest of its service life. In this stage, normal slip no longer occurs, and deformation occurs either by viscous flow or by slipless flow, depending on temperature and imposed load.

Viscous flow is extremely slow. It can and does occur under loadings too small to produce slip. It rather closely resembles plastic flow of a truly amorphous material such as pitch and occurs only in grain boundary regions. Grains are essentially undeformed, and no slip bands are produced. Viscous flow produces a slow permanent deformation by permitting grains to move around and past each other insofar as their irregular shapes and interlocking arrangements permit. Viscous flow is accomplished by directed diffusion of atoms (or small groups of atoms) in regions of high local internal energy in an effort to reduce distortional energy.

Creep rates of most interest, i.e., low enough to be tolerated coupled with loading high enough to be useful, lie between the rapid creep rates produced by slip and the very slow rates produced by viscous flow. The principal mechanism at this level is slipless flow which, despite the name, appears to involve slip, viscous flow, and recovery. Slipless flow does not produce slip bands, yet it produces changes in grains identical to those produced by slip, e.g., strain hardening which can be explained only by slip. Presumably, slip is either less extensive on a greater number of active slip planes, or the active planes occur individually rather than in clusters and thereby do not cause slip bands to appear. Experiments on specimens with mixed grain size frequently show that all grains have generally similar changes in shape and that slip bands develop only on larger grains. It seems probable that slipless flow is a result of viscous flow in highly stressed grain boundaries which reduces stress enough to permit limited slip in adjacent grains.

Available evidence indicates little, if any, correlation between creep properties and mechanical properties at room temperature. Nominally identical alloys show wide variations in creep properties. In fact, "identical" specimens taken from the same bar may have creep rates differing by a factor of two. Creep is extremely structure-sensitive and is much more affected by minor variations in microstructure and prior history than many other characteristics.

Grain size is a major factor in creep. Fine-grained metals generally have higher yield and tensile strengths at room temperature than coarse-grained metals. At elevated temperatures, the reverse is generally true.

Prior cold working has a strong effect on creep which is accelerated in some metals during recrystallization following cold work. This accelerated creep is followed by a period of relatively slow creep in the recrystallized metal. A concentration of 0.1 a/o sounds small, but it implies one solute atom in a cube which has ten atoms on a side. Solute atoms strain or distort the lattice, and we expect this distortion to retard (or decrease) creep.

There is some tendency for solute atoms to concentrate at grain boundaries. This may serve to decrease generation of dislocations in these areas and thus further retard creep. At the same time, this appears to be related to void formation. For example, the void "nuclei" in brass appear to be ZnO particles. If ZnO is removed by remelting, there is practically no void formation under subsequent dezincification. These samples show a considerably reduced tendency for grain boundary cracking during creep and have a greatly increased stress-rupture life.

Although marked changes in creep characteristics can be accomplished by solute additions, these changes are relatively small compared with the effects of a finely dispersed second phase which is strong and metallurgically stable. Finer dispersions give greater strengthening effect, i.e., greater reduction of creep. It is also true that fine dispersions have greater instability which, in turn, tends to accelerate creep by agglomeration. In general, there is an optimum dispersion which is directly related to desired life under service conditions. Materials which are superior in relatively short tests (e.g., a few hundred hours) may be inferior for long-time service (several thousand hours), since optimum dispersion is coarser for higher temperature or for longer times. Precipitation hardening alloys have further complications since the dispersed phase frequently precipitates along grain boundaries and may form a network. Since this second phase is often relatively hard and brittle, such precipitation can make the structure more susceptible to intercrystalline fracture.

EXAMPLE 13-8

Consider a steel with an applied tensile stress at a temperature well above room temperature.

(a) The applied stress can be "high" or "low" and the temperature can be "high" or "low." Given the opportunity, what combination of applied stress and temperature would you select?

(b) A given design of pressure vessel permits a strain of one percent in two years' service. What is the maximum permissible creep rate when stated in terms of %/hr?

Solution:

(a) Use a low applied tensile stress at a low temperature, if possible, to minimize creep.

(b)

$$\text{rate} = 1\%/2 \text{ year} = 0.5\%/\text{year}$$
$$= \frac{1}{(2)(365 \text{ days/year})(24 \text{ hr/day})}$$
$$= 5.7 \times 10^{-5} \%/\text{hr}$$

13-6-3   CERAMICS

One of the outstanding characteristics of ceramics is stability at relatively high temperatures. Many of them retain strength to much higher temperatures than metals. The effect of temperature on the strength/weight ratio of a representative sampling is shown in Fig. 13-15.

Despite the generally excellent high-temperature strength, many of the ceramics are susceptible to thermal shock. Porous ceramics (for thermal insulation) are resistant to thermal shock, whereas the same ceramic in dense form for structural use may be susceptible.

There is no generally accepted standard for thermal shock resistance rating. The parameter $KS/M\alpha$ has been suggested as a quantitative measure since high values of tensile strength at elevated temperature (S), high values of thermal conductivity (K), low values of ductility modulus (M, ratio of stress to strain at fracture), and low values of coefficient of thermal expansion (α) favor resistance to thermal shock.

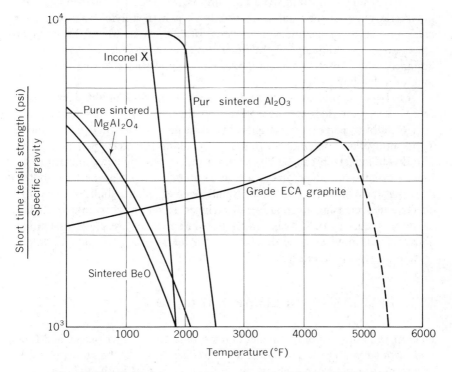

Fig. 13-15. Effect of temperature on strength-weight ratio of various materials.

Although very few of the ceramics will undergo plastic deformation by slip at room temperature, several of them will deform appreciably at high temperatures.

Resistance to high-temperature creep is closely related to composition and microstructure. Many ceramics have one or more crystalline phases, a glass phase and a pore phase. Pores decrease viscosity in approximately direct proportion to the pore volume fraction. The glass phase is often a continuous phase around the crystals and thus tends to control overall creep resistance. Creep resistance in such a ceramic is better than in glass but not as good as for a completely crystalline ceramic.

13-6-4 POLYMERS

Polymers, especially thermoplastics, must be carefully analyzed for creep, or cold flow, at all application temperatures. While the problem may be acute at room temperature, it becomes even more serious at higher temperatures. Both creep (i.e., deformation under constant load) and relaxation (i.e., stress decay under constant strain) are important aspects of the problem. No polymeric material made almost entirely of a high polymer has sufficient creep resistance to sustain high stresses. In composites where the polymer serves mainly as a binder, much higher stresses are possible. Use of properly designed metal inserts can aid greatly in minimizing deformation under long-term loading.

A polymer at a temperature below T_g tends to be brittle but the amount of creep is small and does not increase much with time. In the transition region near T_g, more creep can occur and it increases greatly with time. Creep above T_g in amorphous polymers with no crosslinking is determined mainly by viscosity. Thus creep can be large with a tendency to increase linearly with time. For crosslinked polymers above T_g initial creep may be large but it increases only slightly with time. Small degrees of crosslinking are quite effective in decreasing creep above T_g. Higher degrees of crosslinking further decrease creep. Crystallinity has much the same effect on creep as crosslinking above T_g, although it is detrimental to polymeric strength below T_g where crystallites tend to act as stress concentrators.

13-7 EFFECTS OF NUCLEAR RADIATION

Nuclear radiation is not often encountered outside nuclear reactors and thus is of secondary interest in the context of our discussion. At the same time, it is of general interest since high energy radiation can drastically alter physical, mechanical, and chemical properties.

Although changes induced by radiation may be harmful from the viewpoint of continued use, not all changes are bad. Electron bombardment, for example, improves the temperature stability of polyethylene. An increase in yield strength of metal is not necessarily bad although the accompanying decrease in ductility may be. The principal result of changes is generally considered damage, since, in most cases, the normal condition of material at the start of its service life is essentially an optimum. As yet, a complete and satisfactory understanding of radiation effects has not been reached. In essence, study of radiation effects is a study of defects in materials, i.e., a study of effects of disruption of internal structure.

Interaction of radiation with matter is a complex phenomenon. Each kind of radiation, e.g., fast neutrons, thermal neutrons, gamma rays, alpha particles, beta particles, etc., has its own particular characteristics, and each class of materials, e.g., solids, with different kinds of bonding, i.e., metallic, ionic, covalent, molecular, responds to each type of radiation in its own particular fashion. Pure copper, for example, subjected to a large dose (about 10^{20} nvt) of fast neutrons is still copper but it has some property changes. The same dose applied to diamond destroys crystallinity.

13-7-1 SUMMARY OF EFFECTS

The effects of radiation on properties of materials are principally due to the defects it produces. It is convenient to divide these into three classes: (1) structural and mechanical effects associated with localized strain and change of vibrational frequencies in the vicinity of defects; (2) electronic effects primarily associated with trapping of charge (electrons or holes) by defects; and (3) effects on diffusion-controlled rate processes. Some properties in each class are listed in Table 13-7.

Separation of causes of property changes is an oversimplification since most changes are due to a combination of causes. Some effects of lattice

Table 13-7. Some properties affected by lattice defects

Structural and mechanical	Electronic	Rate processes
Crystal structure	Paramagnetism	Diffusion
Density	Optical absorption	Ionic conductivity
Elastic constants	Photoconductivity	Phase change
Hardness	Dielectric loss	Chemical reactions
Thermal conductivity	Electrical conductivity	
Yield strength		
Tensile strength		
Ductility		

disruption tend to move a system toward thermodynamic equilibrium, while others tend to move it away. Coexistence and interaction of these individual effects often make it difficult to predict the net effect.

The defects discussed above do not constitute a final state, since they have considerable mobility at all but very low temperatures, and can interact with each other and alter total damage. In other words, they can rearrange (anneal) continually during radiation. Radiation must be performed at very low temperatures for all "damage" to be "frozen in" as it occurs. This temperature depends on the property since some effects are relieved at much lower temperatures than others; e.g., effects on electrical conductivity are annealed at very low temperatures. Some defects introduced in common metals are mobile at $-240°C$.

Some indication of the effects of radiation on various materials is given in Fig. 13-16 which gives levels of neutron exposure which: (1) appreciably alter, and (2) render essentially useless some representative materials.

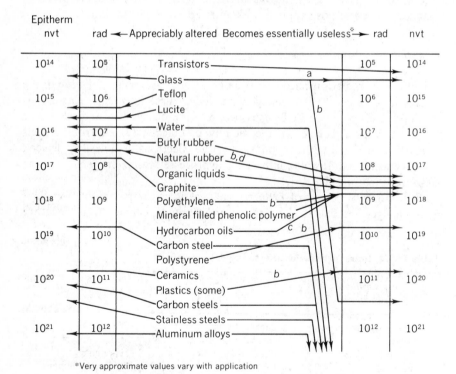

*Very approximate values vary with application

Fig. 13-16. Radiation stability of various substances: (a) Transparent medium; (b) structural material; (c) as lubricant; (d) electrical insulators.

QUESTIONS AND PROBLEMS

1. Define corrosion. Does corrosion differ from erosion? If so, why? If different, is there a relationship between them? In what manner?

2. How may underground steel pipe be protected by magnesium or zinc? Would the same reasoning apply to protection of a ship hull?

3. What is the difference in mechanism, if any, in corrosion protection of iron by galvanizing and by tin plating?

4. Draw a sketch, write the equations, and describe the mechanisms involved in the formation of oxide at the metal-oxide interface. In other words, compare this with Fig. 13-3 and pertinent discussion thereof.

5. The corrosion *rate* of metals in electrochemical corrosion is influenced by a number of factors which can be classed as metallurgical factors and environmental factors. List these and discuss the effect of each factor on corrosion rate.

6. What is the meaning of passivation? Of what significance is it? How may it be accomplished?

7. What is the potential EMF in a galvanic cell if the standard electrodes are Ag and Fe? If a circuit is established, which electrode will be consumed?

8. Of what importance is formation of a film on the surface of a metal exposed to corrosion? What kinds of films may be formed? How does each type of film operate with regard to possible corrosion protection?

9. What is meant by anodizing? Can this be applied to aluminum and its alloys on a commercial basis? If so, is it helpful in protecting against corrosion?

10. The Aluminum Company of America produces Alclad aluminum-base alloys and Reynolds Metal Company produces Pureclad aluminum-base alloys. What is meant by these terms? Of what significance are they from the viewpoint of corrosion protection?

11. How would you interpret the term "internal oxidation"? How could this come about? Of what significance might it be?

12. What is the meaning of "anodic areas"? How do they differ from "cathodic areas"? What determines whether a given area is anodic or cathodic?

13. Discuss the differences and similarities between stress corrosion cracking and intergranular cracking.

14. Discuss the differences and similarities between cathodic protection and mechanical protection. Give examples of each. Is it possible for mechanical protection to "backfire"? If so, how?

15. What is the difference between sensitized and stabilized stainless steel? Of what significance is either (or both) in corrosion resistance?

16. Threaded bronze elbows have been used in connecting steel water pipe. Is corrosion apt to occur? If so, where? By what mechanism? If there is corrosion, how could the joints be made to resist corrosion if steel elbows are not immediately available?

17. What is pitting? What is the general mechanism? How does it occur in stainless steel?

18. A steel fence post will rust most rapidly just above, just below, well above, or well below the soil line?

19. In an oxygen concentration cell, corrosion occurs at the point where the oxygen is least, intermediate, or greatest?

20. Pitting is observed on an automobile bumper under a drop of paving asphalt. What kind of corrosion is involved?

21. "Texas towers" used as platforms for offshore oil well drilling have corrosion problems. Exposure is in three zones: immersed, 0.025–0.030 ipy; splash zone, 0.055–0.065 ipy; and atmospheric zone, 0.005–0.010 ipy. How would you provide corrosion protection for each of the zones?

22. What significance do residual stresses have in corrosion resistance?

23. What is corrosion cracking? How does it operate? What is its practical significance?

24. Explain how oxygen can act as both an inhibitor and stimulator of corrosion.

25. Rationalize the comment that hot water often appears to act simultaneously as a base and an acid.

26. How can susceptibility to stress corrosion be decreased? How does each possible method operate?

27. Discuss the similarities and differences between oxygen concentration cells and metal-ion concentration cells.

28. Austenitic stainless steels with 17% Cr and 8% Ni are susceptible to intergranular attack if cooled slowly from above 1200°C. Explain.

29. Metals such as chromium and platinum are resistant to oxidation. Explain why.

30. Addition of Cr to iron-base or nickel-base alloys increases the oxidation resistance. Explain.

31. In view of the explanation given in question 30, why does the addition of Cr in very small quantities decrease the oxidation resistance?

32. Demonstrate that oxidation controlled by diffusion of ions follows the parabolic law.

33. Compare the oxidation resistance of metals, ceramics, and polymers.

34. What is creep? How does creep take place? What is its practical significance?

35. What is creep limit? Creep strength? What practical significance does either have?

36. Which is better for creep resistance in short-time service, fine- or coarse-grained metal? For long-time service? Why?

37. What is the difference between creep rate and total creep? What is the significance of each in practical design?

38. What are constant-stress, constant-strain-rate, and stress-relaxation tests? What practical significance do they have?

39. What is the distinction between creep and stress rupture? What is the significance of each in practical design? How are these related to tensile testing? Can stress-rupture data be used to predict creep? How and why?

40. What are some metallurgical factors involved in creep? What is the significance of each?

41. What are the effects of temperature, strain rate, microstructural detail, and soluble alloying additions on the creep characteristics of a metal? What is the explanation for each effect?

42. The following data are for creep of a given material under a nominal tensile stress of 20,000 psi.

Temp. (°K)	Creep Rate (%/1000 hr)
700	0.016
755	0.410
810	4.00

(a) Determine if the Arrhenius equation (Eq. 8-2) applies. If so, evaluate the constants.

(b) Assume: Creep Rate $= Ae^{-Q/RT}$, where $A = v_0 \sinh (\sigma/\sigma_0)$.

(Let $Q = 57,000$ cal/g mole,
$R = 1.98$ cal/°K-g mole, $\sigma_0 = 30,000$ psi,
$\ln A = 36.90$ for $\sigma = 20,000$ psi)

What creep rates would you predict at 810°K for stresses of 10,000 and 30,000 psi?

(c) What creep rate would you predict at 865°K for a stress of 20,000 psi? [use data from (b)].

(d) What is the probable validity of the predictions in (b) and (c)?

43. Compare the temperature stability of metals, ceramics, and polymers.

14

MATERIALS DEVELOPMENT

14-1 INTRODUCTION

The history of materials usage is also a story of the development of materials with improved properties, lower cost, etc., for use in preference to an existing material. The aboriginal clothing of animal skins was largely supplanted by cloth woven from animal and vegetable fibers, which in turn have been replaced in many cases by fabrics woven from synthetic fibers. In technical use, for example, stainless steels have better corrosion resistance and elevated-temperature strength than iron while Inconel and Inconel X are superior to nickel in many respects.

If we have a specific application and it is an established fact that no commercial material is available which performs adequately, then we must either: (1) modify the requirements of the application so that an available material is adequate; (2) improve an existing one, or (3) develop a new material. Development of a new material is seldom a simple matter. Many complex and interacting factors are involved. *Successful development may well require the work of several men over many months* (perhaps 3 to 5 years). Obviously, a materials development program is not inexpensive.

14-2 SERVICE CONDITIONS

Before starting a development program, service conditions, i.e., the requirements of the particular application (Table 14-1), must be known. Although only one or two items in Table 14-1 may be paramount, all of them should be considered, especially in any interactions which influence required properties.

Table 14-1. **Principal service conditions**

Operating temperature
Operating period (life)
Operating pressure
Temperature differentials
Pressure differentials
Applied loading
Fluid environment
Radiation exposure

Operating temperature quickly eliminates certain materials, e.g., operation at 500°C obviously eliminates Al- and Mg-base alloys. Temperature differentials induce thermal stresses and influence many reactions.

Operating pressure strongly influences required properties. In some instances, pressure differential can be at least as important. Fluctuations in either pressure or pressure differential can induce fatigue.

Loading, static or dynamic, applied directly or induced, requires strength. For example, a fuel element in a nuclear reactor might be suspended from the top, which is an applied direct loading. Additional loading can be induced by dynamic effects of coolant flow.

Fluid environment has a major influence on necessary properties through corrosion and compatibility requirements.

Radiation exposure can alter physical, mechanical, and chemical properties of "pure" materials. These effects can be more pronounced in alloys, for example, through additional effects on reaction kinetics. For example, radiation decreases corrosion resistance of Zircaloy-2 in HRE fuel by a factor of at least ten.

Desired operating life is a major factor in determining required properties since it appreciably influences the magnitude of effects of other service conditions. For example, extended operation at elevated temperature, even under relatively low loading, can develop creep. Interaction of time and temperature can cause drastic changes in behavior by alteration of reaction rates, allowing transformation reactions (e.g., tempering, overaging, etc.), etc. Most equipments are not expendable and require long operating lives, implying a necessity for inherent stability of materials.

14-3   PROPERTY CATEGORIES

Once service conditions are established, we determine the properties required to meet them. The major property categories are given in Table 14-2. In some cases, only one or two of these groups are truly important whereas other

Table 14-2.   Properties of principal interest

Physical properties
Mechanical properties
Nuclear properties
Corrosion resistance
Castability
Formability
Weldability
Heat treatability
Thermal stability
Radiation stability
Cost

applications may involve several or all categories. We simply emphasize the obvious by stating that physical and mechanical properties must be adequate.

Corrosion resistance must be sufficient for economical operation, and there must be a potential alternative of designing for economical periodic replacement.

If large-scale application of metals is involved, we desire alloys which can be melted and cast by commercial vendors with minimum processing requirements such as inert atmospheres and extremely close temperature control. We further desire ready working (forging, rolling, drawing, pressing, extruding, bending, machining, etc.) into standard shapes such as plate and sheet, rod, tubing, wire, etc. Powder metallurgy can be an alternative to casting and may be the only way to obtain a billet for mechanical working, e.g., beryllium, but this is inherently more expensive.

Fabricated shapes are often joined together metallurgically in preference to mechanical joints. Thus we desire to weld or braze without serious restrictions on atmospheres, procedures, and equipment. Weld deposits and heat-affected zones must not be brittle (as-welded) since lack of ductility often results in fracture of a weld during subsequent welding of some other portion. This places limitations on the amount and type of transformations permitted during welding.

The alloy must be heat-treatable to a microstructure which is stable for long-time operation under possible corrosion and radiation at operating temperature.

14-4   KNOWLEDGE REQUIRED OF FINAL MATERIAL

The material to be developed is, of course, primarily determined by service conditions. A large number of characteristics (Table 14-3) must be determined for the material finally developed. Although some of these charac-

Table 14-3. Knowledge required of final material

Phase diagram	Fabrication procedure effects on:
Transformations	Physical properties
Reactions	Mechanical properties
Kinetics	Radiation stability
Products	Corrosion resistance
Mechanisms	Morphology
Physical properties	Fabrication schedule effects on:
Mechanical properties	Physical properties
Nuclear properties	Mechanical properties
Radiation stability	Radiation stability
Corrosion resistance	Corrosion resistance
In-pile	Morphology
Ex-pile	Texture effects on:
During fabrication	Physical properties
Morphology	Mechanical properties
Different environments	Radiation stability
Effect of composition variations & modi-	Corrosion resistance
fication on:	Morphology
Transformation reactions and	Heat treatment effects on:
mechanisms	Physical properties
Transformation kinetics and products	Mechanical properties
Physical properties	Radiation stability
Mechanical properties	Corrosion resistance (chemical stability)
Nuclear properties	Morphology
Radiation stability	Metallographic procedures
Corrosion resistance	Analytical procedures
Morphology	Importance & effects of impurities
Transformation reactions & kinetics effects	
on:	
Physical properties	
Mechanical properties	
Radiation stability	
Corrosion resistance (chemical stability)	
Morphology	

teristics will predominate and must be determined in considerable detail, some knowledge of each is desirable.

The phase diagram should be determined in the region around the final composition whether the alloy is binary, ternary, or of higher order. The diagram provides a working basis for determining and interpreting transformation reactions, transformation kinetics, and morphology and for designing fabrication schedules, and for quality control requirements.

Transformation reactions and kinetics, along with variations due to compositional changes, must be determined to establish heat treatment and fabrication schedules. These also aid in understanding physical properties, mechanical properties, radiation stability, corrosion resistance, and morphology.

Mechanical properties, as influenced by temperature, compositional effects, fabrication, textures, and heat treatment procedures, must be established. This is necessary for the designer to properly design structural components and for the materials engineer to establish specifications and limitations for commercial production of the material in various forms.

Corrosion (including oxidation) resistance to anticipated environments during manufacture and operation is important. In addition, effects of compositional variations, fabrication methods, heat treatment variables, morphology, and interrelationships of all of these on corrosion resistance must be determined for adequate design and prediction of operating life.

Determination of morphology (macro- and microstructure), as affected by heat treatment and other variables, allows the materials engineer to interpret structures found in manufactured articles. It also gives him sufficient knowledge to specify necessary changes in composition, fabrication schedules, and/or heat treatments to solve manufacturing problems which are commonly encountered. This information also gives the welding engineer sufficient knowledge to interpret structures developed by welding and to make corrective changes in welding procedures.

Metallographic procedures, i.e., etchants and techniques, often must be developed to accomplish the required research work and to give engineers known methods of examination for control and for solution of manufacturing and construction problems. This aspect is most important, for without correct revelation of microstructures little information can be obtained to guide the overall program, or to correlate various effects found in heat treatments, corrosion, or mechanical property studies.

14-5 DEVELOPMENT PROCEDURE

Service requirements and information outlined in Table 14-3 are for the material ultimately developed. They establish the goal but not the path to be trudged in its attainment.

The day when materials can be designed from theory with computing machines is still distant. Until that day, there will probably be nearly as many approaches as there are men working on materials development. Any development program depends on a blending of theory, experience, and intuition. Nevertheless, certain general ideas can be indicated.

A sound starting point is application of theory and known behavior. For example, theory indicates extensive substitutional solid solubility is possible if atomic radii of the elements involved do not differ greatly, provided other factors, e.g., chemical affinity, valence, crystal lattice, etc. are favorable. Experience tells us the difference in radii is limited to about 15%, and the extent of solid solubility decreases with increasing chemical affinity between atoms, disparity of crystal structure, difference in radii, etc.

Knowledge of electron theory, thermodynamics, transformation reactions, and their kinetics, etc., can be profitably applied to reduce the amount of experimental work required. For example, modulus of elasticity of an alloy can be predicted within reasonable limits by using the method of mixtures. Even so, the modulus should be measured (reasonably accurately) before the development program is completed.

The extent of required study is indicated by types of transformations encountered. Plain carbon steels with low carbon content, e.g., AISI 1010, have comparatively few heat treatments which can be applied to modify microstructure and properties. Eutectoid carbon steel (AISI 1080) is more complex, and thus more study is required for understanding. High-speed tool steels, e.g., 18:4:1 (nominal 18% W, 4% Cr, 1% V, 0.65% C, balance Fe) are very complex and are not yet completely understood.

Experience tells us that observed behavior all too often differs drastically from predicted behavior. This means a planned, orderly, and logical program is necessary to determine composition, production, treatment, and properties of the optimum or final material. Standard experimental methods of x-ray diffraction, hardness tests, metallography, resistivity measurements, dilatometry, differential thermal analysis, etc., can be used along with new techniques and tests devised and developed as necessary.

A systematic program requires considerable time and effort to accomplish. A reasonable approach is to use theory in combination with available data to propose a number of compositions. Certain of these will be eliminated by preliminary evaluation. Those remaining can be further developed for final evaluation.

A materials development program, regardless of amount of applied theory, planning, and systemization, will be somewhat empirical and will involve a certain amount of trial and error. Rapid results should not be expected, for materials development is inherently long-term due to extensive interrelationship between the variables and the time required for evaluation. Endurance and persistence are required to execute a program, but there is always the possibility of serendipity.

15

FAILURE ANALYSIS

15-1 INTRODUCTION

Failures in components occur in a very small percentage of the millions of tons of materials fabricated and in service, yet they are of great importance to user, fabricator, and supplier. Failure of only one piece, out of a large number fabricated, can lead to many difficulties including extensive and expensive lawsuits, particularly if serious personal injury or death is a result. It simply is not reasonable, however, in the normal context of engineering practice, to expect absolutely no failures (even if everything has been done properly) when large numbers of a given component are being produced. Since there is some variation in all processes, it is inevitable that a few will be relatively weak. This is commonly recognized in procurement agreements in which the supplier guarantees a maximum failure rate (perhaps not more than one per thousand or per ten thousand) at a price agreed to by both consumer and supplier. When failure rate exceeds the guaranteed maximum or failure leads to a serious, even fatal, accident, it often becomes necessary to thoroughly examine the situation to fix responsibility.

15-2 ELEMENTS OF FAILURE ANALYSIS

When we consider the great variety of materials and the myriad ways they can be used, it hardly seems possible to make an orderly and logical presentation of such a broad and complex topic. Despite the complexity, it is possible to regard four factors as paramount in failure analysis. These are:

1. Material selected.
2. Design of component.

3. Fabrication of component.
4. Operational use of component.

We often think of each of these separately, but this can be misleading since each of the four is very much related to the others.

15-2-1 MATERIAL SELECTION

Everything said in Chapter 2 applies here. Obviously the properties (responses) of the material selected must be able to meet the requirements of the specific application. It should be recognized, after Chapters 3 through 14, that properties can be extensively modified by various treatments.

15-2-2 COMPONENT DESIGN

Component design extends in many directions. One obvious direction, for only one example, is determining the proper geometry of the component with proper regard for minimizing stress concentrations at changes of section, reentrant angles, etc. This is obviously related to choice of material. If a very limited number of suitable materials is available, then the possible sizes are also limited. On the other hand, there are sometimes size limitations which may eliminate a number of materials because they simply do not have sufficient strength in the particular environment.

A variety of design deficiencies are often encountered. The principal one is faulty stress analysis, primarily due to either overlooking or not giving sufficient attention to potential sources of stress concentration. Others are: failure to allow for fabrication and assembly variables, failure to consider effect of possible bending or torsion on components designed to carry tension or compression, failure to consider anisotropic properties (texture), selection of improper material or incompatible combinations of materials, and inadequate or missing specifications.

Related to component design is development of specifications to insure proper performance in service. Another aspect of design in the materials context is the development of fabrication schedules. This includes: (1) procedures for obtaining proper size and shape, and (2) proper treatment procedures for obtaining properties to meet specifications. It should be obvious to us that these two are strongly related since a heat treatment may be necessary after one fabrication step before a following step can be taken.

15-2-3 COMPONENT FABRICATION

It is entirely possible for proper material (of good quality as procured) to be incorporated into an excellent design and yet result in an unacceptable component. There are any number of reasons which can be summarized as failure to follow fabrication schedules, resulting in a component which does not meet specifications. If inspection and quality control are not functioning properly, the producer is quite likely to get into trouble.

Some fabrication deficiencies are: machining errors, poor-quality welding or brazing, surface damage by defective tools, damage by careless use of tools, and inadequate cleaning after internal machining or welding. A number of deficiences are often classified under defective material including: surface decarburization in steels, heat-treating cracks, omission of a heat treatment, forging flaws, overheating during heat treatment, porosity or cracks from casting, and excessive nonmetallic inclusions. While these can be classified as defective material, the deficiencies are normally due to errors or poor control during fabrication.

15-2-4 OPERATIONAL USE

In many situations a component of good quality material is properly fabricated in an excellent design and meets all specifications. Despite this, the component often fails in service. Failure can usually be attributed to abnormal use or improper maintenance (or both).

Abnormal use implies usage of the component in circumstances outside the limits of the original design constraints. This includes: overload (e.g., overspeed in an engine or abnormally high voltage), a change in temperature (normally an increase implying a weakening although temperature reduction can lead to brittle failure in certain alloys), a change in atmosphere increasing corrosion, and uncontrollable events (e.g., a bird striking an aircraft).

Improper maintenance includes: inadequate inspection after installation, failure to replace damaged components, inadequate lubrication, use of inferior replacement parts, improper alteration of components, inferior welding during repairs, inadequate or excessive torque applied to fasteners, failure to replace all fasteners, inadequate cleanup after repairs, insufficient thread engagement, damage from misuse of tools, improper adjustment after repairs, and damage from misuse of inspection equipment. Many of these lead to what is effectively overload under the modified conditions.

15-2-5 SUMMARY COMMENT

It should be obvious that analysis of failure and proper assignment of primary and secondary causes of failure is often a complex (and sometimes unsolvable) problem. This is not said to discourage but to emphasize that care and caution are most important, since the immediately obvious cause is not always the primary cause.

Despite all the possible sources of failure, fatigue is clearly the most common cause. The "beach mark" pattern (Fig. 6-13) which commonly identifies fatigue is not always apparent. High magnifications available in electron microscopy may reveal the crack front at each load cycle. Fatigue fractures in certain metals, such as hard steels, do not usually develop the beach mark pattern but may have a smooth area and a rough area which is typical of fatigue. Study of the pattern of the fracture surface often permits deduction of the point (and orientation of the stresses at the point) at which the failure started.

15-3   FAILURE ANALYSIS OF A PINION

As an example of application of many factors discussed earlier, let us consider a pinion as shown in Fig. 15-1. This pinion was used to drive a large gear where intermittent loading was applied rather rapidly, to the extent of being essentially impact loading on some occasions. The material used was AISI 8620 steel for which the TTT diagram is shown in Fig. 15-2. This pinion had been used extensively with very little difficulty, but the same application in somewhat different circumstances caused an unacceptably large number of failures.

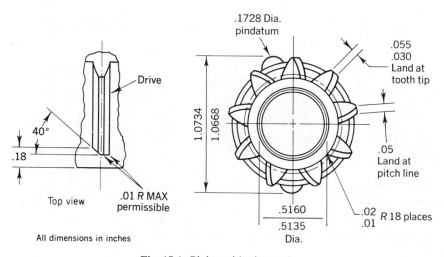

Fig. 15-1. Pinion with nine teeth.

15-3-1   PROCESSING AND SPECIFICATIONS

Material was received as cold-drawn bars about 20 ft. long with a diameter of 1.000 in. (+0.000, −0.002) and with a maximum surface seam of 0.010 in. The bars were sheared to "slugs" of 1 in. length as shown in Fig. 15-3(a).

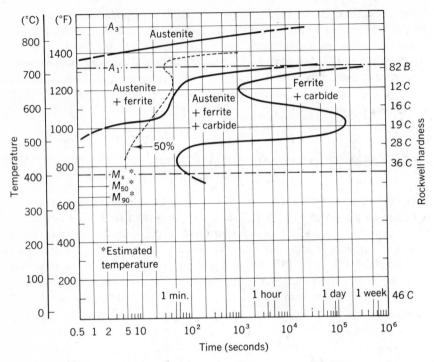

Fig. 15-2. Isothermal transformation diagram for AISI 8620 (C, 0.18; Mn, 0.79; Ni, 0.52; Cr, 0.56; Mo, 0.19); austenized at 1650°F.

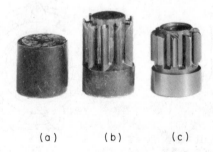

(a)   (b)   (c)

Fig. 15-3. Pinion in various stages of processing: (a) "Slug" as sheared from bar stock; (b) Slug after cold heading and extrusion to form teeth. (c) Finished pinion.

The slugs were normalized at 1310°F in a continuous belt furnace. Belt speed was 4.1 ft/hr with a total cycle of 10 hrs. Specifications permitted a maximum hardness of Rockwell 80B.

Slugs were phosphate coated followed by application of a soap lubricant. A cold heading and finish extrusion on a heavy press produced the shaped teeth as shown in Fig. 15-3(b). A series of machining operations drilled the center hole, rough-finished the hub, and finished the rest of the pinion to final dimensions. All surfaces were copper plated to a minimum thickness of 0.00035 in. and a maximum of 0.001 in. The copper plate was turned off the hub outside diameter. This was followed by processing in a continuous belt carburizing furnace. The pinions had a total heat cycle of 7.5 hr at a furnace temperature of 1650°F in gas atmosphere. Specifications called for a total case depth of 0.035–0.045 in. and a minimum surface carbon content of 1.00%.

The copper plate was stripped in preparation for a second carburizing operation. This was similar to the first but at a furnace temperature of 1560°F. Baskets of pinions moved automatically through the furnace into a quench tank (molten salt at 425°F or oil at 130°F) followed by a spray wash. The case depth on the hub (before grinding) should have been 0.045–0.055 in. The tooth case (at pitch diameter) should have been 0.018–0.024 in.

The quenched pinions were tempered by passing them through a furnace at 350°F at a belt speed of 1.2 in/min for 2 hrs at heat followed by air cooling. Specifications required a Rockwell hardness of 79–82A on the teeth. Tooth structure should have been tempered martensite and a maximum of 10% retained austenite on the case and a combination of martensite and fine pearlite in the core.

Grinding operations on the hub and on the inside diameter produced the finished pinion shown in Fig. 15-3(c).

15-3-2 OBSERVATIONS

The choice of AISI 8620 is entirely reasonable since it is quite possible to develop structures which will meet the stated specifications. It is true that a number of other steels can meet the specifications equally well but AISI 8620 is less expensive than most of the others.

The dimensional tolerances shown in Fig. 15-1 seem reasonable. One critical dimension is not clearly stated. Observe the situation at the end of the tooth at the base of the surface which is at an angle of 40° to the end of the pinion. It is not obvious that a fillet radius is stated although one could infer that it should be 0.01 to 0.02 in., i.e., the same as the fillet radius at the base of the pinion teeth.

The normalizing operation specified a furnace temperature of 1310°F. Examination of Fig. 15-2 indicates that this temperature is marginal. A metal temperature of 1325–1330°F would appear to be better for insuring proper structure before cold forming the teeth. There is no evidence, however, of specific difficulty or failure originating from this marginal temperature.

A number of new, unused, pinions were examined visually with a magnifying glass and at 20X. The drive side surface (Fig. 15-1) of many teeth had long "stringy" defects and general roughness to an even greater extent than shown along the top of the tooth in Fig. 15-4. This general appearance, which is not typical of surfaces which have been cold worked, could be due to small surface seams in the original stock, dirty dies used in forming the teeth, and/or stripping the copper plate.

Fig. 15-4. A new pinion. Defect (on drive side of tooth) extends across root surface and up the back side of the tooth in the foreground. Two small chips were removed from the defect at the root surface. Magnetic particle inspection indicated subsurface cracks along the pitch line and at the fillet at the tooth root. Note general roughness and "strings" of small defects along the surface of the tooth.

The tooth in Fig. 15-4 shows an extreme example of a surface imperfection on a tooth of an unused gear. This defect extended across the root surface and up the back side of the neighboring tooth. Two small chips were removed from the defect in the root surface. Magnetic particle inspection of this pinion indicated the presence of subsurface cracks along the pitch line and at the root of the teeth.

Examination of a number of failed pinions revealed each had one or more teeth with characteristic fatigue failures. One is shown in Fig. 15-5. The fracture started in the fillet at the root of the machined surface which is at 40° to the end of the pinion. This tooth was one of three which failed in apparently the same manner. The second failed tooth appears in the lower part of Fig. 15-5. The third tooth is below the second one in the orientation shown. Figure 15-6 shows the same fractured tooth from above. The characteristic fatigue failure is somewhat more apparent here with the "smooth" portion in the center of the picture, starting at the edge of the 40° machined surface, extending downward, and merging into the "rough" portion of the failure.

Figure 15-7 is a section of this failure at the fillet. This is at the location for which the radius is not clearly specified (see comment on Fig. 15-1, above). If we assume that this radius should be the same as at the root of the teeth (0.010 to 0.020 in.), it is obvious that the assumption is not justified in this case since the radius is approximately 0.002 in.

Fig. 15-5. Fracture of a tooth showing a typical fatigue failure starting at the fillet in the surface at 40° to the end of the pinion. There is some suggestion of a "woody" macrostructure in the failed surface. Visual examination at ×20 with a binocular microscope gives an illusion of an undercut at the fillet.

Fig. 15-6. Same fractured surface as shown in Fig. 15-5, but viewed from the top. The delineation of the "smooth" portion from the "rough" portion is quite distinct.

Figure 15-8 shows a section on the drive (pressure) side of the failed tooth near the fracture. The presence of reformed martensite indicates high localized stress.

Spectrographic analysis confirmed that the material is AISI 8620. A Rockwell hardness of 80A (equivalent to Rockwell 58C) was measured on the hub. A diamond pyramid hardness of 730 (equivalent to Rockwell 61.5C) was found at 0.003 in. below the surface at the pitch line. This is just slightly above the specification of Rockwell 79-82A (56-61C). The total case depth on

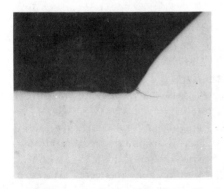

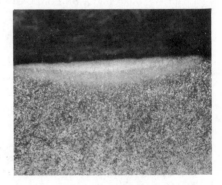

Fig. 15-7. Section through failed tooth through the 40° surface. The measured radius of the fillet is approximately 0.002 in. A crack is apparent at the fillet.

Fig. 15-8. Section on drive (pressure) side of .failed tooth near the fracture showing re-formed martensite, indicating heavy localized stress.

the tooth is 0.022 in. and is thus within specification limits. Tooth core structure is tempered (low-carbon) martensite. The tooth case shows approximately 30–40% retained austenite to a depth of 0.006 in. (Fig. 15-9). This obviously exceeds the specification limit of 10%.

Fig. 15-9. Carburized case on tooth showing 30–40% visible retained austenite to a depth of 0.006 in.

The use of the pinion was essentially the same as previously (when very few failures were observed) except that a somewhat higher loading was applied.

15-3-3   SUMMARY AND CONCLUSIONS OF ANALYSIS

The major factor leading to tooth failure was the high concentration of stress at the root of the surface at 40° to the end of the pinion. This has several contributory factors: stress concentration from the much-too-small fillet radius, an increase in applied load above previous requirement, and the probable existence of subsurface cracks before use.

The general appearance of new pinions and the failure to meet various specifications (e.g., Fig. 15-9) are, in this case, of secondary consideration although these, in the absence of the primary factor of high stress concentration, would lead to excessive wear and pitting along the wear surfaces. This in turn could result in fracture. These secondary aspects are difficult to reconcile with high-quality work in which there has been proper control of processing and proper inspection.

15-4 FAILURE OF A BRASS FIRE EXTINGUISHER

A pressurized water extinguisher was suspected of losing pressure. The manager of the Elks Club in which it was located took steps to have the extinguisher checked. This was being done by an experienced service man using approved testing equipment. He had visually examined the extinguisher and had not observed any "dings," dents, or dampness. He then pressurized the extinguisher to 90 psi with nitrogen (although the normal test pressure was 130 psi) and was uncoupling the extinguisher when it "exploded" and seriously injured him.

15-4-1 OBSERVATIONS

The extinguisher was about 12 years old. It had been kept at different times in a cloak room and the laundry in the Elks Club. It had been checked periodically by trained personnel.

The joint between cylinder wall and end closure was made in the fashion indicated in Fig. 15-10. Both the cylinder and end closure are brass. The "groove" in the joint was probably rolled in after placing the filler metal.

Visual examination of the end closure showed that the upper section in the vicinity of the "groove" had been straightened, but more so in one region than around the rest of the perimeter. Failure presumably started in the region where straightening was more pronounced. Iron oxide (rust) was obvious over a portion of the failure surface in the filler metal between the steel ring and the end closure. Visual examination of the interior of the cylinder indicated the presence of rust on the steel ring on its inner radius, i.e., the surface exposed to the environment. Rust was also observed inside the extinguisher on the surface of the cylinder wall. This latter rust is "separated" from the steel ring by a section of filler metal. The location of rust on the fracture surface (through the filler metal) substantially coincided with the location of rust on the inner surface of the cylinder wall, in terms of location along the perimeter.

Fig. 15-11 shows a section of the inner surface of the cylinder with the steel ring at the bottom of the photograph. The black lines on this photograph indicate the approximate position from which metallographic specimens were taken. The presence of rust on a portion of the fracture surface through the filler metal in this region should be noted.

15-4-2 LABORATORY EXAMINATION

Spectrographic analysis indicates the brass was 70% copper and 30% zinc. The inside of the cylinder and both sides of the end closure were thinly coated

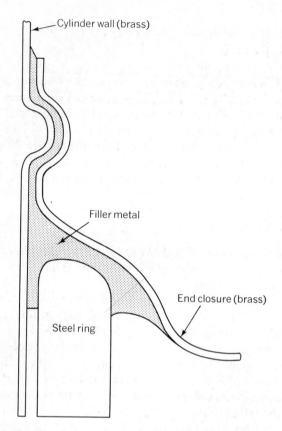

Fig. 15-10. Schematic sketch of joint configuration in brass fire extinguisher (not to scale).

with tin. The steel ring is mild steel which had been annealed or normalized.
The filler metal is about 70% lead and 30% tin.

Fig. 15-12 shows a section of the steel ring with some of the filler metal
adhering to it. There is a portion obviously missing from the steel ring. The
appearance of the other side of the steel ring is quite similar with filler metal
adhering to the steel which has some missing material. The missing portions
are regions which have been removed by corrosive action. All missing portions are in steel which was originally covered by the porous filler metal.

When cuts were made to obtain metallographic specimens, the bond
through the filler metal between the brass cylinder and the steel ring was too
weak to remain integral. Three cuts were made and in each case the cylinder
and ring sections separated.

The surface of the steel ring which interfaced with the brass cylinder wall,
but which was not covered with filler metal, was severely rusted; much more
so than shown in Fig. 15-11.

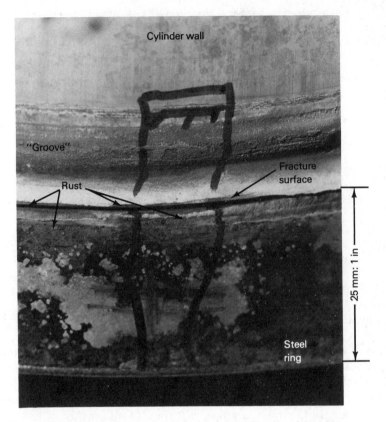

Fig. 15-11. View of failure area inside the brass fire extinguisher. Note there is rust on the inside of the brass cylinder wall above the fracture surface. The black lines indicate approximate positions from which metallographic specimens were taken.

15-4-3   ANALYSIS OF FAILURE

This fire extinguisher had corroded part way through the joint prior to final failure. In effect, there was a significant reduction in thickness in the material which was presumed available to carry the applied load. The remaining intact material was not sufficient to carry the applied load. Failure thus came from a partial failure existing prior to the final failure which started at one region and propagated rapidly around the perimeter.

It appears that the solder was expected to carry a significant portion of the loading. This is questionable since lead-base solders are relatively weak.

The use of the steel ring (presumably for strengthening the base) is highly questionable. Galvanic action, due to the presence of different metals, is almost certain to take place, resulting in corrosion, if there is a humid atmosphere or water present.

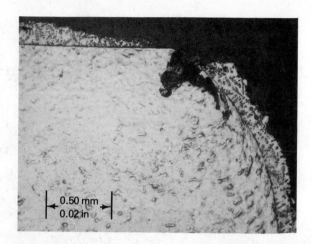

Fig. 15-12. Section of steel ring and filler metal from brass fire extinguisher. Left side of steel ring is on the side of the interface between the ring and cylinder wall. Failure through the filler metal is obvious. Location of a missing portion of steel is equally obvious.

15-4-4 COMMENTS

The National Commission on Materials Policy* said, "One of the most obvious opportunities for materials economy is control of corrosion. It has been estimated that the Nation's annual corrosion cost is $15 billion. One third of this cost can be saved by applying techniques presently developed and available." In this failure, proper attention to information which has been well known for many years, i.e., avoid galvanic cells from dissimilar metals, could have avoided failure and injury to the serviceman.

15-5 FAILURES IN HARD HATS†

A Canadian distributor of safety equipment had a rash of warranty claims due to failures in the harness holders on hard hats imported from Germany. The hat and head harness were received separately and assembled by the distributor to produce the assembly shown in Fig. 15-13. The Canadian Standards Association had tested and approved the headwear as Class B

*"Material Needs and the Environment, Today and Tomorrow," Final Report of The National Commission on Materials Policy, June 1973, Washington, D.C (20402), U.S. Government Printing Office (pg 4C-16).

†Adapted from G. Kardos, "Spontaneous Fracture," ECL-169, Engineering Case Library, Stanford University, Stanford, California.

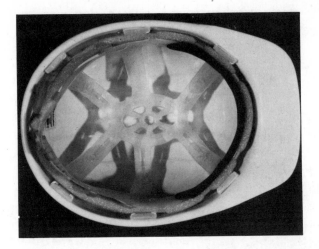

Fig. 15-13. Polyethelene hard hat with head harness installed.

equipment. Failures were reported in approximately 5% of sales over one period of three months. Examination of the failed hats indicated that the majority were failing in the right side holder. Failures occurred in various geographical areas and in various industries. There was no significant variation among the six colors. One noticeable aspect was that a number of hats had failed while in storage.

15-5-1   OBSERVATIONS

A harness insert is shown in Fig. 15-14. Two typical failures are shown in Fig. 15-15. A sketch of the insert-holder assembly is shown in Fig. 15-16. Failures, in all samples examined, originated at the inner corner of the holder (point A, Fig. 15-16).

The nylon insert for the right side was 0.003 to 0.008 in. thicker than the other inserts. This insert also had the date of manufacture on it. Reproducible readings of the holder molded as part of the polyethylene hat could not be made because of the relative flexibility of the material.

15-5-2   TESTING

Failure surfaces showed typical brittle fracture (Fig. 15-15). Attempts to reproduce this type of failure by pulling on the ears of the holder were unsuccessful. The ears could be bent through a large angle without fracture. When fracture did occur, it produced a stringy torn surface with no resemblance to service failure.

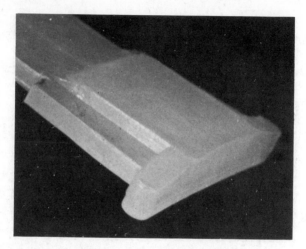

Fig. 15-14. Nylon harness insert.

One hat sample was cycled from room temperature to 25°F, both with the harness inserted during cycling and without the harness inserted. In the case where the harness was not inserted during cycling, the insert was pressed into the holder at the low temperature. No failure occurred. Attempts to pull the ears at 25°F produced failures similar to those observed at room temperature but not like service failures.

A second sample was cycled from room temperature to 0°F without causing failure. Overloading of the holder ear at 0°F produced a brittle fracture similar to service failure, although deflection was much greater than could be accounted for by normal loading.

The geometry of the holder is such that it is expected to carry a bending moment and shear. The holder is thinnest at point A (Fig. 15-16) where there is also a sharp re-entrant corner. The nominal stress at point A from maximum loading and the thin section is greatly compounded by the large stress concentration from the re-entrant corner. Apparently the only reason the element did not fail under static loading is the large compliance of the polyethylene.

15-5-3 CONCLUSIONS

The holders failed in static fatigue, i.e., delayed fracture (Sec. 6-12–1), due to high stresses induced in re-entrant corners. Environmental conditions and low temperature cycling may have contributed to failure but data are too limited to specifically isolate these. In any event, environmental factors are not the prime causes of failure.

Fig. 15-15. Two failed harness holders.

Fig. 15-16. Sketch of harness holder showing interference fit.

15-5-4 COMMENTS

The major factor in failure can be corrected by eliminating the stress concentration. This can be accomplished by introducing a radius at point A so that the section thickness stays constant or increases (Fig. 15-17).

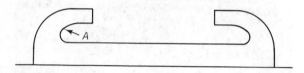

Fig. 15-17. Redesigned harness holder.

QUESTIONS AND PROBLEMS

1. Why is the copper plating used in the first carburizing operation and not in the second on the pinion?

2. Why the difference in temperature of the molten salt and the (alternative) oil used in quenching the pinions after the second carburizing operation?

3. Why would a temperature of 1325–1330°F probably be better than 1310°F in normalizing prior to cold forming?

4. Demonstrate that the heat treatments specified should be proper to develop the specified structures and hardnesses.

5. Why should high localized stress lead to the presence of reformed martensite as shown in Fig. 15-8, i.e., what is the mechanism involved?

6. How would you eliminate the excessive amount of retained austenite shown in Fig. 15-9?

7. Do you concur with the conclusions as to the cause of failure of the pinion? If not, why not?

8. What recommendations would you make for changes in order to obtain satisfactory pinions?

9. Can you recommend any other changes in the design and construction of the fire extinguisher that will improve longevity?

10. Can you suggest any other design changes in geometrical configuration or choice of material to eliminate the static fatigue problem in hard hats?

EPILOGUE

(TO THE STUDENT)

Having worked your way through this book, you surely recognize that there is much more to learn. Whether through self-study or consultation with experts, it is most important to broaden your knowledge. Many designers are not sufficiently familiar with materials and their properties to appreciate the real scope of some problems. As a result, they tend to suspect the unknown and not use various materials which might be applied advantageously. Don't overlook known technology.

Learn from the past. Any mistakes should be your own, not repetitions of those made by others. Design problems have always been with us, but some similar questions keep recurring.

Is this a new problem? *Newsday*, on June 20, 1974, published an article in which a former engineer and senior designer for Litton Industries alleged that the Navy was allowing the firm to build five ships (LHA's) of dangerous and outmoded steel. One quotation says, "These ships could break up in heavy seas and cause the loss of all hands on board." *Metal Progress*, in July 1974, had a letter to the editor that included three pictures of a tugbarge which broke in half in brittle fracture in still water in Port Jefferson, N.Y., on January 10, 1972, at an ambient temperature of 46°F. This vessel was less than one year old.

Is it possible that these are the same problems that were so conspicuous during and after World War II? Of some 4700 welded ships, 250 suffered severe brittle fractures that endangered the vessels (19 actually broke in two or were abandoned) and 1200 suffered potentially dangerous cracks. (The problem is not limited to ships but is found in pressure tanks, pipelines, railroad cars, bridges, power booms, smokestacks, etc.) Since World War II, methods have been developed for selecting materials which permit design and construction of similar structures without invoking the spectre of brittle failure.

Were these World War II ship failures a new problem? There is reason to believe that many brittle failures have occurred without being recognized as such. One case is the failure of a molasses tank in Boston in 1919 which released $2\frac{1}{2}$ million gallons of molasses, killed a dozen people by injury or drowning in molasses, injured 40 others, drowned many horses, damaged many houses, and knocked over a portion of the elevated railway structure.

Was this molasses tank the first indication of trouble? The potential problem was indicated as early as 1879 when a Mr. Kirk, in discussing a paper on "The Use of Steel in Naval Constructions" in the *Journal of the Iron & Steel Institute*, cited a steel plate which "when cold, on being thrown down, split right up. Pieces cut from each side of the split stood all the Admiralty tests. Now given a material capable of standing without breaking an extension of 20% he wanted to know . . . how a plate . . . could split with a very slight extension . . . not to the extent of 1%." Nobody answered his question. Is it possible that too many have been ignoring the question ever since?

Good Luck and a Happy Career!

APPENDIX A

BASIC CONSTANTS AND CONVERSION FACTORS

Speed of light	c	3.00×10^8 m/sec = [1.86×10^5 miles/sec]
Permeability constant	μ_0	1.26×10^{-6} henry/m
Permittivity constant	ϵ_0	8.85×10^{-12} farad/m
Boltzmann's constant	k_B	1.38×10^{-23} joule/molecule °K = [8.617×10^{-5} eV/°K]
Avogadro's constant	N_0	6.02×10^{23} molecules/mole
Universal gas constant	R	8.31 joules/mole °K = [1.99 cal/mole °K] = [0.08231 liter-atm/mole °K]
Planck's constant	h	6.63×10^{-34} joule-sec
Elementary charge	e	1.60×10^{-19} coulomb
Electron rest mass	m	9.11×10^{-31} kg
Bohr radius	α_0	5.29×10^{-11} m = 0.529 A
Electron magnetic moment (Bohr magneton)	β	9.27×10^{-24} joule-m^2/weber = [9.27×10^{-24} amp-m^2]

	mks units	cgs units	English units
Length	1 m	100 cm	39.37 in.
Mass	1 kg	1,000 grams	2.2046 lb.
Energy	1 joule (J)	10^7 ergs	0.73756 ft-lb.
Power	1 watt (W)	10^7 ergs/sec	1/746 h.p.
Force	1 newton (N)	10^5 dynes	0.2247 lb.
Electric Charge	1 coulomb (C)	0.1 emu, 3×10^9 esu	1 coulomb
Electric Potential	1 volt (V)	10^8 emu, $\frac{1}{300}$ esu	1 volt
Magnetic Flux	1 weber (Wb)	10^8 lines	10^8 lines
Magnetic Induction	1 weber/m^2 (tesla)	10,000 gauss	64,500 lines/in.^{2}
Magnetic Field Intensity	1 amp-turn/m^2	$4\pi \times 10^{-3}$ oersted	0.0254 amp-turn/in.

Length	1 A = angstrom = 10^{-10} m
Length	1μ = micron = 10^{-6} m
Energy	1 eV = 1.60×10^{-19} joule
Energy	1 calorie = 4.186 joules = 1.16×10^{-6} kw-hr. = $\frac{1}{252}$ Btu
Pressure	1 atmosphere = 29.9 in.-Hg = 76.0 cm-Hg = 1.01×10^5 N/m^2
Pressure	1 bar = 10^6 dynes/cm^2 = 10^6 N/m^2 = 0.99 atm

APPENDIX B

SOME PROPERTIES OF SELECTED ENGINEERING MATERIALS

Material	Density lb/in.³ (20°C)	Thermal conductivity cal/cm²/°C/cm/sec (25°C)	Thermal expansion µin./in./°C (20 to 100°C)	Electrical resistivity [68°F (20°C)]	Modulus of elasticity psi in units of 10^6	Tensile yield point psi	Tensile yield strength psi	Tensile yield strength % offset	Elongation % in 2 in.	Tensile strength psi	Fatigue limit psi	Fatigue strength psi	Fatigue strength 10^6 cycles
Aluminum (1100-0)	0.098	0.53	23.6	2.9	10.0	—	5,000	0.2	35	13,000	—	5,000	500
(1100-H18)	—	0.53	—	3.0	10.0	—	22,000	0.2	5	24,000	—	9,000	500
(2024-0)	0.100	0.45	22.8	3.45	10.6	—	11,000	0.2	20	26,000	—	13,000	500
(2024-T4)	—	0.29	—	5.75	10.6	—	42,000	0.2	19	64,000	—	20,000	500
(7075-0)	0.101	—	23.2	—	10.4	—	15,000	0.2	17	33,000	—	17,000	500
(7075-T6)	—	0.29	—	5.74	10.4	—	73,000	0.2	11	83,000	—	23,000	500
Copper Annealed (ETP)	0.321	0.934	16.8	1.71	17.0	—	10,000	0.5	45	32,000	—	13,000	100
Hard	—	—	—	—	—	—	45,000	0.5	—	—	—	—	—
Magnesium Annealed (MIA)	0.064	0.33	26.0	5.0	6.5	—	18,000	0.2	17	33,000	—	—	—
Hard	—	—	—	—	—	—	26,000	0.2	7	35,000	—	—	\
Steel Hot rolled (1020)	0.283	0.12	11.7	14.3	29–30	30,000	—	—	25	55,000	25,000	—	—
Cold drawn	—	—	—	—	—	51,000	—	—	15	61,000	28,000	—	—
Hot rolled (1040)	0.283	0.115	11.3	19.0	29–30	42,000	—	—	18	76,000	35,000	—	—
Cold drawn	—	—	—	—	—	71,000	—	—	12	85,000	40,000	—	—
Hot rolled (1080)	0.283	0.110	10.8	18.0	29–30	61,500	—	—	10	112,000	40,000	—	—
SACD*	—	—	—	—	—	75,000	—	—	10	98,000	40,000	—	—

*Spheroidize Anneal Cold Drawn

Material												
Austenitic stainless—Type 304	0.29	0.035	17.3	72.0	28	—	35,000	0.2	55	85,000	34,000	—
Alumina (99% Al_2O_3)	0.138	0.070	6.0	$>10^{14}$	50	—	—	—	—	35,000	—	—
Beryllia (98% BeO)	0.103	0.49	2.4	$>10^{17}$	45	—	—	—	—	—	—	—
Glass												
Polycrystalline	0.095	0.008	5.8	1.6×10^{13}	17.3	—	—	—	—	—	—	—
Silica (quartz)	0.080	0.003	0.54	10^{20}	10.5	—	—	—	—	—	—	—
Graphite (AGOT)	0.062	0.25	2.1	800	1.4	—	—	—	—	1200	—	—
Magnesia (MgO)	0.129	0.082	13.0	10^5	12	—	—	—	—	—	—	—
Rubber (Butyl)	0.033	0.002	575.0	—	—	—	—	—	800	2500	—	—
Polyvinyl chloride	0.056	0.0038	79	10^{16}	0.37	—	—	—	—	7300	—	—
Polytetrafluoroethylene	0.079	0.0006	100	10^{19}	0.05	—	—	—	250–450	2500–6500	—	—
Polyethylene (Med. density)	0.034	0.0008	21	$>10^{15}$	—	—	—	—	200	2000	—	—
Polystyrene (Molded, general purpose)	0.037	0.0003	72	10^{18}	0.5	—	6000–7900	—	1.5–2.3	6000–7900	—	—

APPENDIX C

SOME PROPERTIES OF VARIOUS METALS AND ALLOYS*

Material & Identification†	Resistivity Microhm-cm	Temp. coeff. of Resis./°C	Density g/cm³	Specific Heat cal/g/°C	Thermal Cond. w/cm-°C	Thermal Expansion ppm/°C‡	Melting Point °C**
Aluminum, pure	2.65	0.00429	2.70	0.215	2.22	23.6	660
Aluminum, conductor (99.45 Al)	2.8	0.00403	2.70		2.34	23.6	657
Beryllium	4	0.025	1.85	0.45	1.46	11.6	1279
Bismuth	[2]106.8	0.004	9.80	0.029	0.08	13.3	272
Brass, yellow (65 Cu, 35 Zn)	6.4	0.0016	8.47	0.09	1.17	20.3	931
Bronze, commercial (90 Cu, 10 Zn)	3.9	0.00186	8.80	0.09	1.88	18.4	1046
Bronze, phosphor (95 Cu, 5 Sn)	11	0.004	8.86	0.09	0.71	17.8	1050
Cadmium	[2]6.83	[2]0.0042	8.65	0.055	0.92	29.8	321
Chromium	13.0	[2]0.003	7.19	0.11	0.67	6.2	1878
Cobalt	6.24	0.00604	18.85	0.099	0.69	13.8	1497
Constantan (45 Ni, 55 Cu)	50	±0.00002	8.9	0.094	0.21	14.9	1292
Copper, annealed (ACS)	1.72	0.00393	8.89	0.092	3.94	16.5	1084
Copper, beryllium (97.9 Cu, 1.9 Be)	56.8–7.4	0.0013	8.23	0.1	50.8	17.8	956
Copper, electrolytic (99.95 Cu)	1.71	0.00397	8.89	0.092	3.90	16.8	1084
Gallium	253.4		5.91	0.079	0.33	18	30
Germanium	346		5.32	0.073	0.58	5.7	938
Gold	2.35	[2]0.004	19.32	0.031	2.97	14.2	1064
Inconel 600	98.1	0.0001	8.51	0.109	0.15	11.5	1427
Indium	8.37	0.005	7.31	0.057	0.24	33	156
Invar (64 Fe, 36 Ni)	80	0.0012	8.0	0.123	0.11	<0.1	
Iron, pure	9.71		7.87	0.11	0.75	11.8	1538
Iron, gray (3.16 Tc, 1.54 Si, 0.57 Mn)		0.0065	7.15		0.47	10.5	

APPENDIX C CONTINUED

Material							
Lead	20.65	0.00336	11.36	0.031	0.35	29.3	327
Magnesium	4.45	0.0165	1.74	0.245	1.62	27.1	650
Mercury	798.4		13.55	0.033	0.08		−38
Molybdenum	5.2	0.005	10.22	0.066	1.42	4.9	2615
Monel (67 Ni, 30 Cu)	48.2	0.0020	8.84	[2]0.127	[2]0.26	14.0	1327
Nichrome (80 Ni, 20 Cr)	107.9	0.00009	8.4	0.107	0.134	[4]17.3	1402
Nickel	6.84	[2]0.0069	8.90	[2]0.105	0.92	13.3	1455
Niobium (Columbium)	[2]12.5	0.0040	8.57	0.065	0.52	7.3	2472
Palladium	10.8	0.00377	12.02	0.058	0.70	11.8	1554
Platinum	10.64	[2]0.00393	21.45	0.031	0.69	8.9	1772
Silicon	210[5]		2.33	0.162	0.84	2.8–7.3	1412
Silver	1.59	0.0041	10.49	0.056	4.18	19.7	962
Steel, carbon (0.4–0.5 C, bal. Fe)	7–12		7.8		0.5	11	1482
Steel, silicon (3 Si, oriented)	50	0.001	7.65		0.18		
Steel, stainless, 304	72	0.001	7.9	0.12	0.15	17.3	>1152
Steel, stainless, 347	73		8.0	0.12	0.16	16.7	>1402
Steel, stainless, 410	57	0.0015	7.7	0.11	0.24	11	>1482
Tantalum	12.45	[2]0.0038	16.6	0.034	0.54	6.5	2996
Thorium	13	[2]0.0038	11.66	0.034	0.38	12.5	1753
Tin	11.0	[2]0.0047	7.30	0.054	0.63	23	232
Titanium	42	0.0035	4.51	0.124	[6]0.11	8.4	1670
Tungsten	5.65	0.0045	19.3	0.033	1.66	4.6	3417
Uranium	30	0.0034	19.07	0.028	0.27	7–14	1132
Water, pure	∞		1.00	1.000	0.006		0
Zinc	5.92	[2]0.0042	7.13	0.091	1.13	39.7	419
Zirconium	[2]40.0	0.0044	16.49	0.067	0.88	5.8	1855

*at 20°C or ambient
†pure metals if alloy is not stated
‡parts per million/°C

[1]close packed hexagonal
[2]at 0°C
**International Practical Temperature Scale—1968

[3]ohm-cm
[4]20–1000 C; 13 at 20°C

[5]depends on processing
[6]at −240°C
[7]at 50°C

APPENDIX D

PROPERTIES OF SOME ENGINEERING POLYMERS

NOTE: All properties shown are typical of standard grade resins. In most instances additional grades are available.

Material	Specific gravity	Cost cents/in.³	Impact strength Notched izod (ft.lb./in., 1/8")	Tensile strength psi × 10³	Tensile modulus psi × 10³	Heat distortion temp. °F at 264 psi	Dielectric strength ST 1/8" thick (v/mil)	Water Absorption % in 24 hrs
Thermoplastic Polyester	1.31	3.16	1.2	8.0	340[1]	310[2]	590	0.08
Thermoplastic Polyester Glass Reinforced	1.52	3.84	2.2	17.3	1200	415	750	0.06
Polycarbonates	1.20	3.85	16	8.5	345	270	410	0.15
Nylon 6/6	1.14	3.6	1-2	11.5	425	200	350	1.3
Phenylene Oxide Based Resins								
NORYL® SE-1	1.06	3.44	5.0	9.6	355	265	500	0.066
NORYL® SE-100	1.10	2.74	5.0	7.8	380	212	400	0.07
Acetal	1.42	3.34	1.3	10	520	255	500	0.25
ABS (KJB)	1.21	2.45	4.0	6.0	320	180	350-500	0.4
Polysulfone	1.24	4.50	1.3	10.2	360	345	425	0.22
Phenolic	1.56	1.8	0.3	7.0	1200	360	325	0.25
Acrylic	1.18	1.96	0.5	10	450	180	400	0.30
Polypropylene (G.P.)	0.91	0.68	0.4	5.0	170	140	650	0.02
Vinyl (Type 1)	1.45	1.23	1-2	7.5	415	169	360	0.05
Cellulosics (CAB)	1.2	2.7	3-8	5.0	150	180	350	1.8
Silicone Molding Resins	1.9	15-20	0.04	4.8	—	610	480	0.5
Chlorinated Polyether	1.4	22.8	0.4	3.5	150	200	500	0.01

[1] Flexural Modulus psi × 10³

[2] Heat Distortion Temp °F @ 66 psi

BIBLIOGRAPHY

This bibliography is divided into two sections. The section of general references provides sources of additional background and information pertinent to many areas of the text. The second set of references has particular relevance to individual chapters.

GENERAL REFERENCES

ALFREY, T., and E. F. GURNEE, *Organic Polymers.* Englewood Cliffs, N.J.: Prentice-Hall, Inc., 1967.

AZAROFF, L. V., *Introduction to Solids.* New York: McGraw-Hill Book Company, 1960.

BAER, E., *Engineering Design for Plastics.* New York: Reinhold Publishing Corporation, 1964.

BARRETT, C., and T. B. MASSALSKI, *Structure of Metals.* 2nd Edition. New York: McGraw-Hill, 1966.

BARRETT, C. R., W. D. NIX, and A. S. TETELMAN, *Principles of Engineering Materials.* Englewood Cliffs, N.J.: Prentice-Hall, Inc., 1973.

CAHN, R. W. Ed., *Physical Metallurgy.* Amsterdam, Netherlands: North-Holland Publishing Co., 1970.

DIBENEDETTO, A. T., *The Structure and Properties of Materials.* New York: McGraw-Hill Book Company, 1967.

EDELGLASS, S. M., *Engineering Materials Science.* New York: Ronald Press, 1966.

EISENSTADT, M. M., *Introduction to Mechanical Properties of Materials.* New York: Macmillan Publishing Co., Inc., 1971.

Engineering Alloys. 5th Edition. New York: Van Nostrand Reinhold Company, 1973.

GOLDMAN, J. E., Ed., *The Science of Engineering Materials.* New York: John Wiley and Sons, 1957.

GUY, A. G., *Elements of Physical Metallurgy.* 2nd Edition. Reading, Mass.: Addison-Wesley, 1959.

GUY, A. G., *Essentials of Materials Science.* New York: McGraw-Hill Book Company, 1976.

GUY, A. G., *Introduction to Materials Science*. New York: McGraw-Hill Book Company, 1972.

GUY, A. G., *Physical Metallurgy for Engineers*. Reading, Mass.: Addison-Wesley, 1962.

HANNAY, H. B., "Solid-State Chemistry," *Int. Sci. & Tech.*, No. 22, October 1963, pp. 64–70.

HOVE, J. E., and W. C. RILEY, Ed., *Modern Ceramics, Some Principles and Concepts*. New York: John Wiley and Sons, 1965.

HOVE, J. E., and W. C. RILEY, Ed., *Ceramics for Advanced Technologies*. New York: John Wiley and Sons, 1965.

KEYSER, C. A., *Materials Science in Engineering*. Columbus, Ohio: Merrill Publishing Co., 1968.

KINGERY, W. D., Ed., *Ceramic Fabrication Processes*. New York: John Wiley and Sons, 1956.

KINGERY, W. D., *Introduction to Ceramics*. New York: John Wiley and Sons, 1960.

LEWIS, T. J., and P. E. SECKER, *Science of Materials*. New York: Reinhold Publishing Corporation, 1965.

LYMAN, T., Ed., *Metals Handbook*. 1948 Edition. Cleveland, Ohio: American Society for Metals.

MEARES, P., *Polymers: Structure and Bulk Properties*. New York: Van Nostrand Reinhold Company, 1965.

Metal Progress. Vol. 66, No. 1-A (Supplement of 1948 Edition of *ASM Metals Handbook*), Cleveland, Ohio: American Society for Metals, July 15, 1954.

Metal Progress. Vol. 68, No. 2-A (Supplement of 1948 Edition of *ASM Metals Handbook*), Cleveland, Ohio: American Society for Metals, August 15, 1955.

Metals Handbook. 8th Edition. Metals Park, Ohio: American Society for Metals.
Vol. 1　*Properties & Selection of Metals*, 1961.
Vol. 2　*Heat Treating, Cleaning & Finishing*, 1964.
Vol. 3　*Machining*, 1967.
Vol. 4　*Forming*, 1969.
Vol. 5　*Forging & Casting*, 1970.
Vol. 6　*Welding & Brazing*, 1971.
Vol. 7　*Atlas of Microstructures of Industrial Alloys*, 1972.
Vol. 8　*Metallography, Structures & Phase Diagrams*, 1973.
Vol. 9　*Fractography & Atlas of Photographs*, 1974.
Vol. 10　*Failure Analysis & Prevention*, 1975.

NUTT, M. C., *Principles of Modern Metallurgy*. Columbus, Ohio: Merrill Publishing Co., 1968.

PASK, J. A., Ed., *An Atomistic Approach to the Nature and Properties of Materials*. New York: John Wiley and Sons, 1967.

PECKNER, D., *The Strengthening of Metals*. New York: Reinhold Publishing Corporation 1964.

POLAKOWSKI, N. H., and E. J. RIPLING, *Strength and Structure of Engineering Materials*. Englewood Cliffs, N.J.: Prentice-Hall, Inc., 1966.

REED-HILL, R. E., *Physical Metallurgy Principles*. Princeton, N.J.: Van Nostrand, 1964.

RICHMAN, M. H., *Introduction to the Science of Metals*. Waltham, Mass.: Blaisdell Pub. Co., 1967.

ROSENTHAL, D., *Introduction to Properties of Materials*. Princeton, N.J.: Van Nostrand, 1964.

ROSENTHAL, D. and R. ASIMOW, *Introduction to Properties of Materials*. New York: Van Nostrand Reinhold Company, 1971.

RUOFF, A. L., *Introduction to Materials Science*. Englewood Cliffs, N.J.: Prentice-Hall, Inc., 1972.

SCHULTZ, J. M., *Polymer Materials Science*. Englewood Cliffs, N.J.: Prentice-Hall, Inc., 1974.

"Scientific American," Vol. 217 (Materials Issue), No. 3, September 1967.

SINNOTT, M. J., *The Solid State for Engineers*. New York: John Wiley and Sons, 1958.

Structure and Properties of Materials. New York: John Wiley and Sons, 1964.
    Vol. 1: *Structure.*
    Moffatt, W. G., G. W. Pearsall, and J. Wulff.
    Vol. II: *Thermodynamics of Structure.*
    Brophy, J. H., R. M. Rose, and J. Wulff.
    Vol. III: *Mechanical Behavior.*
    Hayden, W., W. G. Moffatt, and J. Wulff.
    Vol. IV: *Electronic Properties.*
    Rose, R. M., L. A. Shepard, and J. Wulff.

VAN VLACK, L. H., *Elements of Material Science*. 2nd Edition. Reading, Mass.: Addison-Wesley, 1964.

VAN VLACK, L. H., *Materials Science for Engineers*. Reading, Mass.: Addison-Wesley, 1970.

VAN VLACK, L. H., *Physical Ceramics for Engineers*. Reading, Mass.: Addison-Wesley, 1964.

VAN VLACK, L. H., *A Textbook of Materials Technology*. Reading, Mass.: Addison-Wesley, 1973.

WILLIAMS, D. J., *Polymer Science and Engineering*. Englewood Cliffs, N.J.: Prentice-Hall, Inc., 1971.

WINDING, C. C., and G. D. HIATT, *Polymeric Materials*. New York: McGraw-Hill Book Company, 1961.

INDIVIDUAL CHAPTER REFERENCES

CHAPTER 1

SMITH, C. S., "Simplicity and Complexity," *Science & Technology*, No. 77, May 1968, pp. 60–65.

CHAPTER 2

"Characterization . . . Industry's Most Urgent Materials Problem," *Metal Progress*, Vol. 93, No. 4, April 1968, pp. 60–78.

GILLETT, H. W., *The Behavior of Engineering Metals*. New York: John Wiley and Sons, 1951.

RUSKIN, A. M., *Materials Considerations in Design*. Englewood Cliffs, N.J.: Prentice-Hall, Inc., 1967.

SMITH, C. O., "Some Factors in Selecting Materials for Nuclear Reactors," *Naval Engineers Journal*, Vol. 79, No. 6, April 1967, pp. 233–237.

VERINK, E. D., Ed., *Methods of Materials Selection*. New York: Gordon & Breach, 1968.

WAGNER, H. J., and M. EMERSON, "A New Kind of Coin," *Int. Sci. & Tech.*, No. 44, August 1965, pp. 44–50.

WAGNER, H. J., and A. M. HALL, "The New Coinage," *Journal of Metals*, Vol. 18, No. 3, March 1966, pp. 300–307.

CHAPTER 3

"Atomic & Electronic Structure of Metals," The 1966 ASM Seminar Book, Metals Park, Ohio: American Society for Metals, 1967.

BARTLETT, N., "Noble-Gas Compounds," *Int. Sci. & Tech.*, No. 33, September 1964, pp. 56–66.

COHEN, B. L., "Nuclear Orbital Structure," *Int. Sci. & Tech.*, No. 23, November 1963, pp. 65–74.

COMPANION, A., *Chemical Bonding.* New York: McGraw-Hill Book Company, 1964.

DEKKER, A. J., *Solid State Physics.* Englewood Cliffs, N.J.: Prentice-Hall, Inc., 1957.

DROST-HANSEN, W., "The Puzzle of Water," *Int. Sci. & Tech.*, No. 58, October 1966, pp. 86–96.

EVANS, R. C., *Introduction to Crystal Chemistry.* 2nd Edition. New York: Cambridge University Press, 1964.

EYRING, H., J. HILDEBRAND, and S. RICE, "The Liquid State," *Int. Sci. & Tech.*, No. 15, March 1963, pp. 56–66.

GOTTLIEB, M. B., "Plasma—The Fourth State," *Int. Sci. & Tech.*, No. 44, August 1965, pp. 44–50.

HAMMETT, L. P., and H. MASSEY, "Chemical Bonds," *Int. Sci. & Tech.*, No. 25, January 1964, pp. 62–72.

HUME-ROTHERY, W., *Electrons, Atoms, Metals and Alloys.* London: Iliffe and Sons, 1955 (Philosophical Library, N.Y.).

HUME-ROTHERY, W., *Atomic Theory for Students of Metallurgy.* 3rd Edition. London: The Institute of Metals, 1955.

JACKSON, E. A., *Equilibrium Statistical Mechanics.* Englewood Cliffs, N.J.: Prentice-Hall, Inc., 1968.

KAPLAN, I., *Nuclear Physics.* 2nd Edition. Reading, Mass.: Addison-Wesley, 1963.

KITTEL, C., *Introduction to Solid State Physics.* 2nd Edition. New York: John Wiley and Sons, 1956.

KITTEL, C., *Introduction to Solid State Physics.* 3rd Edition. New York: John Wiley and Sons, 1966.

MATTSON, H. W., "Determining Molecular Structure," *Int. Sci. & Tech.*, No. 38, February 1965, pp. 22–31.

McLACHLAN, D., *Statistical Mechanical Analogies.* Englewood Cliffs, N.J.: Prentice-Hall, Inc., 1968.

MOTT, N. F., and I. N. SNEDDON, *Wave Mechanics and Its Applications.* New York: Dover Publications, 1963.

POHL, A., *Quantum Mechanics for Science and Engineering.* Englewood Cliffs, N.J.: Prentice-Hall, Inc., 1967.

SMITH, G. W., "Liquid-Like Solids," *Int. Sci. & Tech.*, No. 61, January 1967, pp. 72–80.

SPEDDING, F. H., "Rare Earths," *Int. Sci., & Tech.*, No. 4, April 1962, pp. 39–46.

SPROULL, R. L., *Modern Physics, A Textbook for Engineers*. New York: John Wiley and Sons, 1956.

STRINGER, J., *An Introduction to the Electron Theory of Solids*. New York: Pergamon Press, 1967.

TEMPERLEY, H. N. V., "Changes of State," *Int. Sci. & Tech.*, No. 46, October 1965, pp. 68–76.

WEINREICH, G., *Solids: Elementary Theory for Advanced Students*. New York: John Wiley and Sons, 1965.

WEISS, R. L., *Solid State Physics for Metallurgists*. Reading, Mass.: Addison-Wesley, 1963.

WEISSKOPF, V., "The Quantum Ladder," *Int. Sci. & Tech.*, No. 18, June 1963, pp. 61–71.

ZIMAN, J. M., "Understanding Solid Matter," *Int. Sci. & Tech.*, No. 67, July 1967, pp. 44–53.

CHAPTER 4

CHALMERS, B., "Crystals," *Int. Sci. & Tech.*, No. 68, August 1967, pp. 51–60.

COHEN, J. B., *Diffraction Methods in Materials Science*. New York: Macmillan Publishing Co., Inc., 1966.

CULLITY, B. D., *Elements of X-Ray Diffraction*. Reading, Mass.: Addison-Wesley, 1956.

GRAY, T. J., Ed., *The Defect Solid State*. New York: Interscience Publishers, 1957.

KIRSCH, H., *Applied Mineralogy*. London, England; Chapman & Hall, New York: Barnes & Noble, 1968.

NEWKIRK, J. B., and J. H. WERNICK, *Direct Observation of Imperfections in Crystals*. New York: Interscience Publishers, 1962.

READ, JR., W. T., *Dislocations in Crystals*. New York: McGraw-Hill Book Company, 1953.

CHAPTER 5

"Atomic & Electronic Structure of Metals," The 1966 ASM Seminar Book, Metals Park, Ohio: American Society for Metals, 1967.

BITTER, F., "Strong Magnets," *Int. Sci. & Tech.*, No. 4, April 1962, pp. 58–64.

BOUWMAN, S., "Magnetic Materials," *Int. Sci. & Tech.*, No. 12, December 1962, pp. 20–32.

BRUGGER, R. M., "Neutrons & Phonons," *Int. Sci. & Tech.*, No. 11, November 1962, pp. 52–59.

FRANKEN, P., "High Energy Lasers," *Int. Sci. & Tech.*, No. 10, October 1962, pp. 62–68.

GARNETT, C. G. B., "Far-Infrared Masers," *Int. Sci. & Tech.*, No. 39, March 1965, pp. 39–44.

HULM, J. K., B. S. CHANDRASEKHAR, and H. RIAMERSMA, "Super-conducting Magnets," *Int. Sci. & Tech.*, No. 17, May 1963, pp. 50–57.

LEWIS, H. R., "Optical Masers," *Int. Sci. & Tech.*, Prototype Issue, pp. 60–70.

MYERS, R. A., "Scanning with Lasers," *Int. Sci. & Tech.*, No. 65, May 1967, pp. 41–50.

NEISMAN, R. R., and R. E. JOHNSON, "Organic Materials in Electronics," *Int. Sci. & Tech.*, No. 29, May 1964, pp. 68–79.

NUSSBAUM, A., *Electronic and Magnetic Behavior of Materials.* Englewood Cliffs, N.J.: Prentice-Hall, Inc., 1967.

QUIST, T. M., "Semiconductor Lasers," *Int. Sci. & Tech.*, No. 26, February 1964, pp. 80–88.

TURRO, N. J., "Light-Excited Molecules," *Int. Sci. & Tech.*, No. 66, June 1967, pp. 42–50.

CHAPTER 6

BARROIS, W. G., Ed., "Manual on Fatigue of Structures," AD 708621, AGARD-MAN-8-70, Langley Field, Virginia: National Aeronautics & Space Administration, 1970.

BIGGS, W. D., *The Mechanical Behavior of Engineering Materials.* New York: Pergamon Press, 1965.

CORTEN, H., and F. PARK, "Fracture," *Int. Sci. & Tech.*, No. 15, March 1963, pp. 24–36.

"Ductility," The 1967 ASM Seminar Book, Metals Park, Ohio: American Society for Metals, 1968.

HEYWOOD, R. B., *Designing Against Fatigue of Metals.* New York: Barnes & Noble, 1962.

HOEPPNER, D. W., Editor, *Fracture Prevention & Control.* Metals Park, Ohio: American Society for Metals, 1974.

KRAVCHENKO, P. Ye., *Fatigue Resistance.* New York: Pergamon Press, 1964.

LYNCH, C. J., "Deformation & Flow," *Int. Sci. & Tech.*, No. 49, January 1966, pp. 72–81.

MACCRONE, R. K., "Plastic Deformation," *Int. Sci. & Tech.*, No. 23, November 1963, pp. 36–43.

McCLINTOCK, F. A., and A. S. ARGON, Ed., *Mechanical Behavior of Materials.* Reading, Mass.: Addison-Wesley, 1966.

TEGART, W. T., *Elements of Mechanical Metallurgy*. New York: Macmillan, 1966.

WEERTMAN, J., and J. R. WEERTMAN, *Elementary Dislocation Theory*. New York: John Wiley and Sons, 1965.

CHAPTER 7

HANSEN, M. and K. ANDERKO, *Constitution of Binary Alloys*. New York, McGraw-Hill Book Company, 1958.

LEVIN, E. M., *et al.*, *Phase Diagrams for Ceramists*. Vol. 1 (1964), Vol. 2 (1969). Columbus, Ohio: American Ceramic Society.

RHINES, F. N., *Phase Diagrams in Metallurgy*. New York: McGraw-Hill Book Company, 1956.

TEMPERLEY, H. N. V., "Changes of State," *Int. Sci. & Tech.*, No. 46, October 1965, pp. 68–76.

"Ultrapure Materials," *Int. Sci. & Tech.*, Prototype Issue, pp. 33–37.

CHAPTER 8

BURKE, J., *Kinetics of Phase Transformation in Metals*. New York: Pergamon Press, 1965.

"Diffusion in Body-Centered Cubic Metals," Metals Park, Ohio: American Society for Metals, 1965.

SHEWMON, P. G., *Diffusion in Solids*. New York: McGraw-Hill Book Company, 1963.

"Solidification," The 1969 ASM Seminar Book, Metals Park, Ohio: American Society for Metals, 1971.

WALTON, A., "Nucleation," *Int. Sci. & Tech.*, No. 60, December 1966, pp. 28–39.

CHAPTER 9

BURKE, J., *Kinetics of Phase Transformation in Metals*. New York: Pergamon Press, 1965.

BYRNE, J. E., *Recovery, Recrystallization and Grain Growth*. New York: Macmillan Publishing Co., Inc., 1965.

CHALMERS, B., "Metallurgy Today," *Int. Sci. & Tech.*, No. 53, May 1966, pp. 18–29.

CHRISTIAN, J. W., *Theory of Transformations in Metals & Alloys*. New York: Pergamon Press, 1965.

FELBECK, D. K., *Introduction to Strengthening Mechanisms*. Englewood Cliffs, N.J.: Prentice-Hall, Inc., 1968.

FINE, M. E., *Phase Transformation in Condensed Systems.* New York: Macmillan Publishing Co., Inc., 1964.

MARTIN, J. W., *Precipitation Hardening.* New York: Pergamon Press, 1968.

"Precipitation from Solid Solution," Metals Park, Ohio: American Society for Metals, 1958.

"Recrystallization, Grain Growth and Textures," The 1965 ASM Seminar Book, Metals Park, Ohio: American Society for Metals, 1966.

SHEWMON, P. G., *Transformations in Metals.* New York: McGraw-Hill Book Company, 1969.

CHAPTER 10

BAIN, E. C., and H. W. PAXTON, *Alloying Elements in Steel.* Metals Park, Ohio: American Society for Metals, 1939, 1961, 1966.

BURKE, J., *Kinetics of Phase Transformation in Metals.* New York: Pergamon Press, 1965.

CHALMERS, B., "Metallurgy Today," *Int. Sci. & Tech.*, No. 53, May 1966, pp. 18–29.

CHRISTIAN, J. W., *Theory of Transformations in Metals & Alloys.* New York: Pergamon Press, 1965.

FELBECK, D. K., *Introduction to Strengthening. Mechanisms.* Englewood Cliffs, N.J.: Prentice-Hall, Inc., 1968.

FINE, M. E., *Phase Transformation in Condensed Systems.* New York: Macmillan Publishing Co., Inc., 1964.

HUME-ROTHERY, W., *Structures of Alloys and Iron.* New York: Pergamon Press, 1966.

PARR, J. G., and A. HANSON, *Introduction to Stainless Steel.* Metals Park, Ohio: American Society for Metals, 1965.

"Phase Transformations," Metals Park, Ohio: American Society for Metals, 1970.

SHEWMON, P. G., *Transformations in Metals,* New York: McGraw-Hill Book Company, 1969.

"Stahlschlussel—Key to Steel," 10th Ed., Published by Verlag Stahlschlussel, 1974, Available at Metals Park, Ohio: American Society for Metals.

WAYMAN, C. M., *Introduction to the Crystallography of Martensitic Transformations.* New York: Macmillan Publishing Co., Inc., 1964.

CHAPTER 11

BAKISH, R., "Vapor-Phase Metallurgy," *Int. Sci. & Tech.*, No. 18, June 1963, pp. 54–60.

BROUTMAN, L. J. and R. H. KROCK, Ed., *Modern Composite Materials*. Reading, Mass.: Addison-Wesley, 1967.

CARLTON, D., and P. CASSIDY, "Not Organic, Not Metallic," *Int. Sci. & Tech.*, No. 68, August 1967, pp. 78–86.

CLAUSER, H. R., "Advanced Composite Materials," *Scientific American*, Vol. 229, No. 1, July 1973, pp. 36–44.

DIETZ, A. G. H., "Fibrous Composite Materials," *Int. Sci. & Tech.*, No. 32, August 1964, pp. 58–69.

FRANK, R. C., "Gases-in-Solids," *Int. Sci. & Tech.*, No. 9, September 1962, pp. 53–59.

GRUNTFEST, I. J., and L. H. SHENKER, "Ablation," *Int. Sci. & Tech.*, No. 19, July 1963, pp. 48–57.

HANSEN, D., and A. H. NISSAN, "Fiber Science," *Int. Sci. & Tech.*, No. 36, December 1964, pp. 74–84.

HEYWOOD, R. B., *Designing Against Fatigue of Metals*. New York: Barnes & Noble, 1962.

KELLY, A., G. C. SMITH, P. J. E. FORSYTH, and A. J. KENNEDY, *Composite Materials*. New York: American Elsevier, 1966.

KLEIN, C. A., "Pyrolytic Graphite," *Int. Sci. & Tech.*, No. 8, August 1962, pp. 60–68.

KROCK, R. H., "Whisker-Strengthened Materials," *Int. Sci. & Tech.*, No. 59, November 1966, pp. 38–48.

MARK, H. F., and S. M. ATLAS, "Heat-Resisting Plastics," *Int. Sci. & Tech.*, Prototype Issue, pp. 44–52.

MARK, H. F., "Tailor-Making Plastics," *Int. Sci. & Tech.*, No. 27, March 1964, pp. 72–79.

MOREY, G. W., *The Properties of Glass*. 2nd Edition. New York: Reinhold Publishing Corporation, 1954.

MORTON, M., "Real Synthetic Rubber," *Int. Sci. & Tech.*, No. 32, August 1964, pp. 70–78.

OUTWATER, J. O., R. A. SCHULTHEISS, F. R. BARNET, S. P. PROSEN, and W. J. McLEAN, "Composite Materials," *Mechanical Engineering*, February 1966, pp. 32–39.

PHILLIPS, C. J., *Glass: The Miracle Maker*. New York: Pitman Publishing Corp., 1941.

RYSHKEWITCH, E., "Metal-Oxide Ceramics," *Int. Sci. & Tech.*, No. 2, February 1962, pp. 54–61.

STOOKEY, S. D., "Modern Glass," *Int. Sci. & Tech.*, No. 7, July 1962, pp. 40–46.

TSURUTA, T., and S. INOUE, "Well-Ordered Polymers," *Int. Sci. & Tec.*, No. 71, November 1967, pp. 66–76.

"Ultra-High-Purity Metals," The 1962 ASM Seminar Book, Metals Park, Ohio, 1962.

CHAPTER 12

ARMAREGO, E. J. A., and R. H. BROWN, "Machining of Metals," Englewood Cliffs, N.J., Prentice-Hall Inc., 1969.

Cast Metals Handbook. Chicago, Illinois: American Foundrymen's Association.

"Casting Design Handbook," Metals Park, Ohio: American Society for Metals, 1963.

COOK, N. H., *Manufacturing Analysis.* Reading, Mass.: Addison-Wesley, 1966.

DEGARMO, E. P., "Materials & Processes in Manufacturing," 4th Edition. New York: Macmillan Publishing Co., Inc., 1974.

DIETER, G., "Powder Fabrication," *Int. Sci. & Tech.*, No. 12, December 1962, pp. 58–65.

EDGAR, C., *Fundamentals of Manufacturing Processes and Materials.* Reading, Mass.: Addison-Wesley, 1965.

GREENBERG, S. A., Ed., *Welding Handbook.* New York: American Welding Society.

LINDBERG, R. A., *Processes and Materials of Manufacture.* Boston: Allyn & Bacon, 1964.

"Machining Data Handbook," Prepared by the Machinability Data Center, Available at American Society for Metals, Metals Park, Ohio: 1972.

MARTIN, D. C., "Modern Welding," *Int. Sci. & Tech.*, No. 66, June 1967, pp. 22–34.

MURRIN, T. J., "The Art of Manufacture," *Int. Sci. & Tech.*, No. 55, July 1966, pp. 82–88.

PARK, F., "High-Energy-Rate Metalworking," *Int. Sci. & Tech.*, No. 6, June 1962, pp. 12–23.

PARK, F., "Nontraditional Machining," *Int. Sci. & Tech.*, No. 23, November 1963, pp. 22–35.

SLAUGHTER, G. M., *Welding and Brazing Techniques for Nuclear Reactor Components.* New York: Rowman and Littlefield, 1964.

Welding Handbook, 6th Edition. Miami, Florida: American Welding Society.
    Sec. 1   *Fundamentals of Welding.*
    Sec. 2   *Welding Processes: Gas, Arc & Resistance.*
    Sec. 3   *Welding, Cutting and Related Processes* (2 volumes).
    Sec. 4   *Metals & Their Weldability.*
    Sec. 5   *Applications of Welding.*

YOUNG, J. F., *Materials and Processes.* 2nd Edition. New York: John Wiley and Sons, 1954.

CHAPTER 13

BOSICH, J. F., *Corrosion Prevention for Practicing Engineers*. New York: Barnes & Noble, 1970.

COOPER, D., "Radiation Processing," *Int. Sci. & Tech.*, No. 27, March 1964, pp. 22–34.

DETTNER, H. W., *Elsevier's Dictionary of Metal Finishing & Corrosion*. New York: American Elsevier Publishing Co., 1971.

EVANS, U. R., *Introduction to Metallic Corrosion*. 2nd Edition. London: Edward Arnold, Ltd., 1963.

FONTANA, M. G., and N. D. GREENE, *Corrosion Engineering*. New York: McGraw-Hill Book Company, 1967.

GARAFALO, F., *Fundamentals of Creep and Creep-Rupture in Metals*. New York: Macmillan Publishing Co., Inc., 1965.

HOAR, T. P., "Corrosion," *Int. Sci. & Tech.*, No. 24, December 1963, pp. 78–85.

KENNEDY, A. J., "Protective Coatings," *Int. Sci. & Tech.*, No. 8, August 1962, pp. 37–43.

LAQUE, F. L. *Marine Corrosion*. New York: John Wiley and Sons, 1975.

MATTSON, H. W., "High-Temperature Materials," *Int. Sci. & Tech.*, No. 30, June 1964, pp. 20–31.

"Oxidation of Metals and Alloys," The 1970 ASM Seminar Book, Metals Park, Ohio, 1971.

SCULLY, J. C., *Fundamentals of Corrosion*. New York: Pergamon Press, 1966.

SMITH, C. O., "Corrosion Mechanisms in Nuclear Systems," *Naval Engineers Journal*, Vol. 78, No. 4, August 1966, pp. 567–584.

SMITH, C. O., "Effects of Nuclear Radiation," *Naval Engineers Journal*, Vol. 78, No. 5, October 1966, pp. 789–804.

UHLIG, H. H., *Corrosion & Corrosion Control*. New York: John Wiley and Sons, 1963.

UHLIG, H., Ed., *Corrosion Handbook*. New York: John Wiley and Sons, 1948.

CHAPTER 15

"Analysis of Service Failures," Cleveland, Ohio: Republic Steel Corporation, 1961.

GILLETT, H. W., *The Behavior of Engineering Metals*. New York: Wiley, 1951.

HEYWOOD, R. B., *Designing Against Fatigue of Metals*. New York: Barnes & Noble, 1962.

RUSKIN, A. M., *Materials Considerations in Design*. Englewood Cliffs, N.J.: Prentice-Hall, Inc., 1967.

"Source Book in Failure Analysis," Metals Park, Ohio: American Society for Metals, 1974.

WIDNER, R. L., and J. O. WOLFE, "Valuable Results from Bearing Analysis Data," *Metal Progress*, Vol. 93, No. 4, April 1968, pp. 79–86.

WULPI, D. J., *How Components Fail*. Metals Park, Ohio: American Society for Metals, 1966.

INDEX

INDEX

MATERIALS

Metals and Alloys

Mechanical

Chap. 6   Chap. 11

Ceramics

Properties

Chap. 2

Electrical

Chap. 5

Polymers

Chemical

Chap. 13